Science for Sustainable S...

Series Editorial Board

Editor in Chief

Kazuhiko Takeuchi, Ph.D., Director and Project Professor, Integrated Research System for Sustainability Science (IR3S), The University of Tokyo Institutes for Advanced Study (UTIAS), Chair of the Board of Directors, Institute for Global Environmental Strategies (IGES), Japan

Series Adviser

Joanne M. Kauffman, Ph.D., Co-editor in Chief, *Handbook of Sustainable Engineering, Springer*, 2013

Scientific Advisory Committee

Sir Partha Dasgupta, Ph.D., Frank Ramsey Professor Emeritus of Economics, University of Cambridge, UK; Volvo Environment Prize, 2002; Blue Planet Prize, 2015

Hiroshi Komiyama, Ph.D., Chairman, Mitsubishi Research Institute, Japan; President Emeritus, The University of Tokyo, Japan

Sander Van der Leeuw, Ph.D., Foundation Professor, School of Human Evolution and Social Change and School of Sustainability, Arizona State University, USA

Hiroyuki Yoshikawa, Dr. Eng., Member of Japan Academy; Chairman, The Japan Prize Foundation; President Emeritus, The University of Tokyo, Japan; Japan Prize 1997

Tan Sri Zakri Abdul Hamid, Ph.D., Science Adviser to the Prime Minister of Malaysia, Malaysia; Founding Chair of the UN Intergovernmental Science-Policy Platform on Biodiversity and Ecosystem Services (IPBES); Zayed International Prize, 2014

Editorial Board

Jean-Louis Armand, Ph.D., Professor, Aix-Marseille Université, France

James Buizer, Professor, University of Arizona, USA

Anantha Duraiappah, Ph.D., Director, UNESCO Mahatma Gandhi Institute of Education for Peace and Sustainable (MGIEP), India

Thomas Elmqvist, Ph.D., Professor, Stockholm Resilience Center and Stockholm University, Sweden

Ken Fukushi, Ph.D., Professor, The University of Tokyo, Japan

Vincenzo Nazo, Ph.D., Professor, The Sapienza University of Rome, Italy

Obijiofor Aginam, Ph.D., United Nations University-International Institute for Global Health (UNU-IIGH), Malaysia

Osamu Saito, Ph.D., Academic Director and Academic Programme Officer, United Nations University Institute for the Advanced Study of Sustainability (UNU-IAS), Japan

Leena Srivastava, Ph.D., Executive Director, The Energy and Resources Institute, India

Jeffrey Steinfeld, Ph.D., Professor Emeritus of Chemistry, Massachusetts Institute of Technology, USA

Scope of the Series

This series aims to provide timely coverage of results of research conducted in accordance with the principles of sustainability science to address impediments to achieving sustainable societies – that is, societies that are low carbon emitters, that live in harmony with nature, and that promote the recycling and re-use of natural resources. Books in the series also address innovative means of advancing sustainability science itself in the development of both research and education models.

The overall goal of the series is to contribute to the development of sustainability science and to its promotion at research institutions worldwide, with a view to furthering knowledge and overcoming the limitations of traditional discipline-based research to address complex problems that afflict humanity and now seem intractable.

Books published in this series will be solicited from scholars working across academic disciplines to address challenges to sustainable development in all areas of human endeavors.

This is an official book series of the Integrated Research System for Sustainability Science (IR3S) of the University of Tokyo.

More information about this series at http://www.springer.com/series/11884

Isabel B. Franco • Tathagata Chatterji
Ellen Derbyshire • James Tracey
Editors

Actioning the Global Goals for Local Impact

Towards Sustainability Science, Policy, Education and Practice

Editors
Isabel B. Franco
Institute for the Advanced Study
of Sustainability
United Nations University Shibuya-ku
Tokyo, Japan

Australian Institute for Business
and Economics
The University of Queensland
Brisbane, Australia

Ellen Derbyshire
Faculty of Business, Economics and Law,
Business School,
The University of Queensland
Brisbane, QLD, Australia

Tathagata Chatterji
Xavier School of Human Settlements
Xavier University Bhubaneswar
Kakudia, Odisha, India

James Tracey
School of Engineering
University of New South Wales
Sydney, NSW, Australia

ISSN 2197-7348   ISSN 2197-7356   (electronic)
Science for Sustainable Societies
ISBN 978-981-32-9929-0   ISBN 978-981-32-9927-6   (eBook)
https://doi.org/10.1007/978-981-32-9927-6

© Springer Nature Singapore Pte Ltd. 2020
This work is subject to copyright. All rights are reserved by the Publisher, whether the whole or part of the material is concerned, specifically the rights of translation, reprinting, reuse of illustrations, recitation, broadcasting, reproduction on microfilms or in any other physical way, and transmission or information storage and retrieval, electronic adaptation, computer software, or by similar or dissimilar methodology now known or hereafter developed.
The use of general descriptive names, registered names, trademarks, service marks, etc. in this publication does not imply, even in the absence of a specific statement, that such names are exempt from the relevant protective laws and regulations and therefore free for general use.
The publisher, the authors, and the editors are safe to assume that the advice and information in this book are believed to be true and accurate at the date of publication. Neither the publisher nor the authors or the editors give a warranty, expressed or implied, with respect to the material contained herein or for any errors or omissions that may have been made. The publisher remains neutral with regard to jurisdictional claims in published maps and institutional affiliations.

This Springer imprint is published by the registered company Springer Nature Singapore Pte Ltd.
The registered company address is: 152 Beach Road, #21-01/04 Gateway East, Singapore 189721, Singapore

Preface

This book explores implementation challenges of the 2030 Sustainable Development Agenda, by specifically focusing on unique operational issues associated with each of the 17 Sustainable Development Goals (SDGs). In doing so, the book draws attention toward sustainability science, education, and community capacity-building needs related to the specific SDG targets and indicators. The target audience of the book are sustainability leaders, namely, policy-makers, sustainable development planning practitioners, academicians, and graduate students in various disciplinary domains associated with sustainability science, education, policy, management, and impact.

The Sustainable Development Agenda, which was adopted by the United Nations (UN) in 2015, is a universal, integrated, and human rights-based program. It underscores links between peace, social justice, and development. Consequently, its associated 17 SDGs are wider and much more multidimensional in scope, compared to its predecessor program, the Millennium Development Goals (MDG) (2000–2015).

The MDG program was the first concerted effort at a global scale to address extreme poverty and basic health-care needs. The eight identified goals were manageable and measurable and, most importantly, could be easily identified by a wide range of stakeholders, across the globe. During the 15-year period, the MDG program was able to achieve certain remarkable outcome – although the progress was uneven. Therefore, there is a need to create a new framework to achieve inclusive sustainable development.

The Sustainable Development Goals encompass the Millennium Development Goals and at the same time incorporate several newer goals, such as building resilient infrastructure, promotion of inclusive and sustainable industrialization, and fostering innovation (SDG 9); reduction of inequality within and among countries (SDG 10); making cities and human settlements inclusive, safe, resilient, and sustainable (SDG 11); ensuring sustainable consumption and production patterns (SDG 12); etc. Table 1 shows a comparison of MDG and SDG targets.

The millennium goals expressed solidarity with the poorest and the most vulnerable. It galvanized the global community to fight poverty and its multiple dimensions.

Table 1 The Millennium Development Goals and the Sustainable Development Goals

Millennium Development Goals	Sustainable Development Goals
MDG1: Eradicate extreme poverty and hunger	SDG 1. End poverty in all its forms everywhere
MDG 2: Achieve universal primary education	SDG 2. End hunger achieve food security and improved nutrition
MDG 3: Promote gender equality and empower women	SDG 3. Ensure healthy lives and promote well-being for all at all ages
MDG 4: Reduce child mortality	SDG 4. Ensure inclusive and equitable quality education
MDG 5: Improve maternal health	SDG 5. Achieve gender equality and empower all women and girls
MDG 6: Combat HIV/AIDS and other diseases	SDG 6. Ensure availability and sustainable management of water and sanitation for all
MDG 7: Ensure environmental sustainability	SDG 7. Ensure access to affordable, reliable, sustainable, and modern energy for all
MDG 8: Develop a global partnership for development	SDG 8. Promote sustained, inclusive, and sustainable economic growth, full and productive employment, and decent work
	SDG 9. Build resilient infrastructure, promote inclusive and sustainable industrialization, and foster innovation
	SDG 10. Reduce inequality within and among countries
	SDG 11. Make cities and human settlements inclusive, safe, resilient, and sustainable
	SDG 12. Ensure sustainable consumption and production patterns
	SDG 13. Take urgent action to combat climate change and its impacts
	SDG 14. Conserve and sustainably use the oceans, seas, and marine resources for sustainable development
	SDG 15. Protect, restore, and promote sustainable use of terrestrial ecosystems, sustainably manage forests, combat desertification, and halt biodiversity loss
	SDG 16. Peace, justice, and strong institutions promote peaceful and inclusive societies for sustainable development
	SDG 17. Strengthen the means of implementation and revitalize the global partnership for sustainable development

Source: prepared by authors based on open-source data available under UN

The 2030 agenda moves on from their targeted action bound programs on a wide array of interlinked developmental concerns.

The 2030 development agenda revolves around the concept of sustainability and also takes a comprehensive system view about the developmental paradigms. Embedded in the concept of sustainability is the idea of striking a balance between meaningful economic growth, environmental well-being, and social justice. Thus, the 17 SDGs are not directed to arrive at a trade-off between competing claims related to progress from multiple ideological standpoints. Rather, they are cross-cutting, are inter-sectoral, and complement each other in many ways (Babier and Burgess 2017). Thus, for example, provision of quality education (SDG 4) and safe drinking water (SDG 6) to deprived areas and slum settlements helps the communities and cities become more sustainable (SDG 11), improves health conditions of the people (SDG 3), and is also simultaneously an antipoverty (SDG 1) measure, as it reduces livelihood vulnerabilities by building community capacity-building. Thus, each of the SDG and targets are multidimensional in scope but also tied with each other.

As the Sustainable Development Agenda now sets the vision for 2030 for global action, its success depends on how far they are localized and integrated with national, subnational, and local plans of various countries. Policy-makers, academics, educators, and practitioners have embarked in activities aimed to integrate SDGs in policy documents, research agenda, and academic course curriculum. Yet, a major problem confronting these actors is a lack of knowledge about the operationalization of SDGs, which compromises their ability to disseminate knowledge in an impactful and contextualized manner. Some researchers and educators have proactively become active participants in the implementation of the SDGs across the world, representing a potential for global change.

Tokyo, Japan   Isabel B. Franco
Brisbane, Australia
Kakudia, Odisha, India   Tathagata Chatterji
Brisbane, QLD, Australia   Ellen Derbyshire
Sydney, NSW, Australia   James Tracey

Contents

1 Towards Impact Sustainability 1
   Isabel B. Franco, Tathagata Chatterji, Ellen Derbyshire,
   and James Tracey

2 SDG 1 No Poverty 5
   Isabel B. Franco and John Minnery

3 SDG 2 Zero Hunger 23
   Emily F. Creegan and Robert Flynn

4 SDG 3 Good Health and Well-Being 39
   María Belén Federico

5 SDG 4 Quality Education 57
   Isabel B. Franco and Ellen Derbyshire

6 SDG 5 Gender Equality 69
   Isabel B. Franco, Paulina Salinas Meruane,
   and Ellen Derbyshire

7 SDG 6 Clean Water and Sanitation 85
   Natalia A. Cano Londoño, Jessi Osorio Velasco,
   Felipe Castañeda García, and Isabel B. Franco

8 SDG 7 Affordable and Clean Energy 105
   Isabel B. Franco, Caitlin Power, and Josh Whereat

9 SDG 8 Decent Work and Economic Growth 117
   Ana Cristina Ribeiro-Duthie

10 SDG 9 Industry, Innovation, and Infrastructure 135
   Isabel B. Franco, Franz Gonzalez Arduz, and Jairo Andres Buitrago

11	**SDG 10 Reducing Inequalities**................................	153
	Caitlin Power	
12	**SDG 11 Sustainable Cities and Communities**	173
	Hitesh Vaidya and Tathagata Chatterji	
13	**SDG 12 Responsible Consumption and Production**	187
	Isabel B. Franco and Lance Newey	
14	**SDG 13 Climate Action**.......................................	219
	Isabel B. Franco, Rosemarie Tapia, and James Tracey	
15	**SDG 14 Life Below Water**....................................	229
	Elisa Palomino	
16	**SDG 15 Life on Land**	247
	Claudia Arana, Isabel B. Franco, Anuska Joshi, and Jyoti Sedhai	
17	**SDG 16 Peace, Justice and Strong Institutions**	265
	Isabel B. Franco and Ellen Derbyshire	
18	**SDG 17 Partnerships for the Goals**............................	275
	Isabel B. Franco and Masato Abe	
19	**Impact Sustainability: Conclusions and Lessons Learned**	295
	Ellen Derbyshire, Isabel B. Franco, Tathagata Chatterji, and James Tracey	

About the Editors

Isabel B. Franco Isabel B. Franco, PhD, is an international leader in sustainability science, policy and practice interested in exploring the role of industry in fostering sustainable, inclusive development. She does this through the development, refinement and application of sustainability assessment, methods and techniques. She has applied those tools to various industries such as extractives (mining, oil and gas), higher education, finance and banking, public sector and international development in Australia, Japan, Thailand, Colombia, Chile, Bolivia, Angola and Zambia. She has been an international advisor for various international organisations, namely, UNDP, UNESCAP, UNU and British Council. She has authored various peer-reviewed publications and has co-authored UN Books *Socially Responsible Business: A Model for a Sustainable Future* and *The Corporate Agenda of Sustainable Development*. With a PhD in Governance and Sustainability from the University of Queensland, Australia, she has managed large multilingual research and consulting projects globally. She has also held academic appointments at the United Nations University – Institute for the Advanced Study of Sustainability (Japan), Keio University (Japan), The University of Queensland (Australia), University of Zambia (Zambia) and Universidad Nacional de Colombia (Colombia). She is the Founder of eWisely (Exceptional Women in Sustainability), the fastest-growing connector of women in sustainability, globally. Her work can be consulted on her Lab DrIsabelFranco.com.

Tathagata Chatterji Tathagata Chatterji is a Professor of Urban Management and Governance at Xavier University, Bhubaneswar. His research interests are: urban sustainability and the political economy of urbanization in developing countries. He had authored two books – *Local Mediation of Global Forces in Transformation of the Urban Fringe and Citadels of Glass – India's New Suburban Landscape*. He reccived the Gerd Albers Award from the ISOCARP for his research on comparative governance. He graduated in Architecture from Calcutta University, holds a postgraduate qualification in Urban Design from Kent State University, and a PhD in Urban Governance and Planning from the University of Queensland. He is a member of the Planning Institute of Australia and a Fellow of the Institute of Urban Designers.

Ellen Derbyshire Ellen Derbyshire is a research analyst with an academic background in Business and Sustainability at the University of Queensland, Australia. Currently, she is a Research Fellow at the Secretary of Mining Antioquia, Colombia. She holds a Bachelor of International Relations, Public Policy and Spanish from the University of Queensland and has worked on several research projects with the United Nations Economic Social Commission Asia Pacific. Her key areas of research are: gender and youth empowerment, strategy for sustainable business, sustainable resource governance and education for sustainable development. Ellen is a member of the Australian Institute of International Affairs and the Centre for Feminist Foreign Policy.

James Tracey James Tracey is a research analyst with an academic background in Environmental Engineering at the University of New South Wales, Sydney, Australia. James has also worked extensively throughout Asia in various education and community capacity-building-related roles. He is passionate about international development, specifically focusing on the role of community capacity building and the importance of environmental integrity.

Chapter 1
Towards Impact Sustainability

Introduction

Isabel B. Franco, Tathagata Chatterji, Ellen Derbyshire, and James Tracey

This book contributes to sustainability studies, as it focuses on local operationalization of all 17 Global Goals in an impactful manner. This book is the result of collaborative and interdisciplinary research work by sustainability leaders from all over the world, namely, scientists, researchers, educators and practitioners. Disconnected educational systems and policy practices from global and local sustainability trends create scepticism about the potential of the research institutions in contributing towards policy debates and issues centring on the question of sustainability, which compromise the wellbeing of all. Preliminary investigations identified that a few reasons for this were limited understanding of the context; lack of an overall approach to sustainability, such as the Sustainable Development Goals (SDGs); and inadequacies in the education sector and collaboration processes amongst academics, educators and practitioners to achieve global sustainability targets (Franco et al. 2018; Franco and Tracey 2019).

I. B. Franco (✉)
Institute for the Advanced Study of Sustainability, United Nations University Shibuya-ku, Tokyo, Japan

Australian Institute for Business and Economics, The University of Queensland, Brisbane, Australia
e-mail: connect@drisabelfranco.com

T. Chatterji (✉)
Xavier School of Human Settlements, Xavier University Bhubaneswar,
Kakudia, Odisha, India
e-mail: tathagata@xub.edu.in

E. Derbyshire
Faculty of Business, Economics and Law, Business School, The University of Queensland, Brisbane, QLD, Australia
e-mail: ellen.derbyshire@uq.net.au

J. Tracey
Faculty of Engineering, University of New South Wales, Sydney, Australia

© Springer Nature Singapore Pte Ltd. 2020
I. B. Franco et al. (eds.), *Actioning the Global Goals for Local Impact*, Science for Sustainable Societies, https://doi.org/10.1007/978-981-32-9927-6_1

In this context, the overarching aims of this book are to provide coverage of results of research conducted in accordance with the SDGs and to better understand the integration of the Global Goals as an integral part of impact research, curriculum and community capacity-building for sustainability. This book also advances sustainability science itself in the development of both research and education models to SDGs integration in an effective manner. It does this through case studies and original research across various topics of sustainability science, education and community capacity-building. Embracing the SDGs as a component of research and education for sustainability may help academics and practitioners gain practical tools to teach students on ways to cope with potential global issues. Understanding the way that academia targets the SDGs and collaborates with external stakeholders for the achievement of the 2030 Agenda is also important in helping plan for future generations. The use of the SDGs as the conceptual framework of this book enables this manuscript to develop a better and more nuanced understanding of global sustainability matters and their impact on the local level. The case studies used for the book present research and best practices that target all SDGs across various locations around the globe currently experiencing significant sustainability challenges.

The SDGs not only meet educators and scientists' expectations, but they also address student's concerns regarding sustainability challenges. Students and the broader community are more often concerned about how global sustainability issues affect them. In response, higher education institutions have engaged in integrating the SDGs, in curriculum and science. This approach creates a higher level of impact amongst education institutions and their external stakeholders. This manuscript is therefore essential for educators, scientists and trainers interested in creating impact by integrating the SDGs in research, educational curriculum and capacity-building.

SDGs integration in curriculum fosters contemporary education practices that go beyond traditional forms of teaching, research and capacity-building and positions education institutions as powerful agents of global change. Embarking on more sustainable education practices is pivotal for helping students and the broader community meet social and environmental demands and for overcoming the challenges of socioeconomic development both at the local and global levels. The SDG approach is rapidly changing the role of education institutions by placing them at the forefront of sustainable development. This approach also provides a competitive advantage to education institutions.

SDGs have been relatively well studied and documented in various fields such as governance and business, yet more research is necessary to understand SDGs integration in education-related sectors. Although there have been many successful SDGs integration initiatives around the globe, most of those do not align with education, research and capacity-building agendas (Franco and Tracey 2019). Consequently, education institutions should facilitate the embrace of SDGs in curriculum, research and capacity-building. Accordingly, this publication aims to increase the knowledge of educators, scientists and trainers on the SDG approach and pathways for its integration. Therefore, this book unpacks the key role this approach can play in promoting impactful education, research and capacity-building practices and, ultimately, in contributing to the SDGs. The book consists of 18 chapters targeting each one of the 17 SDGs in science, curriculum development, policy and capacity-building.

The second chapter, by Franco and Minnery, discusses anti-poverty measures under SDG 1. Applying the sustainable livelihood framework (SLF), they explore community capacity-building projects in resource regions. The research highlights how the SLF shows interconnections between SDG 1 and SDG 11 in tackling multiple dimensions of poverty. The third chapter by Creegan et al. focuses on SDG 2 ending hunger and achieving food security through sustainable agriculture. The chapter highlights the need for organic waste recycling programmes to apply locally available resources, appropriate technologies, and localization measures based on specific community needs assessment. The research also shows how the overlapping nature between SDG 2 and SDG 6 for sustainable waste management and water conservation measures. The fourth chapter by Belen Federico studies health and wellbeing measures under SDG 3. By focusing on effects of ultraviolet radiation on human DNA, the research reported in this chapter shows that DNA is seriously damaged by higher UV radiation caused by ozone layer depletion.

The fifth chapter by Franco and Derbyshire is linked to SDG 4 quality education and connections with SDG 17 partnerships for the goals. The chapter pays nuanced attention to the nature and importance of limiting and fostering factors in a collaborative governance scenario towards the achievement of SDG 4 quality education and in the pursuit of education for sustainable development. Moving on, from there, the next chapter by Franco, Salinas and Derbyshire shows how existing education practices at higher education institutions of Latin America produce patriarchal systems detrimental for the sustainable leadership of women in male-dominated industries. The chapter makes reference to SDG 5 concerns on gender equality. Chapter 7 by Cano et al. is on provision of clean water sanitation under SDG 6. It shows how the mining industry is confronting water scarcity challenges through efficiency in technological applications and through the implementation of different strategies such as the efficiency of resources, the optimization of process efficiency and circular economy.

Chapter 8 by Franco, Powell and Whereat explores how women can boost their assets and capacities to cope with the effects of unsustainable energy consumption. This capacity-building analysis primarily links up with SDG 7 affordable and clean energy but also connects with SDG 5 gender equality and SDG 12 sustainable production systems. Similarly, Chap. 9 although primarily addresses SDG 8 concerns regarding decent work and economic growth but also has great relevance for SDG 12 concerns regarding production systems. Ribeiro-Duthie's research draws attention towards economic opportunities drawn by fair trade certified small producers. Franco, Gonzalez and Buitrago in Chap. 10 show that community capacity-building roadmap can enhance the ability of local communities to foster innovative industrial practices for sustainable resource development and the achievement of SDG 9 Industry, Innovation and Infrastructure. The following chapter by Power investigates how superannuation-related social benefit schemes could further SDG 10 objectives of reducing financial inequalities.

Unlike other sustainability goal which is organized along specific sectoral indicators and targets, SDG 11 objectives of urban sustainability are spatially defined. Chapter 12 by Vaidya and Chatterji focuses on SDG 11 as the analytical framework to explore how the transformative force of urbanization represents opportunity and challenge to meet several sustainability challenges, such as SDG-1, poverty reduction, SDG 4 (education), SDG-5 (gender equality), SDG 6 (clean water and sanitation), SDG 7 (affordable and clean energy), and SDG 8 (economic growth). The following chapter by Franco and Newey looks at responsible production and consumption practices under the SDG 12 umbrella. Based on a case study method, this chapter shows how the corporate sector can contribute to sustainable community development through a multidimensional approach to wellbeing for entrepreneurship. It also provides some conceptual and practical tools towards enhanced accountability for sustainable production and therefore the achievement of SDG 12. Next, Chap. 14 by Franco, Tapia and Tracey provides a capacity-building framework to move from climate education to action and towards the achievement of SDG 13.

Chapter 15 by Palomino discusses SDG 14 target of life below water by looking through utilization of fish skin as a sustainable raw material for fashion industry. Fish skin leather processing could prevent and significantly reduce marine pollution and sustainably protect marine ecosystems in order to achieve healthy and productive oceans. As such this also promotes SDG 12 targets. The fashion industry is also the topic of discussion of Chap. 16 by Arana, Franco, et al. This chapter links to Goal 15 Life on Land as it explores alternative organic materials and recycled processes for the sustainable production of yarns used in the fashion industry in a sustainable manner. Chapter 17 by Franco and Derbyshire makes a strong contribution to SDG 16 as it examines the role of women in sustainable peace in the context of resource regions. Chapter 17 explores the role of global business networks in the pursuit of corporate accountability and sustainability reporting. The book finishes drawing conclusions and impact sustainability recommendations.

The book highlights the value of sustainability science on newly emerging and innovative approach towards research, education, capacity-building and practice in order to transform rhetoric into impact sustainability while encompassing cases from various industries, sectors and geographical contexts. The case studies are collected from different geographical contexts and industries, provide insights into themes that cut across sustainability science and that aid the fulfilment of the SDGs building more resilient, sustainable, equal and inclusive societies and the environment.

References

Franco IB, Tracey J (2019) (accepted) Community capacity-building for sustainable development: effectively striving towards achieving local community sustainability targets: effectively. Int J SustainHigh Educ

Franco I, Saito O, Vaughter P, Whereat J, Kanie N, Takemoto K (2018) Higher education for sustainable development: actioning the global goals in policy, curriculum and practice. Sustain Sci:1–22

Chapter 2
SDG 1 No Poverty

Building Sustainable Communities: A Framework for Supporting Community Livelihoods and Poverty Alleviation in Resource Regions

Isabel B. Franco and John Minnery

Abstract This chapter proposes the use of the sustainable livelihood framework (SLF) as a powerful conceptual approach for research aimed at understanding the interaction between global investment, local livelihoods and poverty reduction in resource regions. The chapter applies the SLF as a tool to develop recommendations for poverty alleviation, showing how it can contribute to SDG 1 No Poverty. The innovative application of the SLF helps us understand the ways in which key areas of research connect and interact as constituent components inherent in the framework. This chapter also argues that this framework helps increase our understanding of the ways communities build capacity to forge sustainable livelihoods in resource regions. It thus presents a justification for the use of the SLF, followed by an examination of the SLF principles, their implications for communities and relevance for empirical research in this field. The chapter also shows the way in which the SLF can be modified for application to local circumstances through case studies conducted in two resource regions of Colombia. Nevertheless, the findings of the research can be applied to other resource locations elsewhere.

Keywords SDG 1 No Poverty · Sustainable livelihoods · Community · Resource regions · Sustainability

---

I. B. Franco (✉)
Institute for the Advanced Study of Sustainability, United Nations University Shibuya-ku, Tokyo, Japan

Australian Institute for Business and Economics, The University of Queensland, Brisbane, Australia
e-mail: connect@drisabelfranco.com

J. Minnery
Griffith University – Cities Research Institute, Brisbane, QLD, Australia

© Springer Nature Singapore Pte Ltd. 2020
I. B. Franco et al. (eds.), *Actioning the Global Goals for Local Impact*, Science for Sustainable Societies, https://doi.org/10.1007/978-981-32-9927-6_2

2.1 Introduction

This chapter proposes the sustainable livelihood framework (SLF) as the governing conceptual approach to research that explores the interaction between global resource development and local livelihoods in resource regions. It also provides recommendations relevant to the achievement of SDG 1 No Poverty through suggestions for poverty alleviation and thus contributes directly to SDG1 No Poverty. The innovative application of the SLF helps us understand the way in which key areas of research connect and interact as inherent constituent components of the SLF. This chapter also argues that this framework increases our understanding of community capacity-building for forging sustainable livelihoods in resource regions. It presents a justification for the use of the SLF, followed by an examination of the SLF principles, their implications for communities and relevance for empirical research in this field. The chapter shows the way in which the SLF can be modified for application to local circumstances.

The SLF is an approach to sustainable development extensively applied by aid agencies to examine poverty issues in the developing world. In the recent years, a wide number of frameworks have served as a platform to better understand community livelihoods in resource regions. The SLF is one such (see Fig. 2.1). Originally developed to help address rural poverty in the Global South (Carney 2003; Rakodi and Lloyd-Jones 2002), here it is adopted to link mining activities and community livelihoods. The chapter also shows the way in which the SLF can be modified for local application, in this case through two case studies conducted in Antioquia and Risaralda, two resource regions of Colombia (Franco 2014). Nevertheless, the findings of the research can be applied to other resource locations.

Resource exploitation provides both strong advantages and powerful difficulties to the communities adjacent to the resource operations (such as mining, oil and gas). This chapter explores the mining case. Communities are often highly dependent on mining for employment and financial support, but this dependence comes with inevitable economic, social and environmental vulnerabilities. If these communities are in the Global South, the difficulties are exacerbated because they experience unequal access to political and economic resources, poor local governance, unbalanced access to the resources and often, low levels of education and skills (Franco 2014). Yet mining can also be a critical component of local, regional and national sustainable development.

This chapter is based on research into mining-dependent communities in resource regions of Colombia. It focuses on community livelihoods in these resource locations and through this develops a framework to show how mining and exploration interventions can help support communities adjacent to mining operations to maintain their livelihoods, meet their own development aspirations and reduce poverty. The framework used in the chapter is based on the sustainable livelihoods approach (Carney 2003, pp. 14–15; Rakodi and Lloyd-Jones 2002), but it modifies this classic approach to make it more appropriate to resource regions where mining forms a major component of the local and regional economy.

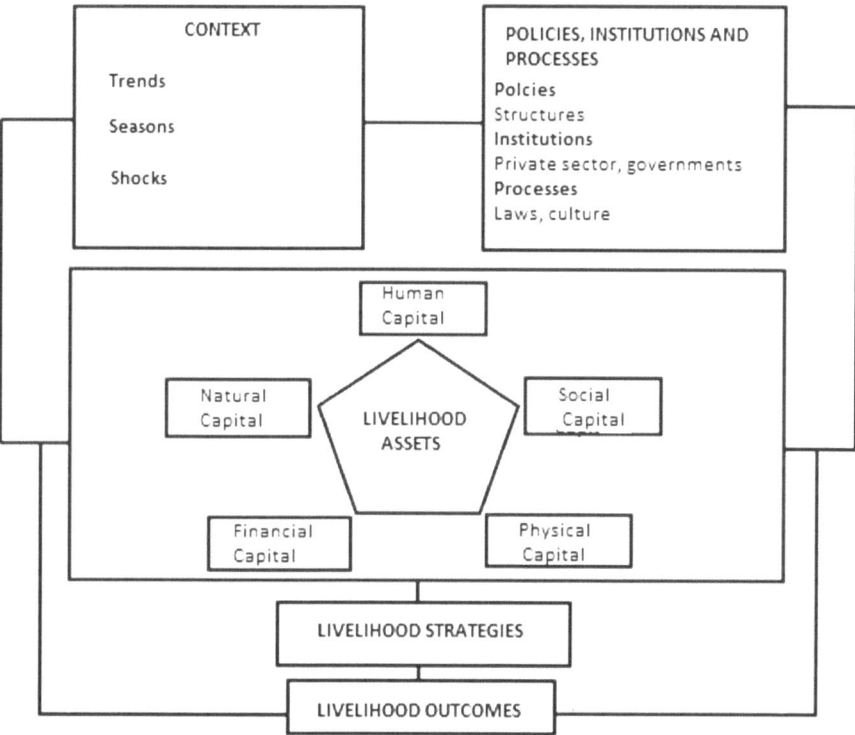

Fig. 2.1 Modified sustainable livelihood framework. (Modified from Mandke 2007 and Rakodi and Lloyd-Jones 2002)

The chapter provides (below) an overview of the research and introduces the case study area. It then examines the implications of the principles underpinning the sustainable livelihoods framework for communities adjacent to mining and thus demonstrates how the original principles and structure of the classic SLF need to be modified to be applied in this context. This is followed by a presentation of the research findings, concentrating on three components of the sustainable livelihoods framework: context, governance and livelihood assets. The original SLF structure is shown in Fig. 2.1, and the details of how the original SLF are modified to apply to the subject matter of the research are shown in Fig. 2.2. The ways in which it is modified are explained more fully in the remainder of this chapter, but in summary the contextual component and governance component of the modified SLF deal principally with aspects particularly relevant to communities dependent on mining, and the focus is on the community asset of human capital rather than all five forms of asset (human capital, plus natural capital, social capital, financial capital and physical capital) usually considered. The final section of the chapter provides recommendations for ways of improving social sustainability for settlements and communities in mining regions based on the key findings identified in the paper.

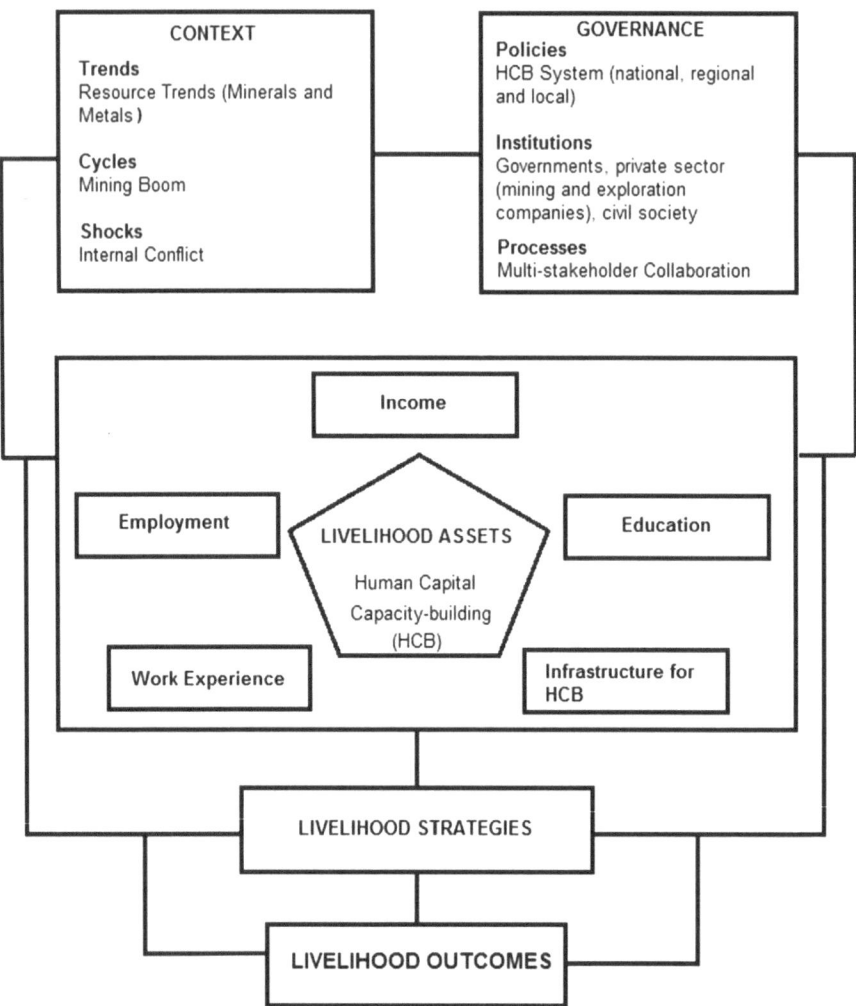

Fig. 2.2 Modified sustainable livelihood framework. (Modified from Mandke 2007 and Rakodi and Lloyd-Jones 2002 (see Franco 2014))

2.2 Methodology

The research findings came from document analysis (policy analysis), face-to-face semi-structured interviews with community, mining and government stakeholders, field work and participant observation and group interviews in two case study areas in Colombia. The research deals with diverse and multiple sets of data requiring the application of the case study method. The case studies allow for detailed and comprehensive information to be collected about a more focused issue. The research was also organised to follow the structure of the SLF.

2.2.1 Research Overview and Case Study Areas

Colombia is located in the north of South America and has experienced an escalating mining growth over the last three decades. It is the main producer of coal in Latin America and the twelfth largest globally, the third major producer of nickel after Cuba and the Dominican Republic and is also known as a leading exporter of gold and emeralds (Idarraga et al. 2010; Torres 2001; Vilora de la Hoz 2009). Mining along with oil extraction represented 4.6% of the national GDP in 2005 (UPME 2006, p. 19). The resource boom is not particular to a specific region, with exploration and mining operations being spread throughout the country.

The research presented here was undertaken in Risaralda and Antioquia, two resource regions, located in the Colombian Andes mountain range. These geographical areas are some of the most active producers of minerals and metals in Colombia. Mining and exploration projects operated in Risaralda impact on communities and human settlements and their livelihoods in both urban and non-urban areas. Risaralda is highly urbanised, with 665,104 (77.4%) of the total population of 859,666 living in urban areas in 2005 (DANE 2005). With a total of 230,532 households, the average household size was approximately 3.7 persons in 2005.

The State of Antioquia is located on the North Pacific Coast of Colombia. Antioquia has an active mining industry, as it holds the largest reserves of gold, silver, coal, platinum and construction materials in Colombia. Mining projects currently operated by domestic and multinational companies have significant implications for local communities (Cámara de Comercio de Medellín para Antioquia 2010; Sistema de Información Minero Energético Colombiano 2010). Antioquia is also highly urbanised with 4,340,744 (77.5%) of its population of 5,601,507 living in urban areas in 2005 (DANE 2005). The average household size, at 3.8 persons, was similar to that in Risaralda.

2.2.2 The Sustainable Livelihoods Framework (SLF)

The structure used to guide the research presented in this chapter was based on what is variously called the 'sustainable livelihoods framework' (Hostetler 2006) or the 'sustainable livelihoods approach' (Mazibuko 2013). As Mazibuko (2013, pp. 174–5) makes clear, it is not a theory or a model but 'as a framework ... [it] helps in considering the phenomenon and recognizing patterns'. It was used in this way in this research. One of the core components of the framework is the notion of 'livelihood' itself. A livelihood comprises the capabilities and activities required for the means of living, so that 'a livelihood is sustainable when it can cope with and recover from stresses and shocks and maintain or enhance its capabilities and assets both now and in the future' for both households and communities (DFID 1999, p. 1). The SLF was conceived as a way of thinking about the objectives, scope and priorities for development (Carney 2003, pp. 14–15; Rakodi and Lloyd-Jones 2002).

In 1998, the British Government's Department for International Development (DFID) adopted the SLF as an approach to assessing and evaluating developmental projects funded by it. Since then several other international organisations, like the United Nations Development Programme and the NGO and CARE, have also adopted the SLF to undertake their projects (Carney 2003). Its potency as a framework lies in the fact that it 'recognizes that people have many capabilities, have various assets and engage in many activities to earn their living' as well as recognising that 'institutions and processes should be clearly understood' (Mazibuko 2013, p. 175). The focus on assets is critical (Mitlin 2003). The framework also resonates with Sen's (1979) approach to strengthening communities' capabilities, for which they need rights and opportunities.

However, despite its widespread use, it has a number of weaknesses. One is that it neither provides an adequate role for the private sector nor provides for broader ideas about governance dynamics (Carney 2003). The study reported in this chapter created a modified version of the SLF that incorporated and expanded these two elements: how private corporations take part in creating sustainable livelihoods, and a broader understanding of governance arrangements. Researchers such as Stead (2015) have linked governance capacity (as broadly defined) to a multinational understanding of urban prosperity, but in this research, governance was explored within the tighter framework of one of the elements of the SLF, namely, human capital.

This modified version of the SLF (see Fig. 2.2) structured the research reported here. The revised framework still follows the key components proposed by Rakodi and Lloyd-Jones (2002, p. 9), namely, the external environment; vulnerability context; policies, institutions and processes; livelihood assets; livelihood strategies; and livelihood outcomes. These key components are linked as shown in Figs. 2.1 and 2.2, but the modifications identified in Fig. 2.2 create a framework that is more suited to understanding the Colombian mining case and the research focus specifically on human capital as an intangible asset. Such modifications to suit the relevant context have been used by other researchers. For example, Shen, Hughey and Simmons (2003, p. 20), in their review of the use of the sustainable livelihoods approach in the tourism industry, referring to Cahn's (2006) application of it to her Samoan case studies, noted that 'a "one size fits all" SLF approach is neither possible nor appropriate – context is important'.

The modified framework shown in Fig. 2.2 identifies the trends, cycles and shocks that affect mining and human settlements in Colombia. Governance processes constitute the core component of the policies, institutions and processes box, but in the context of mining in Colombia, these processes incorporate the activities of multinational and domestic private sector mining and exploration companies, as well as informal mining actors. Human capital and its development are examined as the principal focus of the livelihood assets box. Although the standard SLF normally comprises five forms of asset (or capital), this research focuses only on human capital in order to reduce the potential complexity of the analysis. The modified SLF constituted a guide to organise the research reported in this paper and also served as a means to review existing activities, understand cause and effect relationships, and provide a structure for analysis and a checklist of ideas (Rakodi and Lloyd-Jones 2002).

2.3 Discussion

2.3.1 SLF Application and Research Findings

This section provides a brief explanation of the elements of the SLF that were modified to enable it to be applied to the mine-based communities in Colombia. It also presents the major research findings derived from this application. The main modifications are shown in the boxes labelled 'Context', 'Governance' and 'Livelihood assets' in Fig. 2.2.

2.3.1.1 Context SLF Component

The broader term 'Context' is used here rather than the original term 'Vulnerability context' found in the SLF literature. Trends are the principal contextual factor for mining in Colombia. Colombian settlements and communities are diverse. Hence, mining impacts on local communities and the benefit they get from the industry are quite different across its various regions. There are some locations where the compensation for natural resource extraction has positively impacted on communities (Warhust 2001); therefore, they cannot always be categorised as vulnerable.

According to the SLF literature, the trends can be positive or negative or national or international and have a strong influence on community livelihoods. Trends in mining activities include increased globalisation of production and sales, greater need for highly capitalised production as easily exploited reserves are depleted and changes in consumption patterns that are built on particular mineral resources, such as demand for gold in India and increasing energy demand in the developing world (Surborg 2012). On the other hand, cycles (referred to as 'seasonality' in the classic SLF literature, such as Rakodi and Lloyd-Jones 2002, p. 14) pertain to recurrent economic shifts and employment opportunities. In a mining context, they also refer to the stages of the mine life cycle (Hitch et al. 2014). The current mining boom is fostering important global mining and exploration projects across Colombia which have aspects of both longer-term trends and short-term cycles. These resource fluctuations have had implications for the domestic economy and local communities' livelihoods, particularly in terms of regular fluctuations in the value of minerals and livelihood and employment opportunities.

The 'Context' SLF component also includes the concept of shocks. These shocks constitute factors that might destroy community assets such as floods and droughts but also resulting in or from conflicts or wars (Rakodi and Lloyd-Jones 2002, p. 14). The revisited version of the SLF proposes the Colombian internal conflict (or civil war) as a shock. It is also argued that internal conflict and local impacts have had serious implications for community livelihoods. The application of the SLF to the Colombian mining context highlights that shocks, in the form of civil war and the involvement of illegal groups, are perhaps the most sensitive issues that settlements and communities in resource regions have to deal with and therefore deserve major attention. The involvement of illegal groups in resource regions through informal small-scale mining has exacerbated historical conflicts and adversely affected stakeholder interactions. The interviews conducted for this research showed it has also diminished the potential of enhancing community assets as a driver for social sustainability (Franco 2014):

> Why do you think there is violence in the world? What do you do when you have a family to support and you do not have a job? ... you do whatever to get some income to feed your children, right? Well, there are many people experiencing this situation in this region, even though this is a mining region ... The situation is very complex in these towns. At least 3 to 4 people are killed every day ... This has got worse during the last three months. (NGO Director from Antioquia, Interview, November, 2012)

Civil society actors and government representatives both agree that relationships between companies and other stakeholders have deteriorated in the two regions. In addition, the escalation of global mining interests in the regions is causing resentment amongst legitimate small-scale miners. These groups are highly dependent on mining which they perceive as the only employment opportunity and therefore their only livelihood option. Consequently, legitimate small-scale miners regard the arrival of multinational exploration and mining companies as a threat to their livelihoods. This situation has increased resistance from these groups against multinational companies. The whole situation has also been intensified by armed conflict.

Internal conflict dynamics have been exacerbated by illegal groups including guerrillas, paramilitary and *bacrim* (criminal bands). These illegitimate actors own small-scale mines to launder money or support their illegal businesses. In addition, they have found the current mining boom to be the best opportunity to extort money from companies operating in the two regions. Community livelihoods have been heavily impacted by the indirect benefits of mineral extraction to these groups. Very often communities and human settlements cannot actively engage in human capital/asset enhancement initiatives as they feel threatened by these groups. On the other hand, governments and companies are reluctant to further engage with some small-scale miners and community members due to their suspicions of links between locals and illegal groups. This has not only aggravated the level of discontent and violence in remote resource areas but has also resulted in destruction of livelihoods, since communities reap very little benefit from collaborative endeavours in these circumstances. These limiting contextual factors have also been detrimental for the role of the private sector in the governance environment in resource regions.

2.3.1.2 Governance: Policies, Institutions and Processes

A broader examination of the governance environment was carried out in the light of the important component of the classic SLF of policies, institutions and processes (PIP) (Fig. 2.1). This component deals with the governance environment in which livelihoods are constructed (Rakodi and Lloyd-Jones 2002, pp. 15–16). For the purpose of the research reported in this chapter, governance processes include the interactions between three stakeholders: government, the private sector (particularly mining and exploration companies) and civil society (Minnery 2007). Thus, in Fig. 2.2, this is labelled as the 'Governance' box.

The mainstream presentations of the SLF exhaustively discuss government and community roles in developing sustainable livelihoods; however, they have little to say about the private sector's role (see, e.g. Rakodi and Lloyd-Jones 2002). The research reported here emphasises the critical potential importance of the private sector in the design of sustainable livelihood options for communities and in the opportunities to reduce poverty in resource regions. The literature on mining emphasises that companies have a responsibility to contribute to other economic futures in addition to mineral extraction that in fact they have a social responsibility (Hitch et al. 2014). Some of the incentives for private mining companies to do this include, but are not limited to, obtaining a 'social licence to operate', responding to international standards and regulatory frameworks and being accountable to their shareholders but also to their wider stakeholders (Freeman 1984). Companies' contributions to community livelihoods differ according to the context and governance dynamics in which they are immersed (Franco and Robertson 2014; Franco 2014):

> We do not implement initiatives different from mining because we do not have direct relationship with other sectors…we think that tourism or agriculture are sectors in which we do not fit. For this reason we focus on education for mining. (Senior Corporate Representative, Interview, October, 2012)

In addition, it was also found that if the improvement of local livelihoods is identified as a policy goal, then companies need both to be more active in support of this and to be more accountable to communities. This will help communities enhance their coping capacities, which will then enable them to overcome imminent challenges posited by the expansion of mining and exploration projects. Research shows that while corporate social responsibility agendas are vital to help communities design livelihoods relevant to their development aspirations, in practice such agendas have serious constraints that challenge corporate efforts (Franco 2014). In most cases these agendas for human capital/asset enhancement are top-down approaches intended mainly to enhance human capital to attract and retain skilled workers for the industry.

Nevertheless, mining and exploration companies can still play a stronger role with the implementation of bottom-up and locally driven social responsibility agendas. Such approaches can become some of the main enablers for local development, poverty mitigation and overall sustainability. Developing bottom-up (economically diversified) as opposed to top-down (mining focused) social responsibility agendas could have positive implications for communities. In addition, the research indicates that those agendas need to go beyond just mining practices and provide local communities with alternative non-mining livelihood options, enabling an expansion of their life plans and the enhancement of the human capital assets required to become more sustainable during and after the life of the mine (Buitrago-Franco and Robertson 2014).

2.3.1.3 Livelihood Assets SLF Component

Human capital is one of the core livelihood assets in the SLF. According to DFID (1999, p. 1), a livelihood comprises the capacities required for the means of living. This perspective positions people as active participants with a positive contribution to make in their development rather than as passive or deprived actors starved of assets. Community assets are diverse and vary across individuals, households and communities. Colombian communities are often starved of financial and skilled human capital (Cardenas 2011). Strengthening community assets can help locals forge more sustainable livelihoods (Rakodi and Lloyd-Jones 2002, p. 10). The classic SLF involves five forms of capital, which are human, social, physical, financial and natural capitals (as shown in Fig. 2.1). However, the research reported in this chapter focuses specifically on human capital. This section identifies the forms of human capital that need to be further enhanced in resource regions of Colombia to help communities forge more sustainable livelihoods. These are shown in Fig. 2.2, but the justification for using them is contained in the remainder of this section.

The investigation showed that there are five main priority areas for human capacity-building for local communities in mining resource regions. Developing these will hopefully assist locals in forging sustainable livelihoods and so reducing poverty. These areas are employment, education, work experience, income and infrastructure for human capital development (Franco 2014):

> I am the only miner in my family. I have three children and want them to study. I am aware of the risks of working at the mine and I did not want them to be part of the industry. This is a mining town... a person who does not go to school becomes a miner. Most of us do not know anything else than mining, hence, we have to work here and sustain our family members with the little income we get. (Community Member, Focus Group, October, 2012)

Although other components of human capital such as health and leadership are important, improving the five identified forms of human capital seems from the investigation to be the most effective way to strengthen the ability of communities to cope with mining-induced changes over time. An important finding from the research here reported was that priority areas which are the most valuable for communities are those that help them achieve their own sustainable development goals. Because of this, bottom-up human capital capacity-building approaches structured into social responsibility agendas are more likely to create value for communities than are top-down approaches. In those cases in which companies had embraced community-oriented and community-driven agendas, locals claim to have become more resilient. Communities that have been properly consulted about their capacity-building priorities have been able to strengthen key assets, becoming more capable of coping with mining-induced changes. However, such approaches need to be included as constituents of both corporate and government policies. It is not solely the companies' responsibility to make community capacity-building available for communities; governments also need to share responsibilities with companies in this regard, and governments also need to consult communities. All consultation should be genuine and not be merely the token provision of information; it should be such that communities are involved in decision-making about matters such as priorities for the allocation of funds.

At the corporate level, a more people-centred approach to corporate social responsibility for community capacity-building has helped companies in mining locations play a strong role in enhancing human capital and allowed them to contribute strongly to the development of sustainable livelihoods. The implementation of non-mining capacity-building initiatives involving the broader community was found in the research to be more valuable for locals. There are a few cases of companies driving community capacity-building initiatives that place local people at the centre of development. These actions involve the broader community regardless of their engagement with the mining industry. For example, there is active participation from local entrepreneurs and vulnerable groups, but with the assistance of mining companies, so that they can access better business opportunities and so develop livelihoods that do not involve mining. These activities can also have the potential to reduce poverty.

These are examples where both corporations and governments have embarked on socially responsible programmes based on the aspirations and expectations of communities. This has had positive impacts on locals, as the resultant efforts at community capacity-building are more attuned to the context in which they are embedded. In this approach the needs and expectations of the community are highly valued in the implementation of community capacity-building activities. These conditions have made the existing community capacity-building approaches meaningful for all stakeholders but especially for communities, all of which could serve as a valuable model for actors in other regions to follow.

As discussed earlier, the current approach in Antioquia is mainly intended to attract and retain human capital that is skilled or trained for the mining industry. Although there are some wider entrepreneurship development initiatives in place, they were seen by respondents to be insufficient to face the potential challenges created by the mining sector. In spite of corporate investments in community capacity-building programmes and activities, those initiatives have not yet tackled issues that were seen to be critical by the community. Development aspirations have been partially achieved, but powerful mining stakeholders are not playing a sufficiently strong role in protecting community assets or in helping to forge sustainable livelihoods. In addition, the government and private sector actors involved need to take community capacity-building more seriously and devote more resources and attention to the enhancement of community human capital assets, especially by delivering relevant training.

The research identified that there are key forms of training that can be delivered to provide communities and human settlements with more sustainable livelihood options. This was the case in both case study locations. Community training needs to be implemented according to gender and context variables. For example, participants feel there is a potential for dressmaking training for females in both regions. Funded by a Canadian company, some women in Risaralda are learning about and participating in dressmaking. This initiative has the potential to create sustainable value for the company and the community itself. Women are currently designing and making company and school uniforms, so creating a valuable supply chain for the local economy. A similar initiative is taking place in Antioquia; however, active corporate engagement is needed. Women from Segovia, Antioquia, have obtained some funding from the government, but there is minimal support from the large-scale mining companies, so the initiative is under threat. Other women more actively involved in the small-scale mining sector have been partially engaged in community capacity-building initiatives in mining, but this is limiting their possibilities to further expand their livelihood options beyond mining. In this context, women need to become active participants and benefit from a range of training in both case study locations.

In addition, small-scale miners participating in existing community capacity-building in mining need to be trained in other areas different from mineral processing. Lessons can be learned from the current small-scale licensing process in Antioquia. This process involves community capacity-building in areas like finance, accounting, administration and literacy. However, this community capacity-building approach should not be exclusive for small-scale miners but should be applied to the broader community. Key findings also show that there are other forms of training that are being delivered to the whole community, particularly in Risaralda, and that need to be enhanced and replicated in other resource regions in Colombia. For example, training in jewellery design not only adds value to mineral extraction but creates alternative livelihoods for locals. Similarly, major attention needs to be paid to existing agriculture community associations. Agriculture is a key sector in Colombia's economy, and so stakeholders, particularly the private sector, should support agriculture-based livelihood options. Companies operating in Risaralda contribute to fostering both the mining and agriculture sectors. This approach has assisted locals in keeping their traditional skills and knowledge alive. For example, coffee and blackberry production training has been provided by the private sector in the Risaralda case.

2.3.1.4 Sustainable Livelihood Strategies Component

The assets available or stock of capital at the community level can be accumulated, restored, exchanged or put to work to generate income and prevent poverty. Or they can be depleted. The transformation of assets in any of these positive forms constitutes the livelihood strategies in the light of the SLF. Such strategies are more likely to be effective if communities become active participants in their own development. However, this argument needs to be examined carefully, as communities might not have the capacities or skills to develop their own sustainable livelihood strategies. Hence, strategies in place might be temporary rather than sustainable in the long-term. They may also be of varying degrees of relevance for locals. Following Mandke (2007), livelihood strategies can be examined in consideration of four aspects: combination, substitution, sequencing and clustering. These strategies will be examined here in relation to community capacity-building to recommend alternative ways of creating more sustainable livelihoods for mining-based communities and human settlements in resource regions in Colombia.

2.3.1.5 Combination and Substitution

Combination and substitution are two main strategies that can be implemented to foster sustainable livelihoods. It is advisable to apply these two strategies according to the context and governance environment in which the communities are embedded. Combination entails examining how mining and other livelihoods coexist and the implications of this coexistence for communities. However, assets and livelihoods can also be replaced to help communities adjacent to mining operations develop livelihoods in tune with their development aspirations in what the SLF refers to as substitution. The Risaralda case study illustrates the successful implementation of a combination strategy in which both mining and other traditional livelihoods like agriculture are combined to help communities become more resilient to potential mining impacts.

In Risaralda, examples can be seen where mining has also been combined with jewellery design and trade, a livelihood option that is currently adding value to resource extraction in the region. Conversely, mining training in Antioquia is often substituting for traditional livelihood options such as agriculture, which is compromising community sustainability and even exacerbating poverty.

In this context, the implementation of an economic diversification approach coupled with other initiatives such as food security and farming programmes is highly recommended due to the likelihood that mining impacts will have many adverse implications for livelihoods. This will allow activities like mining to coexist with other industries and activities relevant to the local economy and will add value to other livelihood options relevant to communities, as well as helping to ensure that livelihood options remain open after mines have closed.

2.3.1.6 Sequencing and Clustering

Both sequencing and clustering are examined here. Sequencing relates to the resources that need to be allocated one after the other to build livelihoods, while clustering relates to the group of livelihood assets associated with specific livelihood strategies (DFID 1999). In other words, an examination of these two concepts accounts for existing community assets and potential strategies to further employ these assets. Based on the key findings of the research, communities need to develop alternative livelihoods to mining. In doing so, there are specific assets that should be enhanced to help provide them with meaningful livelihood opportunities. As discussed earlier, five forms of human capital (income, employment, education, work experience and infrastructure for community capacity-building) need to be further enhanced so that locals can achieve their development aspirations.

Existing income rates are not enough to meet community members' basic needs or those of their families. Governments and companies (with input from communities) need to partner to formulate income generation strategies. Up-skilling communities so that they are helped to obtain higher salary rates is highly recommended, but as shown in the research, this needs to be combined with increased opportunities to earn incomes. This will not only create value for the community but also for the company itself in terms of goodwill and support for its community service obligations.

Employment is an asset that needs to be revisited as it can cause tensions at the community level. Current processes intended to up-skill informal miners and the broader community need further development in order to help locals enhance their capacity to get gainful employment and therefore develop more resilient livelihoods. The research showed that these need to be both within and outside the mining sector. The participation of tertiary institutions and other educational organisations is vital to increase both education and job opportunities and therefore help reduce poverty.

Education is an essential asset to enhance other human capital such as income and employment. Despite the implementation of educational policies at the regional level, education does not always reach the broader, more localised, community. The research showed that locals who benefit the most from educational initiatives are those with the financial capital to be able to afford it. The scarcity of economic resources to enable access to tertiary education is threatening community opportunities to access the educational system. In addition, current vocational community capacity-building initiatives are mainly mining-orientated, particularly in the Antioquia case, which is becoming a limitation for locals whose livelihood options and development aspirations are not always directly linked to the mining industry. Thus, it is recommended that education, particularly at the tertiary level, be secured for locals in non-mining subjects. These need to be subjects that will enhance their ability to take up opportunities that are offered outside the mining sector.

The dearth of both gainful employment options and appropriate education is hindering locals from gaining relevant work experience. This is leading to diminished opportunities for income generation and preventing them from employing other human capital assets that they may have access to. Research shows that issues relating to work experience arose from the Colombian community capacity-building approach and the educational system itself. A more effective educational system implemented to overcome the obstacles of limited prior learning or prior experience faced by community members is needed. Giving credits for work experience is essential. Tertiary education institutions need to embrace such education approaches to help vulnerable communities from resource regions achieve their development aspirations and reduce poverty through community capacity-building in the form of education. Therefore, a review of existing community capacity-building approaches needs to be carried out by national level organisations. The review should be carried out with input from both companies and the community. It also should be followed by targeted capacity-building at the government level, so that government organisations can effectively implement community capacity-building approaches and help other stakeholders like companies to get actively involved in its implementation.

Stakeholders have faced serious challenges in community capacity-building implementation due to infrastructure issues. Resource allocation for infrastructure for community capacity-building as well as stakeholders' responsibilities in this domain remains unclear for some parties. Evidence shows that stakeholders are often unaware of their potential roles and even of their legal responsibilities for infrastructure provision. It is recommended that communities play an active role in advising on the allocation of resources. It is equally important that both governments and companies further engage and commit to the delivery of infrastructure for community capacity-building actions, especially where these have been identified by communities as being of high priority, thereby increasing community opportunities to meaningfully benefit from these initiatives.

2.4 Impact Sustainability: Final Remarks

The strengthening or misuse of capacities or assets might bring either positive or adverse livelihood opportunities or outcomes for communities. For instance, an increase in economic activities can foster labour market opportunities. If communities are well equipped in terms of capacities or assets, they can gain valuable benefits resulting in significant livelihood outcomes. These outcomes derive from the combination of livelihood opportunities and community assets. Therefore, if a community's capacities match gainful livelihood opportunities, they might be able to increase income, improve their existing conditions and reduce poverty.

Enhancing community assets will bring positive livelihood outcomes for local communities. In doing so and following the SLF, it is necessary to identify and use clear links between livelihood strategies and the community assets they utilise. Economic diversification was recommended as a combination livelihood strategy. However, such a strategy needs to be accompanied by labour market opportunities to enhance other assets, such as employment and income, which will improve the community's existing conditions. If Colombian communities (and those adjacent to mining and exploration projects in Latin America and elsewhere) are well equipped in terms of assets (education, employment, income, work experience and infrastructure), they are more likely to achieve livelihood outcomes that are sustainable for the longer term and so reduce poverty across the region.

As the sustainable livelihood framework – a modified version of which was the governing framework of the research here reported – is widely used across the developing world, the sustainability recommendations given in this chapter are likely to help in achieving the aims of SDG 1 No Poverty in other developing locations and elsewhere.

References

Cahn M (2006) Sustainable rural livelihoods, micro-enterprise and culture in the Pacific Islands: case studies from Samoa. PhD dissertation, Development Studies, Massey University, New Zealand. Retrieved 22 January, 2015, from http://muir.massey.ac.nz/bitstream/handle/10179/1532/02_whole.pdf

Cámara de Comercio de Medellín para Antioquia (2010) Minería: Potencial para Iniciativas Cluster en Antioquia Documento Comunidad. Cámara de Comercio de Medellín para Antioquia, Medellín

Cardenas M (2011) Población Guajira, Pobreza, Desarrollo Humano y Oportunidades Humanas para los Niños en La Guajira. Master dissertation, Universidad Nacional de Colombia. Retrieved from http://www.bdigital.unal.edu.co/3573/1/Tesis__Mauricio__Cardenas.pdf

Carney D (2003) Sustainable livelihoods approaches. Retrieved 25 March, 2014, from http://www.eldis.org/vfile/upload/1/document/0812/SLA_Progress.pdf

DANE (Departamento Administrativo Nacional de Estadísticas) (2005) Perfil Risaralda. DANE, Bogota

DFID (Department for International Development) (1999) Sustainable livelihoods guidance sheets. DFID, London

Franco I (2014) Building sustainable communities: enhancing human capital in resource regions. PhD Dissertation, The University of Queensland, Brisbane

Franco BI, Robertson S (2014) Mine life cycle planning – creating lasting value for communities. Paper presented at the conference life-of-mine 2014, Brisbane

Freeman RE (1984) Strategic management: a stakeholder approach. Cambridge University Press, New York

Hitch M, Ravichandran AK, Mishra V (2014) The real options approach to implementing corporate social responsibility polices at different stages of the mining process. Corp Gov 14(1):45–57

Hostetler M (2006) Enhancing local livelihood options: capacity development and participatory project monitoring in Caribbean Nicaragua. PhD thesis, Graduate Program in Geography, York University, Toronto

Idarraga A, Munoz A, Vélez H (2010) Conflictos Socio-Ambientales por la Extracción Minera en Colombia: Casos de la Inversión Británica. Censat Agua Viva, Bogota

Mandke PR (2007) Understanding the linkages between tourism and urban poverty reduction using a sustainable livelihoods framework. PhD dissertation, The University of Queensland, Brisbane

Mazibuko S (2013) Understanding underdevelopment through the sustainable livelihoods approach. Community Dev 44(2):173–187

Minnery J (2007) Stars and their supporting cast: state, market and community as actors in urban governance. Urban Policy Res 25(3):325–345

Mitlin D (2003) Addressing urban poverty through strengthening assets. Habitat Int 27:393–406

Rakodi C, Lloyd-Jones T (2002) Urban livelihoods: a people centred approach to reducing poverty. Earthscan Publications, London

Sen A (1979) Equality of what? Stanford University, Oxford

Shen F, Hughey KFD, Simmons DG (2003) Connecting the sustainable livelihoods approach and tourism: a review of the literature. J Tour Hosp Manag 15:19–31

Sistema de Información Minero Energético Colombiano (2010) Distritos Mineros. Retrieved March 26, 2014, from http://190.90.10.157/Distritos%20Mineros/

Stead D (2015) What does the quality of governance imply for urban prosperity? Habitat Int 45:64–69

Surborg B (2012) The production of the world city: extractive industries in a global urban economy. PhD dissertation, The University of British Columbia. Retrieved from https://circle.ubc.ca/bitstream/handle/2429/40719/ubc_2012_spring_surborg_bjoern.pdf?sequence=1

Torres I (2001) The mineral industry of Colombia. Retrieved March 26, 2014, from http://minerals.usgs.gov/minerals/pubs/country/2001/comyb01r.pdf

UPME (Unidad de Planeación Minero Energética) (2006) Colombia País Minero 2019. Unidad de Planeación Minero Energética, Bogota

Vilora de la Hoz J (2009) Cerro Matoso y la Economía del Ferroníquel en el Alto San Jorge (Córdoba). Banco de la Republica: Documentos de Trabajo sobre Economía Regional 119:41–58

Warhust A (2001) Corporate citizenship and corporate social investment. Corp Citiz 28:57–73

Chapter 3
SDG 2 Zero Hunger

Organic Waste-to-Resource Compost Program Development: Cultivating Circular Sustainable Systems

Emily F. Creegan and Robert Flynn

Abstract The backbone of society and the basis of self-sufficiency is the agricultural advancement of food, fiber, fuel, and industry. However, in many non-industrialized countries, self-sufficiency has been eroded by several factors, including environmental impacts, lack of educational infrastructure, and inequities in resource management and distribution. Additionally, with the rise in human populations and climate pressures, the need to increase food production security and water conservation measures is imminent. By emulating the productivity of natural ecosystems and returning carbon-based materials to the soil, agricultural production may be enhanced with a reduced reliance on potentially water-polluting and often prohibitively expensive synthetic fertilizers. Through various case studies showcasing the effectiveness of educational tools, we work to demonstrate the need for more financial investment in biomass utilization program development in order to cultivate emerging circular economies. Furthermore, we highlight the importance of carbon and ecosystem cycling curriculum in school settings. In situ soil application of organic materials engenders soil carbon sequestration – a climate change mitigation mechanism – and overall soil health and soil water conservation. Compost processing and program development ameliorates potential organic materials pathogen transference while mineralizing nutrients for plant uptake.

Keywords Education · Biomass utilization · Compost · Food security · Nutrition · SDG 2 Zero Hunger

---

E. F. Creegan (✉)
Department of Plant & Environmental Sciences, New Mexico State University (NMSU), New Mexico, NM, USA
e-mail: ecreegan@nmsu.edu

R. Flynn (✉)
Extension Plant Sciences Department, New Mexico State University (NMSU), New Mexico, NM, USA
e-mail: rflynn@nmsu.edu

3.1 Introduction

A revival of nurturing a full cycle organic waste-to-resource system, with educator, community, nonprofit, public and private entities, and farmer collaborations, is crucial to sustainable community and agricultural development. However, all stakeholders should be sensitive to the desires of the local people, their culture, and society and cognizant of available and applicable resources and technologies. Furthermore, academics should communicate with the public and extend their knowledge through outreach services. Based on various case studies showcasing the effectiveness of curriculum and educational tools, more financial investment in biomass utilization education and carbon source availability is needed in developing circular economies. The United Nations Sustainable Development Goal (SDG) 2 encompasses the need to "end hunger, achieve food security and improved nutrition and promote sustainable agriculture." Additionally, both the UN Millennium Development Goals (MDGs) and SDG 6 highlight the need to develop more holistic waste management and water conservation measures. These goals can be fostered by both improved soil organic matter (SOM) practices and effective compost program development.

Compost processing is the controlled decomposition of organic materials; by harnessing naturally occurring bacteria, thermophilic or "hot compost" reduces organic materials pathogen potential. Compost processing of locally produced organic waste materials provides safe and optimal handling and processing of organic waste streams. Compost case studies, outlining substrate and system feasibility analyses, highlight several interrelated food-security and environmental benefits.

Living or once living materials, known as *organic matter*, *organic materials*, or *biomass*, serve as a nexus for nutrient cycling, water conservation, moderation of climate change, and food production. Human excrement, animal manures, and food waste pose deleterious effects to watersheds and overall community health, particularly in many non-industrialized regions where waste management and sewage treatment infrastructure may be lacking. The utilization of organic waste materials via compost processing has been shown to enhance soil and plant productivity, increase soil water retention, sequester carbon, and decrease external synthetic fertilizer and chemical inputs. This chapter provides an educational model for organic waste-to-resource initiatives associated with food production for long-term sustainability in alignment with SDG 2 Zero Hunger.

3.2 Literature Review

3.2.1 Agricultural Nutrient Sources

The soil naturally cycles and recycles organic materials in the form of leaves, twigs, microorganisms, bird droppings, and other biota. The soil encompasses biological, physical, and chemical components. Oxygen fills soil pore spaces among soil particles which provide oxygen to soil microorganisms and plant roots. Soil pore spaces also contain water and decomposed soil organic matter adheres water molecules, providing a water source to soil microorganisms that facilitate organic matter decomposition, and nutrient and water availability for plant production. Around the world, the soil plays vast and critically dynamic roles, including food, fuel, and fiber production; water and air, contaminant filtering and conversion; a complex and diverse biological habitat; and a building and structure platform (Brady and Weil 2008).

Societal organic waste materials include food, landscape, animal manure, and biosolids, which are comprised of many macro- and micronutrient sources. The natural microbial cycling of soil organic matter improves nitrogen levels in soil, which is often the limiting factor in non-legume plant systems. Furthermore, cycling of organic material can also effectively decrease soil nitrogen leaching potential (Hepperly et al. 2009). However, unlike carbon, nitrogen is not easily accessible from the atmosphere by plants; consequently, nitrogen is synthetically produced in mass quantities (Finn et al. 2015; Fortier et al. 2010). Carbon and nitrogen are vital components of soil microbial functions and associated plant nutrient availabilities (Albaladejo et al. 2008). Increasing soil organic material has been shown to increase soil biota, facilitating plant nutrient uptake and more diverse and resilient soil and plant systems (Brady and Weil 2008).

In many industrialized countries, macronutrients are often supplied to the crop via synthetic fertilizer application (Creegan 2017). Synthetic fertilizers typically provide a more immediate nutrient source to the plant (Kallestad et al. 2008). However, synthetic nitrogen fertilizers have been shown to potentially leach out of the soil system (Omer et al. 2018). Although synthetic fertilizers supply essential nutrients to the crop, synthetic fertilizer use can pose deleterious effects to aquatic ecosystems (Creegan 2017). Eutrophication of surface water is due to the prolific growth of algal blooms following increased levels of aquatic nutrient source from landscape and agricultural runoff, and other sources (Sharpley et al. 2003). Algal blooms deplete the water of oxygen, often resulting in high rates of aquatic life mortality. Globally, the high-volume use of synthetic fertilizers may have potentially harmful effects and may not be sustainable or economically feasible in the long-term for non-industrialized countries.

3.2.2 Environmental Factors and Food Production as Related to Organic Materials

In 2015, organic waste, including paper and paperboard, yard trimmings, and food products, accounted for the largest single waste stream – over 50% of all thrown away waste – in the USA (US Environmental Protection Agency 2018). Organic waste materials are often landfilled or incinerated, posing harmful effects to local and global watersheds and air quality. The phrase "organic waste-to-resource" challenges the predilections of today's world for landfilling and incinerating "waste" by cultivating a more approach. All ecosystems utilize materials; there is no waste in nature. Mimicking the natural environment and the regenerative processes of the planet, organic materials can be harnessed to foster soil regeneration, promoting sustainable agricultural practices and increased crop productivity (Rosenani et al. 2016). Organic material and healthy soils play a critical role in soil water retention and both environmental and economic stability, particularly during drought and in productive arid regions (Lepsch et al. 2019). Employment of soil organic matter (SOM) can facilitate production and other associated benefits for plants, including tree, vegetable crops, and potted plants (Brady and Weil 2008).

Organic materials (OM) utilization often promotes greater soil microbiota and more resilient soil and plant systems (El-Gawad 2008). Healthier plants are more resistant to pest invasion, promoting less use and reliance on pesticides and herbicides. Being less reliant on external inputs may have economic benefits and promotes localized community self-resiliency. An important part of this resiliency and renewal process is safely returning nutrients to the source (soil) in the form of compost, as a cycle of continuous regeneration (Hepperly et al. 2009).

All life is made of carbon components. Fossil fuels are derived from highly decayed carbon-based material. When burned for fuel, carbon is oxidized to create carbon dioxide – the most prolific greenhouse gas on Earth. Coastal, island, and non-industrialized countries are most vulnerable to the effects of climate change. A climate change mitigation strategy is to sequester carbon in the soil by returning carbon-based organic materials to the source. With this, a new way of "carbon-farming" practices can be cultivated (Velasquez-Manoff 2018).

Global methane is approximately 23 times more potent greenhouse gas than carbon dioxide. Methane's annual contribution to climate change causing gases is approximately 1/3 of carbon dioxide emissions (Henson 2014, 32). Methane gas is produced by the anaerobic decomposition of organic materials and comes from various sources, including landfills. Compost program development requires the local utilization of heavy organic waste materials, decreasing both financial and environmental costs associated with waste transportation.

3.3 Methodology

3.3.1 Case Study 1: On-Farm Organic Waste-to-Resource

The top two pecan growing regions in the USA are currently New Mexico and Georgia. Increases in the cost of synthetic fertilizers have dramatically diminished the profitability of growing pecans (Huang 2009; Wells 2012). Additionally, as the pruning practice of pecan trees has become more common, the amount of carbon-based organic waste materials has increased in the form of woody biomass. Historically, the woody biomass was burned as the primary form of waste management; however shredding machinery has enabled many farmers to utilize and recycle the wood biomass back into their orchards. Wood shreddings are often combined with manure, field green waste, food waste, or composted and incorporated into the soil.

A 4-ha block from a 647-ha pecan farm in Roswell, New Mexico, USA, was selected to evaluate the effects of soil water, soil carbon sequestration, and nutrient availabilities on soil properties from co-composted tree trimmings and locally derived manure, along with direct in pecan field soil in situ applications. The farm's pecan biomass (a carbon source) and local dairy manure (a nitrogen source) were incorporated in an outdoor aerobic windrow composting system. A tractor pulls a compost turner which turns (aerates) and waters the shredded pecan biomass and manure, facilitating microbial decomposition (Fig. 3.1). Windrow (a row of organic material) temperatures, carbon dioxide (as a measure of microbial respiration rates), and moisture levels were consistently monitored throughout the composting process. The finished compost products were then incorporated into commercial pecan orchard soil. An economic assessment and feasibility of the on-farm compost program and finished compost product application is being assessed.

Fig. 3.1 Windrow turning and watering

3.3.1.1 The Benefits of Composting: Safely Utilizing Organic Materials

Composting employs microorganisms naturally present in OM to decompose and mineralize organic materials. Finished compost is recognized as a soil conditioner that provides an organic nutrient source for soil microbial populations that assist with plant productivity (Brinton et al. 2012). Compost is comprised of decomposed material containing primary, secondary, and micro nutrients (Egrinya et al. 2008). Proper composting techniques are dependent upon many factors, including OM substrate, substrate carbon to nitrogen ratios, oxygen and moisture levels, surface area of the OM, and the form and duration of compost processing (UMass Amherst 2018). Shredding OM substrates increases the available surface area for improved microbial decomposition.

Microorganisms are naturally present in all life; the human body, for example, is abundant with living bacteria. Diverse soil microbiology facilitates plant nutrient uptake and associated plant growth parameters (Molina-Romero et al. 2017). Plant matter and human and animal excrement is also naturally rich with microorganisms. The compost process mimics most natural ecosystems and requires greater carbon-based materials to lesser nitrogen-based materials for proper microbial decomposition rates (an approximate 30:1 carbon-nitrogen ratio). Examples of common carbon-based materials include dried leaves, shredded paper and newspaper, paper towels, and agricultural organic waste materials such as sugarcane bagasse. More common nitrogen-based materials include "fresh" green grass, animal and human excrement, and most fresh plant detritus, including food waste.

The more prevalent forms of compost processing include thermophilic/"hot"/aerobic composting, anaerobic (without oxygen) composting, enclosed aerated systems (which often encompasses ventilation and/or forced aeration systems), and vermicomposting (the employment of worms, primarily red wigglers "*Eisenia fetida*" as the primary OM decomposers). Simplified compost processes with minimal mechanization that may be more suitable for non-industrialized countries include smaller-scale boxed compost systems and/or "covered" systems that encase nitrogen-based organic materials with a locally available shredded carbon source.

Thermophilic ("heat loving") bacteria proliferate in properly established and maintained compost systems. Effective aerobic composting produces high temperature pathogen and weed seed killing heat created by organic exudates and action from the rapid microbial decomposition of OM. The finished compost product is a stable, humus-like material that can be safely handled, stockpiled, and utilized. Conversely, simply applying organic materials to a soil, such as animal manures, will not ameliorate potential pathogen loads or mitigate weed seed transference. To be noted, although the finished compost product typically contains a multitude of nutrients, it is not considered a fertilizer as the exact nutrient proportions cannot be consistently determined.

Composting can be accomplished in small- or large-scale systems; however a minimum of a cubic yard of initial organic substrate material is recommended for thermophilic processing. Larger on-farm composting or compost facilities, found in several municipalities in California (USA) and elsewhere, have been established throughout the world. To ensure proper compost processing and reduction of pathogen potential, the California Integrated Waste Management Act of 1989 has decreed a minimum threshold compost processing temperature of 3 consecutive days of 55 degrees Celsius or higher.

3.3.1.2 Aerobic (Hot) Composting

The materials below, produced by the New Mexico State University (NMSU, Las Cruces, New Mexico, USA) Skeen Compost Club, highlight the more common forms of smaller-scale compost processing, compost utilization, and "troubleshooting" tips. Other Guides from NMSU include Backyard Composting Guide H-110 by John Allen and Circular CR637- Managing Organic Matter in Farm and Garden Soils by Robert Flynn and John Idowu.

Aerobic composting is the controlled decomposition of organic materials typically by bacteria. For small-scale compost systems, basic recipes include incorporating (by turning with a shovel) nitrogen-based non-meat/non-dairy kitchen food scraps or fresh green landscaping material in addition to a shredded carbon source such as shredded newspaper, torn up paper egg cartons, etc. The material in your bin should always be about as moist as a wrung-out sponge (add water to your compost bin as needed).

Tip: Shoot for 30:1 carbon to nitrogen (incorporate much more carbon-based material than nitrogen)

3.3.1.3 Vermicompost

Vermicomposting is a method of composting that utilizes worms (specifically red wigglers) as the primary decomposers. Worm castings (poop!) are an excellent plant food and fertilizer for indoor/outdoor plants and landscaping.

3.3.1.4 How to Maintain a Vermicompost System

1. Place the bin in a shady area (not too hot and not too cold, e.g., the bin can be kept under the kitchen sink; the worms won't leave the bin unless they are not getting something they need).
2. Feed them a Vegan diet (non-dairy/non-meat kitchen scraps, including coffee grounds, paper towels, tea bags – less any metal, fruit and vegetables scraps, etc.).
3. Ensure the worms have access to sufficient air (check the bin has enough holes for air circulation).

3.3.1.5 Harvesting and Using Compost

Harvest the finished compost by using a shovel. Harvest worm castings by hand separating the worms from their castings. Incorporate all finished compost into landscaping or garden spaces by top dressing and watering in or shovel-incorporating into the soil.

3.3.1.6 Compost Bin Troubleshooting Tips

Issue	Solution
Malodorous bin or abundant flies	Cut down the amount of nitrogen-based organic material the worms are receiving, increase the organic materials surface area by shredding or chopping the material, and/or add more shredded carbon-based materials such as paper towels
Worms exiting the bin	The bin is either too hot or too cold, or alternatively the worms too much or too little food

3.3.2 Case Study 2: A City Model

The City of San Francisco (California, USA) boasts an approximate 80% materials diversion rate from landfills. In 2017, approximately 3 million tons of waste was produced in the City of San Francisco, and approximately 2.5 million tons was diverted from landfills (this material was either recycled or composted). Much of the City's success can be attributed to a government assembly bill that mandated promoting waste diversion. This is a fee-based program with applied enforcement. Of these applied mandates, This includes the "door-to-door" outreach program in multiple languages to all city sectors and educational initiatives, bin placement, and resident access of more than 99%" of the waste diversion. The "Fantastic Three" three bin (recycling, composting, landfilling) system for residential and waste collection was initiated in 1999. The green-colored bin is the organic materials collection bin; the blue-colored bin is the recycling collection bin; and the smallest of the three bins, the black-colored bin, is the landfill bin.

SF Environment School Education Program (SFESEP): A department of the City and County of San Francisco credits a large part of their diversion rate success to widespread education initiatives, including citywide public and private kindergarten-12th grade curriculum and educational signage. The SFESEP reaches approximately 20,000 students annually throughout San Francisco and provides online composting curriculum. The no-cost curriculum can be found on this site: https://sfenvironment.org/k-12.

SF Environment also provides a no-cost customizable online "Signmaker Tool" in multiple languages; specialized (specific to materials generated in a given region) recycling, composting, and landfilling, sign templates are available on this website: https://sfenvironment.org/recyclingcomposting-and-landfill-signs.

3.3.3 Case Study 3: A University Model

The New Mexico State University (NMSU, Las Cruces New Mexico, USA) Skeen Compost Club (SCC) compost program and associated organic waste-diversion educational materials was developed in 2017 by Dr. Ivette Guzman and the author and several past professors and current students.

SCC developed educational signage in the department kitchen that demonstrated what can be placed in the organic materials collection bins, including paper towels, grains, breads, coffee grounds, tea bags, and vegetable and fruit scraps. As is common in smaller-scale compost systems, this no-meat no-dairy system was established to deter pathogen potential and potential compost processing vectors (i.e., rats, mice, and insects). Additional educational materials were created for the department green house organic waste collection bins. Approximately 112 kg of organic waste was collected and diverted from the local landfill during the in-school months of September, October, and November in 2017, alone.

3.3.3.1 Small-Scale Waste Audit

Waste audits are important in determining total approximate waste generation and volume estimates and can help determine the needed compost system size. Smaller-scale waste audits can be easily conducted with the following steps:

1. Lay down three small-medium size tarps to separate organics, recyclables, and landfill materials.
2. As a safety precaution, wear thick gloves, closed toed shoes, and long pants while conducting the audit.
3. Separate the source material (from the waste bins), and weigh all materials on at least 2 separate days to get a relative volume percentile for the organics, recyclables and landfill materials.

3.3.3.2 SCC Small-Scale Compost Program Development Learning Lessons

A 1-gallon organic materials collection bin system was initiated in the department kitchen area; however, patrons were often continuing to throw-away coffee grounds and paper towels, in particular. Two other 1-gallon organic materials collection bins with labeled "coffee grounds for compost" and "used paper towels – to be composted" were placed in strategic locations where patron bin-use was improved. This simple step increased the overall organic materials diversion rates.

Collected organic materials bins are weighed, and the organic material is placed in an outdoor compost processing bin. The outdoor bin was constructed from recycled lumber and corrugated tin tops and metal handles. The compost system temperatures are monitored, the organic material is aerated, moisture levels are maintained by incorporating water into the system, and the substrate surface area is increased by chopping the material with a shovel.

3.3.4 Case Study 4: For-Profit Community Collaborations Models

Sodexo Inc. is a Fortune 500 global food vendor company and provides the food vending services for Eastern New Mexico University (Portales, New Mexico, USA). Sodexo's Better Tomorrow Plan is a "year-round commitment to the environment, including waste reduction, water efficiency, energy efficiency and sustainable sourcing." A partnership has been established between Sodexo and a local pecan farmer in creating a pre-consumer organic-waste-to-farm-resource compost program. The University kitchen staff place all pre-consumer kitchen organic waste materials into strategically located and labeled 5-gallon buckets with secure lids. The full buckets are kept in the kitchen freezer and then transported to the farm for windrow compost processing.

In 2016, Sodexo announced a commitment to "Zero Food Waste to Landfills," in collaboration with other businesses, government, and nonprofit organizations, in reducing food waste in the USA by 50% in year 2030 (established by the Obama Administration in 2015). Sodexo cites "freeing up landfill space" and "reducing ozone-depleting methane gas emissions" as beneficial effects of reducing food waste in landfills.

In an effort to minimize landfilled waste many restaurants are converting from non-compostable to compostable bowls, plates, and other utensil-ware. Rubio's Coastal Grill has converted to offering compostable ware for much of its take-out products, as a part of Rubio's brand ethos and strategy. Compostable and biodegradable to-go containers that are made from molded fibers prevent more than approximately 300 tons of their past alternative Styrofoam option from being landfilled, yearly. In collaboration with compost program development companies, compost facilities and educational initiatives, this compostable ware organic material can be captured, diverted from the landfills, and utilized in a local community.

3.3.5 Case Study 5: GiveLove – Going Global with Community Sanitation

GiveLove (www.givelove.org) is a skills training nonprofit organization committed to the teaching and promotion of ecological sanitation *(EcoSan)/humanure/biosolid* (human excrement) compost program development. Much of the information cited here was provided by Alisa Keesey, GiveLove Program Director, and Samuel Souza, GiveLove Research and Training Coordinator (Keesey and Souza 2018).

As stated in the 2015 UN Sustainable Development Goals (SDGs), the achievement of total sanitation is to develop solid waste containment, disposal, and reuse alternatives. GiveLove develops container/bucket-based compost sanitation toilet systems in communities that typically practice open defecation or the use of pit latrines. GiveLove works with the given community in identifying ongoing, easily accessible and locally available shredded carbon sources as the carbon component and bulking agent of the compost system. GiveLove collaborates with government entities, communities, schools, and other nonprofit organizations to provide simple, low-cost, extremely low technology, non-urine diverting compost toilets and on-site compost processing (Keesey and Souza 2018).

GiveLove initiated the *Green School* model with the objective of developing school-based compost toilets and hygiene programs as an entry point of more expansive community humanure compost program development. GiveLove also provides training on using the finished compost products to grow vegetables, trees, and medicinal plants for consumption and sale - working to foster more resilient and self-sufficient circular economy communities (Keesey and Souza 2018).

At schools, students, teachers, and administrators must learn how to use and manage the compost toilets. With this, everyone learns the science of composting as GiveLove works with all participants in on-site compost system establishment. Projects are introduced step-by-step and initially focused on small pilot programs. GiveLove notes widespread skepticism at the onset of the compost toilet introduction and application of the finished compost product, but full adoption is often quickly achieved once users experience the benefits of this full cycle biosolid waste management alternative (Keesey and Souza 2018).

The compost systems are a connecting tool for sustainable community development and a learning-by-doing model in addressing community health impacts and environmental challenges. The programs are designed for illiterate or low-literacy people and for community information exchange and education dissemination. The organization encourages hands-on, peer-to-peer learning and ongoing problem-solving. With pathogen killing effective humanure compost program development, the individual becomes an empowered agent of change, cultivating water purifying, healthy soil food producing, climate change mitigating, self-reliant and resilient communities" (Keesey and Souza 2018).

3.3.6 Case Study 6: The Need and Potential for Ghana

Ghana is a coastal nation in Central Africa, with a population of approximately 27 million, high poverty rates, and relatively low gross domestic product (GDP). With a large portion of the population in the agricultural sector and limited agricultural technologies and resources, the country is particularly vulnerable to climate change impacts. Conventional agricultural practices, including the use of heavy machinery, prolific synthetic fertilizer application, and lack of soil organic matter practices, exacerbate climate change effects. Compost program development and finished compost product application could facilitate climate change mitigation and resiliency measures.

Ghana is diverse geographically and ecologically. Much of the larger flora has been cleared for agricultural or construction purposes and has largely not been revegetated. Agricultural crops, including cocoa, yams, grains, timber, nuts, and palm oil, constitute the base of agriculture of Ghana's economy. In 2013, agriculture employed approximately half of the total labor force in Ghana (*Agriculture in Ghana: Facts and Figures* 2011). A Danish government *National Climate Change Adaptation Strategy (2016)* reports that while many governments are taking initiative to reduce the effects of climate change, the extreme challenges faced by the more highly vulnerable populations are largely not being addressed (CC DARE 2016). The report cites the need for governments to be more proactive in their resiliency strategies rather than reactive, citing the former as far less costly.

With the growing Ghanaian population and continued food and environmental insecurities and pressures, soil health and soil health resiliency strategies in relation to crop productivity are paramount. Remediating and cultivating healthy soil systems begin with the renewal and regeneration of soil biomass, and the utilization of locally sourced organic materials. This begins with available organic waste materials assessment and identification. However, according to one article, reliable national data on waste generation that will guide effective waste management, policy, and sustainable practices is "absent" in Ghana (Miezah et al. 2015). The research comprised a national analysis of the rate of waste generated, physical composition of the waste, and the sorting and separation efficiency of the waste. The results showed that as of year 2015, the rate of waste generated in Ghana was approximately 0.47 kg/person/day, equating to 12,710 tons of waste per day per the current national population. Furthermore, the study identified a significantly larger percentage of the total waste as biodegradable - approximately 61% of the total waste stream (Miezah et al. 2015). Identification and subsequent composting of locally sourced carbon-based materials, initially human and animal manures that pose deleterious impacts to watersheds, warrant attention.

As of year 2014, in two of the largest Ghanaian cities, over 4000 tons of solid waste was generated per day (Monney 2014). According to Monney, this is partly due to an "erroneous perception about solid waste" and the Ghanaian people generally viewing waste as disposable, instead of a valued potential resource (Monney 2014). The article also cites other countries, such as the Philippines and Sweden, "making money" out of solid waste. Monney (2014) cites the overall lack in waste management and materials utilization infrastructure, emphasizing the need to focus on "the control of generation, storage, collection, transfer and transport, processing, and disposal of solid wastes in a manner that conforms to the best principles of public health, economics, engineering, conservation, aesthetics, and other environmental considerations" (Monney 2014).

Additionally, there may be a "huge" gap in Ghana between research and policy (Monney 2014). Academia should be communicating with the public and offering extension/outreach services in improved and sustainable waste management and agriculture techniques. Academic communication with policy makers in developing climate change resiliency goals and regulations is also lacking. Monney (2014) states the need for initial seed funding and entrepreneurial leaders to develop waste-to-resource programs and community member biomass recycling education, biodegradable products, facilities infrastructure, and transportation of community-produced organic materials to the compost facility or farm. Developing these models to increase soil carbon sequestration will increase soil tilth and associated crop productivity, creating more resilient, self-sufficient, and climatically and financially stable systems (Agriculture in Ghana: *Facts and Figures* 2011).

3.4 Impact Sustainability: Final Remarks

3.4.1 *Fostering Circular Sustainable Economies*

Closing the carbon loop, organic materials serve as a connecting link between food production, recycling and waste management, water conservation, climate change resiliency, self-sufficiency, and sustainability. However, in many non-industrialized countries, self-sufficiency has been eroded by several factors, including environmental impacts, lack of education and dissemination, and globalization. Additionally, with the continued exponential rise in human populations, the need to increase food production and water conservation measures is imminent.

By imitating the productivity of natural ecosystems and returning carbon-based materials to the soil source, biomass utilization promotes a reduced reliance on potentially water-polluting and costly synthetic fertilizers. There is a need for global biomass waste volume assessments and associated waste audits. Proper compost processing provides a safe means to utilize organic waste materials, fostering sustainable landscape and agricultural practices and circular economies. Developing educational tools and models and requiring carbon and ecosystem cycling curriculum in public and private school settings are necessary components of effective compost program development. Although advanced machinery and technologies may be more readily available in industrialized countries, compost program development employing locally available organic materials can and must translate to non-industrialized countries.

A revival of nurturing a full cycle organic waste-to-resource system, with community, academia, public and private entities, and farmer collaborations, is crucial in sustainable agricultural development. However, these measures and goals should be culturally sensitive and predicated on available resources and technologies. As emphasized in the Ghanaian case study, sound research must support policy development. Furthermore, academics should be communicating with the public and offering extension and outreach services. Based on various case studies showcasing the effectiveness of curriculum and educational tools, more financial investment in biomass utilization education and carbon source availabilities is needed in cultivating emerging circular, rather than linear, economies.

The United Nations Sustainable Development Goal (SDG) 2 encompasses the need to "end hunger, achieve food security and improved nutrition and promote sustainable agriculture." Interrelated, the UN Millennium Development Goals (MDGs) and SDG 6 highlight the need to develop more holistic waste management and water conservation measures. Compost program development and effective humanure program development, in particular, foster this.

Acknowledgements This research is funded by the USDA National Needs Fellow program, Grant: #2015-38420-23706. Thank you to Bruce Haley and staff.

References

Agriculture in Ghana: facts and figures (2011) Statistics, Research and Information Directorate, Ministry of Food and Agriculture, May 2011. Retrieved from: http://mofa.gov.gh/site/wp-content/uploads/2011/10/AGRICULTURE-IN-GHANA-FF-2010.pdf

Albaladejo J, Lopez J, Boix-Fayos C, Barbera GG, Martinez-Mena M (2008) Long-term effect of a single application of organic refuse on carbon sequestration and soil physical properties. J Environ Qual 37:2093–2099

Brady NC, Weil RR (2008) The nature and properties of soils, 14th edn. Pearson Prentice Hall, Upper Saddle River

Brinton WF, Bonhotal J, Fiesinger T (2012) Compost sampling for nutrient and quality parameters: variability of sampler, timing and pile depth. Compost Sci Util 20(3):141–149

CC DARE (2016) Climate change and development – adapting by reducing vulnerability. A Joint UNEP/UMDP Programme for Sub-Saharan Africa Funded by the Danish Ministry of Foreign Affairs: Ghana National Climate Change Adaptation Strategy. Retrieved from: http://adaptation-undp.org/sites/default/files/downloads/ghana_national_climate_change_adaptation_strategy_nccas.pdf

Creegan E (2017) Effects of compost and biochar application on turf soil nitrate and carbon content under water deficit conditions. Unpublished Master's thesis, California State Polytechnic University, Pomona, California. Retrieved from: Cal Poly Pomona University Library Bronco Scholar database: http://broncoscholar.library.cpp.edu/discover?query=Emily+Creegan&submit=Go&scope=

Egrinya A, Inanaga S, Li X, An P, Li J, Duan L, Li Z (2008) Effectiveness of mulching vs. incorporation of composted cattle manure in soil water conservation for wheat based on ecophysiological parameters. J Agron Crop Sci 194(1):26–33

El-Gawad AA (2008) Employment of bioorganic agriculture technology for Zea mays cultivation in some desert soils. Res J Agric Biol Sci 4(5):553–565

Finn D, Page K, Catton K, Strounina E, Kienzle M, Robertson F, Armstrong R, Dalal R (2015) Effect of added nitrogen on plant litter decomposition depends on initial soil carbon and nitrogen stoichiometry. Soil Biol Biochem 91:160–168

Fortier J, Gagnon D, Truax B, Lambert F (2010) Nutrient accumulation and carbon sequestration in 6-year old hybrid poplars in multiclonal agricultural riparian buffer strips. Agric Ecosyst Environ 137:276–287

Henson R (2014) The thinking person's guide to climate change. American Meteorological Society, Boston

Hepperly P, Lotter D, Ulsh CZ, Seidel R, Reider C (2009) Compost, manure and synthetic fertilizer influences crop yields, soil properties, nitrate leaching and crop nutrient content. Compost Sci Util 17(2):117–126

Huang W (2009) Factors contributing to the recent increase in U.S. fertilizer prices, 2002–2008. USDA, Washington, DC

Kallestad JC, Mexal JG, Sammi TW (2008) Mesilla Valley pecan orchard pruning residues: biomass estimates and value-added opportunities research report 764

Keesey A, Souza S (2018) Container-based sanitation as a low-cost solution in high-need areas: a case study review of community-based compost sanitation in six countries. Dry Toilet Conference full paper. Partially. Retrieved from: www.givelove.org

Lepsch H, Brown P, Peterson C, Gaudin A, Khalsa SD (2019) Impact of organic matter amendments on soil and tree water status in a California orchard. Agric Water Manag 222:204–212

Miezah K, Obiri-Danso K, Kádár Z, Fei-Baffoe B, Mensah MY (2015) Municipal solid waste characterization and quantification as a measure towards effective waste management in Ghana. Waste Manag 46:15–27. https://doi.org/10.1016/j.wasman.2015.09.009

Molina-Romero D, Baez A, Quintero-Hernández V, Castañeda-Lucio M, Fuentes-Ramírez LE, Bustillos-Cristales MR et al (2017) Compatible bacterial mixture, tolerant to desiccation, improves maize plant growth. PLoS One 12(11):1–21. https://doi.org/10.1371/journal.pone.0187913

Monney, I (2014) Ghana's solid waste management problems: the contributing factors and the way forward. Modern Ghana, May 27. Retrieved from: https://www.modernghana.com/news/544185/1/ghanas-solid-waste-management-problems-the-contrib.html

Omer M, Idowu OJ, Ulery AL, Van Leeuwen D, Guldan SJ (2018) Seasonal changes of soil quality indicators in selected arid cropping systems. Agriculture 8(8):1–12

Rosenani AB, Rovica R, Cheah PM, Lim CT (2016) Growth performance and nutrient uptake of oil palm seedling in prenursery stage as influenced by oil palm waste compost in growing media. Int J Agron:1–8. https://doi.org/10.1155/2016/6930735

Sharpley AN, Daniel T, Sims T, Lemunyon J, Stevens R, Parry R (2003) Agricultural phosphorus and eutrophication, 2nd edn. United States Department of Agriculture Agricultural Research Service ARS–149. Retrieved from: https://www.ars.usda.gov/ARSUserFiles/oc/np/phosandeutro2/agphoseutro2ed.pdf

UMassAmherst (The University of Massachusetts, Amherst) (2018) Organic waste management. The Center for Agriculture, Food, and the Environment. Retrieved from: https://ag.umass.edu/greenhouse-floriculture/greenhouse-best-management-practices-bmp-manual/organic-waste-management

US Environmental Protection Agency (EPA) (2018) Facts and figures about materials, waste and recycling, August 1. Retrieved from: https://www.epa.gov/facts-and-figures-about-materials-waste-and-recycling/guide-facts-and-figures-report-about-materials#NationalOverview

Velasquez-Manoff M (2018) Can dirt save the earth? The New York Times Magazine, April 18. Retrieved from: https://www.nytimes.com/2018/04/18/magazine/dirt-save-earth-carbon-farming-climate-change.html

Wells ML (2012) Pecan tree productivity, fruit quality, and nutrient element status using clover and poultry litter as alternative nitrogen fertilizer sources. HortScience 47(7):927–931

Chapter 4
SDG 3 Good Health and Well-Being

Effects of Ultraviolet Radiation on Human DNA: A Point of View from Sustainable Healthcare

María Belén Federico

Abstract Skin cancer incidence is increasing. The WHO reports between 2 and 3 million non-melanoma skin cancers and 132,000 melanoma skin cancers globally each year, while 1 in every 3 cancers diagnosed is a skin cancer. Several factors are responsible for skin cancer incidence, and some of them are more easily treated than others. Furthermore, as social and contextual factors within communities can often hinder UV exposure reduction (e.g., the societal promotion of tanning), primary prevention is not always sufficient. Early detection and treatments can be remarkably improved through a better understanding of the molecular events activated after UV radiation reaches human cells. As such, this chapter aims to evidence how basic research regarding the effects of UV radiation on the human genetic material works to improve diagnostic tests and the treatment of skin cancer, thus improving the patient's quality of life and reducing fatalities.

Keywords SDG 3 health and well-being · Skin care · DNA damage · UV radiation · Community

4.1 Introduction

Skin cancer is challenging even in countries of high incidence. Policy efforts to reduce the toll of skin cancer now compete with other major behavioral health issues including obesity, harmful alcohol consumption, and smoking (Gordon and Rowell 2015). Furthermore, in the last decades, social factors have heavily contributed to the increased rate of skin cancer cases. Lifestyle and fashion trends have driven the desire for tan skin, especially in adolescents and adults. The pursuit of a "fashionable" tan through deliberate sun exposure or the use of sunbeds has made significant

M. B. Federico (✉)
Leloir Institute Foundation, Buenos Aires, Argentina
e-mail: belefederico@gmail.com

contributions to increasing skin cancer rates around the world (Chang et al. 2014). On the other hand, incidence will continue to increase as nations with aging populations enter the prime ages for onset of skin cancers (i.e., squamous and basal cell carcinomas). Malignant melanoma, the most serious of all skin cancers, is also increasing in incidence and not only arises in older adults but also in adults (Gordon and Rowell 2015). Beyond the facts mentioned previously, increasing incidence of UV rays reaching Earth's surface as a consequence of the ozone hole is also increasing the number of skin cancer patients. The United Nations Environment Programme (UNEP 1998) has estimated every further 10% depletion of the ozone layer results in an additional 300,000 non-melanoma and 4500 melanoma skin cancer cases. Today, skin cancer is the most common type of cancer in fair-skinned populations in many parts of the world (Narayanan et al. 2010).

The increase of skin cancer incidence represents a big challenge for public health, and primary prevention is deemed insufficient to reduce that number. Because of that situation, secondary prevention should be reinforced in order to improve early detection and treatment and tertiary prevention to reduce the impact of the ongoing illness. While several good diagnostic tests and skin cancer treatments are available in the sanitary system today, more research is needed in order to create better and more accuracy tools to fight against this type of cancer. Being UV radiation the main cause of skin cancer, one of the main challenges for current and future skin cancer research is to gain a profound understanding regarding the molecular and cellular events activated in the human body as a consequence of the UV rays. The aim of this chapter is to show how basic research regarding the effects of UV radiation on the human genetic material can help to improve diagnostic tests and the treatment of skin cancer, thus improving the patient's quality of life and reducing fatalities.

4.2 Literature Review

4.2.1 The Impact of Ultraviolet (UV) Radiation on Human Life

4.2.1.1 Solar UV Radiation

The sun is the natural and principal source of UV radiation (Fig. 4.1). UV radiation can be divided, according to its wavelengths, into three categories; UV-A (320–400 nm), UV-B (280–320 nm), and UV-C (100–280 nm). About 95% of the total UV radiation reaching Earth's surface is categorized as UV-A, while UV-B radiation accounts for the final 5% (Britt 2002). UV-A rays are absorbed by the skin and are as such responsible for its aging and wrinkling, while UV-B radiation produces erythema, burns, and skin cancer (Ohnaka 1993). On the other hand, UV-C rays are absorbed by the ozone layer and therefore do not reach Earth's surface.

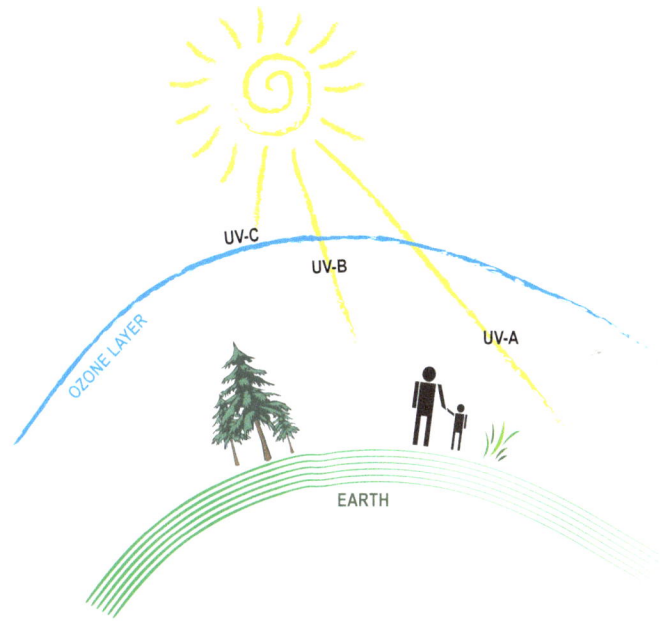

Fig. 4.1 Solar ultraviolet (UV) radiation. UV radiation can be divided, according to its wavelengths, into three categories: UV-A (320–400 nm), UV-B (280–320 nm), and UV-C (100–280 nm). While UV-C rays are absorbed by the ozone layer, UV-A and UV-B rays reach Earth's surface

4.2.1.2 UV Radiation as an Issue for Human Health

Although sun exposure has positive effects on mood and stimulates production of vitamin D, prolonged UV radiation exposure has proven to be extremely harmful for humans. The World Environment Conference, held in Rio de Janeiro in 1992, specifically recommended to "undertake, as a matter of urgency, research on the effects on human health of the increasing UV radiation reaching the earth's surface as a consequence of depletion of the stratospheric ozone layer." Epidemiological and scientific evidence strongly connects UV ray's exposure to several skin cancers (Griffiths et al. 1998; Norval et al. 2011), and that connection has been extensively evaluated during the last decades. Overexposure to UV radiation is the main preventable cause of skin cancers – both melanoma (the most serious type) and non-melanoma skin cancers (NMSC) (Brown et al. 2018). Sixty-five percent of melanoma cases are associated with exposure to UV rays, and the amount is worse for NMSC, reaching 90%, including both basal cell carcinoma (BCC) and squamous cell carcinoma (SCC). NMSC has increased and is becoming more prevalent even in younger age groups; it currently accounts for nearly 15,000 deaths and 3.5 million new cases in the United States alone, representing a major public health concern (Kim and He 2014). A 10 % ozone depletion was found to give rise to a 16–18% increase in the incidence rate of SCC (men and women), a 19% increase in the incidence rate of malignant melanoma for men, and a 32% increase in the incidence rate of melanoma for women (Moan and Dahlback 1992). The difference between the numbers for men and women is almost significant and may be related to the different intermittent exposure pattern to sunlight between the two sexes. Getting a sunburn just once every 2 years can triple the risk of melanoma (Dennis et al. 2008). Moreover, sunburn during childhood or adolescence can increase the risk of skin cancer later on in life (Dennis et al. 2008; Gandini et al. 2005).

Extensive research using animal models has revealed a more active role of UV-B in the induction of skin cancer than UV-A (Roy 2017). Although solar UV exposure is known to be associated with several health problems beyond tumor generation (e.g., accelerated skin aging, eye diseases, and a defective *immune* response), this chapter will focus on the relationship between UV rays and cancer.

4.2.1.3 Skin Cancer: UV Radiation as a Mutagenic and Carcinogenic Agent

The deoxyribonucleic acid molecule, better known as DNA, is a critical target for UV radiation damage. DNA is a central component of all living organisms as it is the genetic material responsible for guiding cell functions. DNA duplication is necessary to transfer the genetic information from the mother cell to its progeny, which means this molecule must remain unaltered to allow next cell generations to preserve the information as accurately as possible. Unfortunately, there are several DNA damaging agents which cause DNA lesions; these DNA damaging agents include UV radiation which can generate 100,000 DNA lesions per cell per day when exposed to peak hour sunlight (Ciccia and Elledge 2010).

The outcomes of DNA lesions differ according to their type, relative position in the DNA molecule, and their quantity, but most are eventually repaired by one of the multiple DNA repair mechanisms. However, damaged DNA is not always repaired as DNA repair mechanisms are often inefficient or saturated. If a lesion goes unrepaired, it could change the DNA's sequence in a permanent way, which results in a mutation. The sources responsible for the generation of mutations, such as UV radiation, are called mutagenic agents (Ikehata and Ono 2011). UV radiation as a mutagenic agent has been broadly investigated, and the lesions generated by UV radiation on the DNA molecule will be explained in detail within Sect. 4.2.2.1. UV rays are also classified as carcinogenic agent, i.e., they promote cancer due to the transformation of a normal cell to a cancerous cell. The International Agency for Research on Cancer (IARC) has extensively analyzed the carcinogenicity of UV and concluded: "There is sufficient evidence in humans for the carcinogenicity of solar radiation. Solar radiation causes cutaneous melanoma and non-melanocytic skin cancer." The increased rates of skin cancer in patients with xeroderma pigmentosum, who have a deficiency in their capacity to repair UV-induced DNA damage, suggests that direct UV DNA damage may be one of the key steps in the causation of these cancers (World Health Organization 1995). Cancer's development is a complex multistage process that implies the alteration of cellular DNA either as a result of changes to the DNA sequence or in its structure (Griffiths et al. 1998). Thus, the mutations generated by UV radiation (mutagenic agent) can be responsible for the development of cancer (carcinogenic agent). Skin carcinogenesis through DNA damage is thus considered a predominant paradigm for UV toxicity (Kim and He 2014).

4.2.1.4 The Necessity to Improve Skin Cancer Diagnosis and Treatments

Assessments by UNEP show increases in skin cancer incidence and sunburn severity due to stratospheric ozone depletion for at least the first half of the twenty-first century. On the other hand, insufficient strategies regarding primary prevention increase skin cancer incidence. Primarily, social and cultural behaviors directly inhibit the success of skin cancer awareness campaigns. Secondly, primary prevention represents the most cost-effective approach to reducing cancer and other NCDs, and it is no clear its impact in the future years (Adami et al. 2001). Besides, despite cancer being a global public health problem, many governments have not yet included cancer prevention in their agendas. In fact, only a few countries, such as Scotland, Belgium, Germany, Spain, France, Brazil, Canada, or Australia, have regulations regarding the use of tanning beds (Mitchell 2018; "From Australia to Brazil: sun worshippers beware" 2009; Stark 2009). This situation heightens the significance of secondary prevention in the fight against this type of cancer. As a result, early detection and treatment are key components of skin cancer control, particularly with regard to reducing morbidity and mortality in the short term (Iannacone and Green 2014). Strategies to go deeper into the etiology of skin cancer are needed in order to improve cancer diagnosis and treatments.

In the last few years, a strong effort has been made to elucidate the cellular mechanisms responsible for protecting cells from cancer development after UV radiation. UV rays affect the integrity of the genome, creating genomic instability, a major driving force for the generation of tumors and cancer. Genomic instability refers to alterations in the DNA molecules as a consequence of 1) errors produced during DNA duplication, 2) lesions generated by several mutagenic agents and 3) alterations in metabolic pathways, among others. While a considerable amount of knowledge is available concerning the interaction of UV rays with the DNA, controversy exists as to which UV-lesion constitutes the most important type of premutagenic damage (UNEP 1987). Therefore, in-depth knowledge on the underlying mechanisms involved in the cellular response following detection and processing of UV-induced lesions in the genome has become one of the important areas of biomedical research in the context of human health (Roy 2017). This chapter will introduce the latest advances in some of these molecular pathways and demonstrates the contribution of these scientific breakthroughs to the improvement of human health, specifically in terms of secondary prevention.

4.2.2 Ultraviolet Radiation and DNA

4.2.2.1 UV Radiation Generates Lesions in the DNA Molecule

To better understand the effects of UV rays on DNA, it is necessary to delve into the DNA structure. The elemental unit of the DNA molecule is called nucleotide, and it has three components: a five-carbon sugar, a phosphate molecule, and a nitrogen-containing base (Fig. 4.2a). DNA contains four different nitrogenous bases: the pyrimidine bases, thymine (T) and cytosine (C), and the purine bases, adenine (A) and guanine (G). Nucleotides bind each other through the phosphate group generating two independent nucleotide chains held together through electrostatic forces (hydrogen bonds) between complementary nitrogen bases (Fig. 4.2b). Thymine is complementary to adenine and cytosine to guanine. The order of nucleotides along the chains encodes the genetic information carried by DNA. As one simple DNA molecule has tens of millions of nucleotides long, the four-letter nucleotide alphabet can encode nearly unlimited information. Because of its conformational structure, the DNA is a double-helix molecule (Pray 2008).

As it was mentioned previously, UV rays damage our DNA. When cells receive the radiation, the main lesion created is a strong aberrant link (covalent link) between two adjacent thymines (Fig. 4.3). Two different links can be formed between those thymines generating a cyclobutane pyrimidine dimer (CPD) or 6–4 pyrimidine-pyrimidone photoproduct (6,4PP) (Ikehata and Ono 2011). The CPDs alone constitute up to 75% of the total UV-induced photoproducts (Roy 2017). Due to this abnormal bind, those two thymines lose their interaction with the

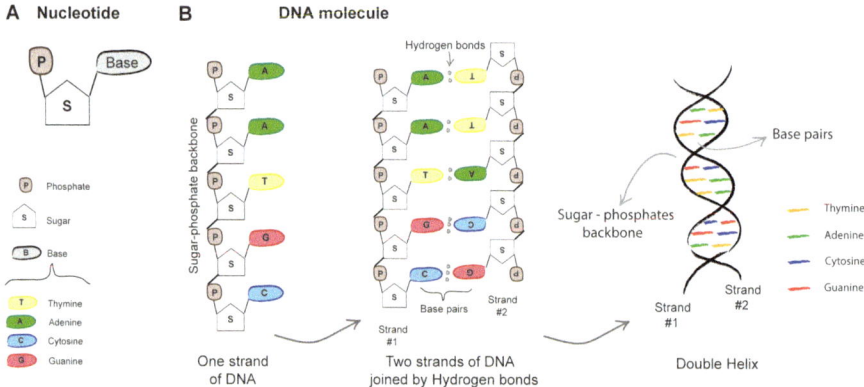

Fig. 4.2 DNA structure. (**a**) The elemental unit of the DNA molecule is called nucleotide, and it has three components: a five-carbon sugar, a phosphate group, and a nitrogen-containing base. (**b**) Nucleotides bind each other through the phosphate molecule generating two independent chains that hold together through electrostatic forces (hydrogen bonds) between complementary nitrogen bases. Because of its conformational structure, the DNA forms a double-helix molecule. There are many ways to represent the DNA double-helix; the far-right diagram simplifies the DNA model, in order to focus on the base pairing

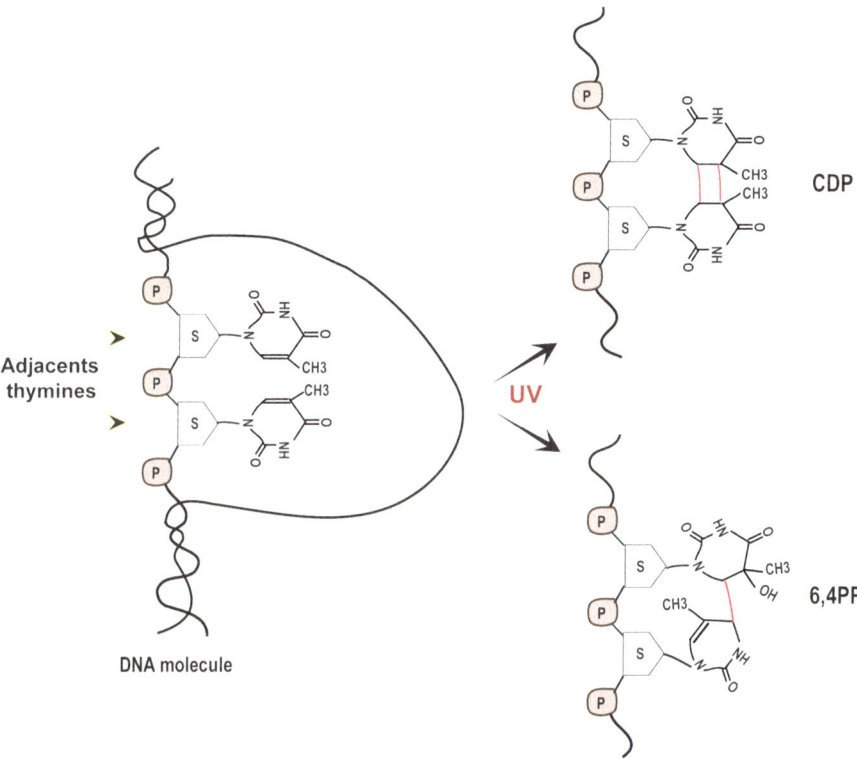

Fig. 4.3 Thymine dimers. UV radiation induces a link (covalent link) between two adjacent thymines present in the same strand of the DNA molecule. According to the position of the link, two different dimers can be formed: cyclobutane pyrimidine dimers (CPDs) or 6–4 pyrimidine-pyrimidone adducts (6–4 PPs)

complementary bases (in the opposite strand), and a conformational and abnormal change occurs to the DNA molecule. This distortion in the DNA molecule can inhibit DNA synthesis (process called DNA replication) and other important cellular events (Roy 2017). In addition, UV radiation can cause DNA damage indirectly, by generating high intracellular concentrations of reactive oxygen species as a consequence of the cellular stress response (Roy 2017; Griffiths et al. 1998).

4.2.2.2 DNA Damage Response After UV Radiation

Aforementioned, the DNA's sequence must be kept unchanged in order to guarantee accurate transfer of genetic information to cell progeny. Therefore, DNA lesions generated by UV rays are a serious threat for the cell and must be repaired in order to prevent defective genetic material. Fortunately, cells have developed a complex and very dynamic network known as DNA damage response (DDR) which deal with that situation. DDR is formed by several mechanisms which are activated to detect DNA damage, signal its presence, and mediate its repair (Jackson and Bartek 2009). Several DNA repair mechanisms, damage tolerance processes, and pathways to check the quality of the DNA (called cell-cycle checkpoints) are members of this sophisticated DDR. Animal models (including genetically engineered mice and human skin xenografts) and cell lines have been widely used to investigate the key role of the DDR mechanisms in UV-induced skin cancer.

4.2.2.2.1 Repair of Photoproducts in Human Cells

The main mechanism responsible for the elimination of thymine dimers is called nucleotide excision repair (NER). This pathway involves three steps: (1) the recognition of the lesion (actually, the recognition of the distortion generated in the DNA structure by the lesion); (2) the removal of a small area of the damaged DNA strand; and (3) the resynthesis of DNA over the deleted region using the undamaged strand as a template (Spivak 2015). NER reduces the number of mutations, by removing most of the UV-lesions before DNA duplication begins. In fact, failures in this repair pathway generate xeroderma pigmentosum which is associated with an increased (>1000 fold) cancer incidence for all types of skin cancers (Griffiths et al. 1998). NER has been studied for many years, and its functions and components are known in detail.

4.2.2.2.2 DNA Damage Tolerance Pathway

Although NER is a very efficient mechanism, excessive DNA damage can cause a saturation of this pathway initiating DNA duplication with unrepaired DNA lesions, jeopardizing the DNA sequence's accuracy and cell survival. When the DNA duplication machinery abuts a photoproduct, the continuity of DNA synthesis is challenged and, if not resolved, can induce cell death. To avoid this situation, cells trigger DNA damage tolerance pathways. Translesion DNA synthesis (TLS) is an alternative DNA replication pathway (Bertolin et al. 2015). TLS can use damaged DNA as template to allow the progression of DNA duplication over the lesion. TLS is not a repair mechanism, as the lesion is kept in the DNA molecule to be eliminated later (Fig. 4.4). Polymerase eta, one of the principal proteins of TLS, can recognize a thymine dimer and add the correct bases (two adenines), unblocking DNA replication. When polymerase eta is absent, a set of alternative TLS polymerases can recognize and bypass UV-lesions, but they do not respect the bases complementarity (i.e., these backup polymerases incorporate random nucleotides, lacking base complementarity). As a result, the TLS process avoids cell death but can induce mutations on the DNA; thus cell viability is preserved at the expense of increased genomic instability (Bertolin et al. 2015). The relevance of polymerase eta after UV radiation is evince in xeroderma pigmentosum variant (XPV) patients which do not express this protein. XPV is an inherited genetic disorder characterized by extreme

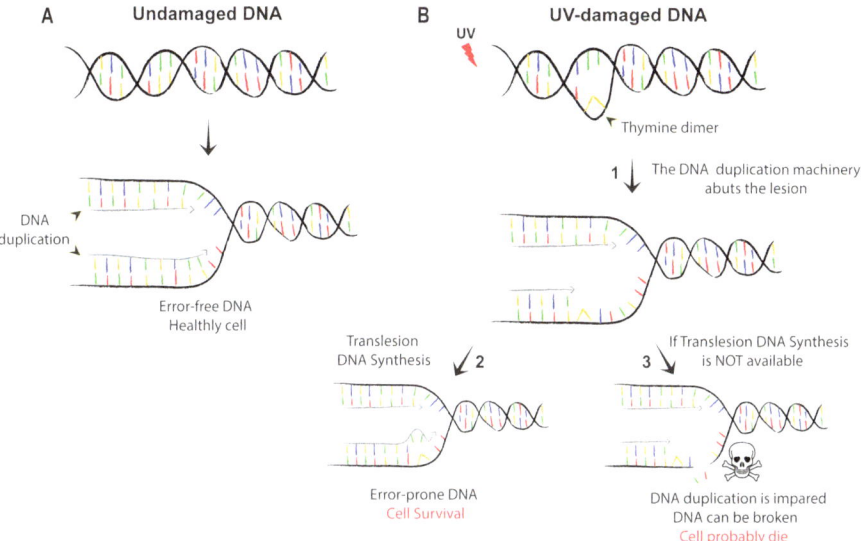

Fig. 4.4 Translesion DNA synthesis (TLS). (**a**) DNA duplication machinery proceeds normally when no lesions are presented on the molecule. (**b**) When DNA is damaged by UV rays, (1) the machinery responsible for the duplication of this molecule cannot work properly across a thymine dimer, and the system is halted. (2) To deal with this situation, cells activate an alternative pathway called translesion DNA synthesis (TLS) that recognizes the lesion and continuous replication across it. TLS induce mutations on the genetic material but allow cell survival after UV radiation. (3) When TLS is not available, DNA duplication does not restart, DNA can be broken, and cell will probably die

Fig. 4.5 Double-strand break (DSB) created as a consequence of a UV-lesion. DSBs can arise during DNA synthesis as a consequence of the stalling of the DNA duplication machinery at a pyrimidine dimer. These kinds of lesions are extremely toxic for the cell and can induce cell death or genomic instability

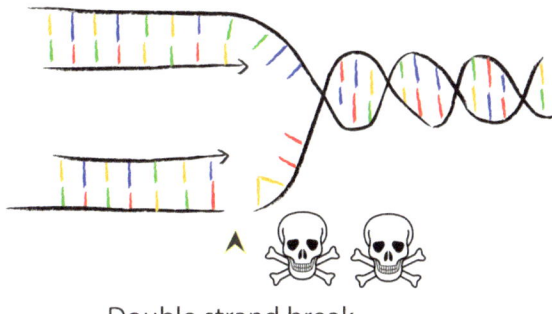

Double strand break created as a consequence of an UV lesion

sensitivity to the sun and increased susceptibility to skin cancers than the general population (Cruet-Hennequart et al. 2010).

DNA duplication after UV irradiation does not uniquely depend on TLS components. Unrepaired UV-lesions can also induce double-strand breaks (DSBs) in the DNA molecule (Fig. 4.5). DSBs are very lethal lesions for a cell as both DNA strands are cut, and genetic material can be lost (Figueroa-González and

Pérez-Plasencia 2017). DSBs are not caused directly by UV rays (Yajima et al. 2009), but they have been reported, even after low doses of radiation (Federico et al. 2016). Studies on human cell cultures showed that DSBs after UV radiation are created when the machinery responsible for the DNA replication abuts a photoproduct (Federico et al. 2016). Supporting the appearance of DSBs, the levels of a protein that are activated when DSBs are formed (the histone variant H2AX) increased in a manner dependent on UV dose. H2AX plays an essential role in the recruitment and accumulation of DNA repair proteins to sites of DSB damage (Fernandez-Capetillo et al. 2003). Moreover, another marker of DSBs as the phosphorylation of ATM kinase (pATM) showed a similar pattern. During DNA synthesis, DSBs are resolved by an error-free mechanism (i.e., a mechanism that does not alter DNA sequence) called homologous recombination repair (HRR). The protein FANCD2, member of a DNA repair mechanism called Fanconi pathway, promotes HRR of UV-triggered DSBs by recruiting Rad51 into the damaged site. When FANCD2 or Rad51 are absent, HRR is not available, and an alternative and very mutagenic pathway takes charge of DSBs resolution (Federico et al. 2016). This alternative pathway, called non-homologous end joining, induces genomic instability by the elimination of some DNA fragments, the addition of extra nucleotides to the sequence, and/or the fusion of one DNA molecule with a second one.

Fanconi anemia patients (which have failure on the Fanconi pathway) present several skin-associated defects including hypo-/hyperpigmentation and café au lait spots (Federico et al. 2016). These characteristics evince a central role of the Fanconi pathway on the DDR after UV radiation.

The DNA molecule can be degraded after prolonged DNA duplication stalling, i.e., the nucleotides of the DNA sequence are separated from each other, and the DNA molecule is shortened (Schlacher et al. 2011). This event has also been described after UV radiation, and Rad51 has a central role in this process. Rad51 protects DNA from degradation and avoids excessive elongation of nascent DNA after UV rays (Vallerga et al. 2015). This mechanism is an example of the complexity of DNA synthesis regulation across UV damage. Thus, more research is required to elucidate the complex network activated when DNA is damaged by UV rays.

4.2.2.2.3 DNA Damage Checkpoint

The cell cycle is the event through which a cell duplicates its genetic material and all its components and divides into two identical daughter cells. In order to allow a proper cell cycle progression, cells have developed "checkpoints." In response to DNA damage, the cells stop cycling by activating checkpoint machinery, thus allowing DNA repair systems to correct replication errors. If the DNA errors are repaired, checkpoint signals will disappear, and the cell cycle will be restarted. If the DNA damage cannot be properly repaired, cell fate includes cell death or replication of aberrant DNA into subsequent cell progeny (Wang et al. 2015). Therefore, the checkpoint system is an essential cellular component guarding the integrity of essential genetic information. The DNA damage checkpoint network contains sensors, transducers, and effectors. The proteins ATR and ATM are the main transducer of DNA damage. ATR responds to a broad spectrum of DNA damage, including DSBs and a variety of DNA lesions that interfere with replication. ATR is the main transducer after UV radiation and Chk1 is its downstream transducer. Chk1 is exclusively associated to the maintenance of active DNA synthesis after UV radiation in a manner in which Chk1 release from DNA molecule prompts TLS to avoid replication stalling (Speroni et al. 2012). On the other hand, ATM is primarily activated by DSBs being Chk2 and p53 its principal effectors. p53 gene is one of the most frequently mutated gene in human tumors and especially in skin cancers like SCC, BCC, and precancerous tissues. Convincing models for the initial steps of UV carcinogenesis, especially for SCC, already exist, based on findings that p53 mutations in NMSC are detected at higher frequency (50–90%) (Benjamin and Ananthaswamy 2007).

4.2.3 Markers for Skin Cancer Diagnostic, Progression, and Treatment

With rates of skin cancer rising in many parts of the world, SDG Target 3.4 "by 2030, reduce by one third premature mortality from non-communicable diseases through prevention and treatment and promote mental health and well-being" is ever more salient. As it is known that early stages of skin cancer can be cured with good prognosis, skin cancer can be treated by means of secondary prevention as well as tertiary prevention, which already includes therapy and rehabilitation (Greinert 2009). The more sensitive and accurate the diagnostics and treatments, the better results can be observed.

Traditional epidemiology uses environmental and/or lifestyle factors as markers to evaluate the risk of some diseases or the early detection of them. For skin cancer, the duration of exposure to sun, intensity, frequency, age at exposure's time, skin type, and number of nevi are considered important factors to evaluate the potential risk of disease development (Greinert 2009). But these methods are often biased by subjective failures to remember exact time points of appearance of, for example, preclinical symptoms of a disease, individual behavior in the environment, and other individual components of lifestyle (Greinert 2009). While these factors have been used for a long time and they are already quite specific, they lack the precision to find biological plausibility as an important criterion to assigning causality (Greinert 2009). Thus, these markers are not the ideal tools for increasing test's sensitivity and specificity on the early detection stages.

In order to design better diagnosis tests and treatments, it is extremely necessary to use "more objective" markers. In 1998, the National Institute of Health Biomarkers Definitions Working Group defined a biological marker or biomarker as "a characteristic that is objectively measured and evaluated as an indicator of normal biological processes, pathogenic processes, or pharmacologic responses to a therapeutic intervention." In the last 25 years, the epidemiology area has incorporated new tools in order to improve the diagnostic, prognosis, and treatments available today for many diseases. The use of molecular and cellular biology techniques in epidemiologic research has initiated a new field called molecular epidemiology. Molecular epidemiology studies the biological events that connect an environmental carcinogen with the occurrence of cancer by using biomarkers. Biomarkers allow to assess exposure, internal dosage, biological effective dosage, altered structure/function, invasive cancer diagnosis, tumor metastasis and prognosis, as well as susceptibility (Chen and Hunter 2005). Currently, there are poor biomarkers for skin cancer diagnostics, progression, prognosis, and metastasis, factors which are of main interest in secondary prevention (Greinert 2009). Few biomarkers are available to follow progression of cancer in patient populations and/or to guide decision-making with respect to dose and administration schedule (Kirsch 2015). Thus, one of the benefits of deepening the knowledge on the molecular events that are activated during the DDR after UV radiation is to find better biomarkers to improve diagnosis tests, predicting tumor response to both radiation and chemotherapy, and to develop more precise skin cancer treatments.

4.2.3.1 DNA Repair Proteins as Markers for Skin Cancer

As it was explained in Sect. 4.2.2.2, following induction of lesions on the DNA, cells activate a DDR in order to detect the lesions and repair them to allow the normal progression of the cell cycle. Thus, the expression of proteins involved in the DDR may change during progression of healthy cells to malignancy (cancerous cells) turning them into excellent biomarkers for diagnostic or prognosis of patients. Researchers from dermatological and oncological departments have joined forces to set up novel and more accurate skin cancer biomarkers based on the DDR activated after UV radiation.

H2AX, a protein activated after DSBs formation – see Sect. 4.2.2.2 – has been evaluated as possible biomarker. Whereas normal human melanocytes (skin cells which produce the protective skin-darkening pigment melanin) displayed low-level expression of activated H2AX (γH2AX), melanoma cell lines showed significantly increased levels of this marker (Warters et al. 2005). Automated quantitative analysis (AQUA) and Tissue Studio are commercial technologies that use digitized immunofluorescence microscopy images to quantify protein expression in defined tissue compartments. In a work published in 2014, these algorithms were used to quantify H2AX activation in several cells and tissues from the skin. After standardizing the system with normal human fibroblast cells, 40 melanoma cells lines and 22 metastatic melanoma tissues were analyzed. This software showed a high intensity of γH2AX on melanoma cells. AQUA also revealed high γH2AX expression in metastatic melanoma samples similar to that seen for the melanoma cell lines, whereas the normal surrounding tissue had significantly lower γH2AX levels (Nikolaishvilli-Feinberg et al. 2013). Another work reported positive γH2AX and Chk2 stain for dysplastic nevi and positive γH2AX stain for melanomas, whereas the surrounding normal skin did not (Gorgoulis et al. 2005). Thus, it is possible that this kind of analysis may be useful in discerning the stages of tumor progression based on H2AX activation levels.

Another key component of the DDR that has been evaluated in the context of skin cancer is the protein ATM, which becomes activated (pATM) upon contact with UV radiation. The pATM expression patterns were evaluated in cultured keratinocytes, skin explants, and a spectrum of premalignant to malignant keratinocyte skin lesions. pATM was observed in the nucleus of a normal skin cell, 30 min after UV radiation, and it returned to its basal localization after 2 h. On the other hand, pATM expression in precancerous keratinocyte lesions was greater when compared to the invasive lesions where pATM was weaker (Ismail et al. 2011). These results showed a more active DDR in the first stages of this type of cancer, which are consistent with the hypothesis that the DDR acts as a barrier to cutaneous tumor formation. In fact, early precursor lesions expressed markers of an activated DDR in several types of tumor, with a diminishing response in more advanced cancers. This highlights the importance of detecting and quantifying DNA damage (Figueroa-González and Pérez-Plasencia 2017). Although these experiments are very promising, there remain many unanswered questions to be addressed in future studies. Beyond the DDR proteins mentioned above, other markers involved in several cell metabolic pathways have been characterized as biomarkers for skin cancer. The utility of all these proteins as biomarkers needs to be validated in future clinical trials, and more research is needed in order to introduce them in intervention programs of secondary prevention.

4.3 Impact Sustainability: Final Remarks

Skin cancer incidence is increasing. The WHO reports between 2 and 3 million non-melanoma skin cancers and 132,000 melanoma skin cancers globally each year; 1 in every 3 cancers diagnosed is a skin cancer. According to Skin Cancer Foundation Statistics, one in every five US citizens will develop skin cancer in their lifetime, and more people are diagnosed with skin cancer each year in the United States than all other cancers combined.

Several factors are responsible for skin cancer incidence, and some of them are more easily treated than others. Primary prevention is not always sufficiently effective. Without adequate widespread, comprehensive implementation and primary prevention strategies may have little effect on attitudes toward sun protection and sunburn prevention at the community level. Social and contextual factors within communities can also create barriers to reducing UV exposure. Moreover, better politics to reduce the risk of skin cancer among outdoor workers is needed. Fortunately, after WHO's designation of indoor tanning devices as Class 1 human carcinogens (the highest-risk level), several countries have banned indoor tanning, specifically, Brazil and Australia. In addition, France, Spain, Portugal, Germany, Austria, Belgium, the United Kingdom, Australia, Iceland, Italy, Finland, and Norway prohibit indoor tanning for youth younger than 18 years old (Mitchell 2018; "From Australia to Brazil: sun worshippers beware" 2009; Stark 2009). Although important strides have been made in skin cancer prevention across several countries, they have not been sufficient to curb the rising rates of skin cancer incidence.

The increase in harmful UV rays as a consequence of ozone hole expansion is another key factor responsible for the rise of skin cancer incidence. Numerous actions have therefore been taken in order to diminish this ozone depletion in the stratosphere. While the application of the Montreal Protocol has allowed the ozone hole to reduce and remain stable, no clear data is available regarding a reduction on the levels of UV rays reaching the Earth's surface. Moreover, the total reduction of the ozone hole is expected to happen slowly due to the long half-life of ozone depleting molecules. The WHO estimates that a 10% decrease in ozone levels will result in additional 300,000 non-melanoma and 4500 melanoma skin cancer cases.

Beyond the serious consequences for life, skin cancer also represents a huge problem for public health. It is clear that management of skin cancer exerts a sizable burden on health systems. Economists in Europe believe that the current growth in healthcare spending is not sustainable (Gordon and Rowell 2015). Skin cancer prevention programs require exorbitant amounts of money, while diagnostics and treatments are also expensive. In most studies, doctor visits, biopsies, surgical excisions, therapies, and hospital stays are the major direct treatment costs related to skin cancers. Although the costs of squamous or basal cell carcinomas per lesion are not high, the high numbers of cancers being treated produce high aggregate costs (Gordon and Rowell 2015).

In order to reduce the number of patients reaching advance stages of cancer, and therefore, reduce the number of fatalities, better diagnosis and prognosis tests are required. Early detection and treatments can be improved by a better understanding of the events activated after UV radiation reaches human cells. Particularly relevant is the field of the DDR after UV radiation as it helps to gain better understanding regarding the genetic regulation and/or dysregulation which leads to skin cancer. Research on this field has helped in the development of new biomarkers which can be identified by modern molecular genetic methods. Until recently these biomarkers have not been thought to be involved in progression and metastasis of skin tumors or to be used as prognostic markers (Greinert 2009). Although these new biomarkers predict future and successful tests, some points need to be evaluated in greater detail. Future research should determine (1) all the lesions generated by UV on the DNA molecule (e.g., pyrimidine dimers, DSBs, oxidative lesion) and characterize their role as carcinogenic agents; (2) the different effects of UV-A, UV-B, and UV-C on the DNA molecule, malignant transformation, and skin tumor development; and (3) biomarkers of UV exposure on early effects of the skin cancer. Also, it is important to keep in mind that successful implementation of these diagnostic/treatment tools will depend not only on the basic research but also on the precision value of the test, the cost, and speed of the assay.

Acknowledgments I would like to thank Dr. Vanesa Gottifredi, Head of the Cell Cycle and Genomic Stability Laboratory at Leloir Institute, for being an excellent mentor and transmitting me her knowledge about DNA damage response and UV radiation.

References

Adami H, Day N, Trichopoulos D, Willett W (2001) Primary and secondary prevention in the reduction of cancer morbidity and mortality. Eur J Cancer 37:118–127. https://doi.org/10.1016/s0959-8049(01)00262-3

Benjamin C, Ananthaswamy H (2007) p53 and the pathogenesis of skin cancer. Toxicol Appl Pharmacol 224(3):241–248. https://doi.org/10.1016/j.taap.2006.12.006

Bertolin A, Mansilla S, Gottifredi V (2015) The identification of translesion DNA synthesis regulators: inhibitors in the spotlight. DNA Repair 32:158–164. https://doi.org/10.1016/j.dnarep.2015.04.027

Britt A (2002) Repair of damaged bases. Arabidopsis Book 1:e0005. https://doi.org/10.1199/tab.005

Brown K, Rumgay H, Dunlop C, Ryan M, Quartly F, Cox A et al (2018) The fraction of cancer attributable to modifiable risk factors in England, Wales, Scotland, Northern Ireland, and the United Kingdom in 2015. Br J Cancer 118(8):1130–1141. https://doi.org/10.1038/s41416-018-0029-6

Chang C, Murzaku E, Penn L, Abbasi N, Davis P, Berwick M, Polsky D (2014) More skin, more sun, more tan, more melanoma. Am J Public Health 104(11):e92–e99. https://doi.org/10.2105/ajph.2014.302185

Chen Y, Hunter D (2005) Molecular epidemiology of cancer. CA Cancer J Clin 55(1):45–54. https://doi.org/10.3322/canjclin.55.1.45

Ciccia A, Elledge S (2010) The DNA damage response: making it safe to play with knives. Mol Cell 40(2):179–204. https://doi.org/10.1016/j.molcel.2010.09.019

Cruet-Hennequart S, Gallagher K, Sokòl AM, Villalan S, Prendergast ÁM, Carty MP (2010) DNA polymerase η, a key protein in translesion synthesis in human cells. In: Genome stability and human diseases. Springer, Dordrecht, pp 189–209

Dennis L, Vanbeek M, Beane Freeman L, Smith B, Dawson D, Coughlin J (2008) Sunburns and risk of cutaneous melanoma: does age matter? A comprehensive meta-analysis. Ann Epidemiol 18(8):614–627. https://doi.org/10.1016/j.annepidem.2008.04.006

Federico M, Vallerga M, Radl A, Paviolo N, Bocco J, Di Giorgio M et al (2016) Chromosomal integrity after UV irradiation requires FANCD2-mediated repair of double strand breaks. PLoS Genet 12(1):e1005792. https://doi.org/10.1371/journal.pgen.1005792

Fernandez-Capetillo O, Mahadevaiah S, Celeste A, Romanienko P, Camerini-Otero R, Bonner W et al (2003) H2AX is required for chromatin remodeling and inactivation of sex chromosomes in male mouse meiosis. Dev Cell 4(4):497–508. https://doi.org/10.1016/s1534-5807(03)00093-5

Figueroa-González G, Pérez-Plasencia C (2017) Strategies for the evaluation of DNA damage and repair mechanisms in cancer. Oncol Lett 13(6):3982–3988. https://doi.org/10.3892/ol.2017.6002

From Australia to Brazil: sun worshippers beware (2009) Bull World Health Organ 87(8):574–576. https://doi.org/10.2471/blt.09.030809

Gandini S, Sera F, Cattaruzza M, Pasquini P, Picconi O, Boyle P, Melchi C (2005) Meta-analysis of risk factors for cutaneous melanoma: II. Sun exposure. Eur J Cancer 41(1):45–60. https://doi.org/10.1016/j.ejca.2004.10.016

Gordon L, Rowell D (2015) Health system costs of skin cancer and cost-effectiveness of skin cancer prevention and screening. Eur J Cancer Prev 24(2):141–149. https://doi.org/10.1097/cej.0000000000000056

Gorgoulis V, Vassiliou L, Karakaidos P, Zacharatos P, Kotsinas A, Liloglou T et al (2005) Activation of the DNA damage checkpoint and genomic instability in human precancerous lesions. Nature 434(7035):907–913. https://doi.org/10.1038/nature03485

Greinert R (2009) Skin cancer: new markers for better prevention. Pathobiology 76(2):64–81. https://doi.org/10.1159/000201675

Griffiths H, Mistry P, Herbert K, Lunec J (1998) Molecular and cellular effects of ultraviolet light-induced genotoxicity. Crit Rev Clin Lab Sci 35(3):189–237. https://doi.org/10.1080/10408369891234192

Iannacone M, Green A (2014) Towards skin cancer prevention and early detection: evolution of skin cancer awareness campaigns in Australia. Melanoma Manag 1(1):75–84. https://doi.org/10.2217/mmt.14.6

Ikehata H, Ono T (2011) The mechanisms of UV mutagenesis. J Radiat Res 52(2):115–125. https://doi.org/10.1269/jrr.10175

Ismail F, Ikram M, Purdie K, Harwood C, Leigh I, Storey A (2011) Cutaneous squamous cell carcinoma (SCC) and the DNA damage response: pATM expression patterns in pre-malignant and malignant keratinocyte skin lesions. PLoS One 6(7):e21271. https://doi.org/10.1371/journal.pone.0021271

Jackson S, Bartek J (2009) The DNA-damage response in human biology and disease. Nature 461(7267):1071–1078. https://doi.org/10.1038/nature08467

Kim I, He Y (2014) Ultraviolet radiation-induced non-melanoma skin cancer: regulation of DNA damage repair and inflammation. Genes Dis 1(2):188–198. https://doi.org/10.1016/j.gendis.2014.08.005

Kirsch D (2015) Biomarkers for predicting radiation response. Semin Radiat Oncol 25(4):225–226. https://doi.org/10.1016/j.semradonc.2015.05.011

Mitchell D (2018) Canada moves to ban indoor tanning in kids under 18. Emax Health, White Plains

Moan J, Dahlback A (1992) The relationship between skin cancers, solar radiation and ozone depletion. Br J Cancer 65(6):916–921. https://doi.org/10.1038/bjc.1992.192

Narayanan D, Saladi R, Fox J (2010) Review: ultraviolet radiation and skin cancer. Int J Dermatol 49(9):978–986. https://doi.org/10.1111/j.1365-4632.2010.04474.x

Nikolaishvilli-Feinberg N, Cohen S, Midkiff B, Zhou Y, Olorvida M, Ibrahim J et al (2013) Development of DNA damage response signaling biomarkers using automated, quantitative image analysis. J Histochem Cytochem 62(3):185–196. https://doi.org/10.1369/0022155413516469

Norval M, Lucas R, Cullen A, de Gruijl F, Longstreth J, Takizawa Y, van der Leun J (2011) The human health effects of ozone depletion and interactions with climate change. Photochem Photobiol Sci 10(2):199. https://doi.org/10.1039/c0pp90044c

Ohnaka T (1993) Health effects of ultraviolet radiation. Ann Physiol Anthropol 12(1):1–10. https://doi.org/10.2114/ahs1983.12.1

Pray L (2008) Discovery of DNA structure and function: Watson and Crick. Nat Educ 1:1

Roy S (2017) Impact of UV radiation on genome stability and human health. In: Ultraviolet light in human health, diseases and environment. Springer, Cham, pp 207–219

Schlacher K, Christ N, Siaud N, Egashira A, Wu H, Jasin M (2011) Double-strand break repair-independent role for BRCA2 in blocking stalled replication fork degradation by MRE11. Cell 145(4):529–542. https://doi.org/10.1016/j.cell.2011.03.041

Speroni J, Federico M, Mansilla S, Soria G, Gottifredi V (2012) Kinase-independent function of checkpoint kinase 1 (Chk1) in the replication of damaged DNA. Proc Natl Acad Sci 109(19):7344–7349. https://doi.org/10.1073/pnas.1116345109

Spivak G (2015) Nucleotide excision repair in humans. DNA Repair 36:13–18. https://doi.org/10.1016/j.dnarep.2015.09.003

Stark J (2009) Tanning salons are fading fast. THE AGE. theage.com.au

UNEP (1998) Ozone layer-climate change interactions. Influence on UV levels and UV related effects. Environmental effects of ozone depletion: 1998 assessment

UNEP/WG 160 (1987) Fifth meeting of the Working Group for Scientific and Technical Co-operation for MED POL

Vallerga M, Mansilla S, Federico M, Bertolin A, Gottifredi V (2015) Rad51 recombinase prevents Mre11 nuclease-dependent degradation and excessive PrimPol-mediated elongation of nascent DNA after UV radiation. Proc Natl Acad Sci 112(48):E6624–E6633. https://doi.org/10.1073/pnas.1508543112

Wang H, Zhang X, Teng L, Legerski R (2015) DNA damage checkpoint recovery and cancer development. Exp Cell Res 334(2):350–358. https://doi.org/10.1016/j.yexcr.2015.03.011

Warters R, Adamson P, Pond C, Leachman S (2005) Melanoma cells express elevated levels of phosphorylated histone H2AX foci. J Investig Dermatol 124(4):807–817. https://doi.org/10.1111/j.0022-202x.2005.23674.x

World Health Organization. Office of Global and Integrated Environmental Health (1995) Health and environmental effects of ultraviolet radiation: a summary of environmental health criteria 160, ultraviolet radiation. World Health Organization, Geneva

Yajima H, Lee K, Zhang S, Kobayashi J, Chen B (2009) DNA double-strand break formation upon UV-induced replication stress activates ATM and DNA-PKcs kinases. J Mol Biol 385(3):800–810. https://doi.org/10.1016/j.jmb.2008.11.036

Chapter 5
SDG 4 Quality Education

Governing Education for Sustainable Development: Towards Inclusive and Equitable Quality Education

Isabel B. Franco and Ellen Derbyshire

Abstract This chapter provides a contribution to the achievement of the Sustainable Development Goal 4, Quality Education (SDG 4), through analysing stakeholder partnerships and collaboration for the implementation of Education for Sustainable Development (ESD). It argues that a better understanding of limiting and fostering factors preventing greater collaboration for ESD is required in order to meet the targets set by SDG 4 which advocates for Inclusive and Equitable Quality Education. This chapter provides a multilevel governance analysis (global, national and institutional) on ESD and key stakeholders' perceptions on limiting and fostering factors in collaborative governance for ESD. This manuscript also increases our understanding of whether ESD policy developed at the international level can be imagined at the national, institutional and local levels and highlights the factors involved in collaboration for ESD. This hopefully helps us understand how to develop rhetoric that can advance from policy to impact. Conclusions drawn from these analyses highlight key inconsistencies and gaps that resulted from incompatible perspectives and motivations for realizing ESD at all levels of governance.

I. B. Franco (✉)
Institute for the Advanced Study of Sustainability, United Nations University Shibuya-ku, Tokyo, Japan

Australian Institute for Business and Economics, The University of Queensland, Brisbane, Australia
e-mail: connect@drisabelfranco.com

E. Derbyshire
Faculty of Business, Economics and Law, Business School, The University of Queensland, Brisbane, QLD, Australia
e-mail: ellen.derbyshire@uq.net.au

5.1 Introduction

Education lies at the heart of Sustainable Development (SD) and is the fundamental medium through which Sustainable Development is imagined and implemented. Although education impacts stakeholders at the global, regional, local and individual levels, it has proven to be the most difficult to legislate, ratify, implement and manage. The challenge does not lie in the promotion and implementation of key education sustainability strategies. The challenge lies in securing investment, interest and motivation to implement ESD. This requires a better understanding of limiting and fostering factors that prevent key stakeholders from collaborating towards education for sustainable development. Following Healey (2006), Minnery (2007) and Franco (2014), stakeholder networks involve both collaboration and conflict. These two aspects impact the effectiveness of sustainability initiatives and therefore the level of relevance and success of these initiatives. Therefore, this chapter raises the question of stakeholder collaboration for Education for Sustainable Development as a constituent component of collaborative governance (Minnery 2007; Franco 2014). Qualitative analysis of key policies and strategies for implementation and complexities in collaboration processes highlight key gaps and inconsistencies in imagining global policies at the regional and local levels. Thus, this chapter proposes key changes for indicators that measure and monitor the implementation of ESD at the regional level. It also addresses the key systemic and structural limitations that prevent the SDG 4 targets from being implemented by key actors in a multi-stakeholders collaboration scenario.

This chapter is structured as follows: The first section presents a literature review of key ideas and perspectives on ESD since 2005. This is followed by a policy review of ESD strategies at the global, regional, local and individual level. This section also presents a timeline of evolving strategies developed to manage changes in politics, society, environment and the economy. It also looks at the role of stakeholders in actioning the SDG 4 targets with respect to ESD. The third section delves into the key challenges that ESD has faced since 2005 and presents limiting and fostering factors in multi-stakeholder collaboration for ESD. The chapter finishes with recommendations for impact sustainability.

5.2 The Governance of ESD: A Review of the Literature

ESD is a fundamental entity which must be ingrained into the mentality of key actors engaged in collaborative governance for ESD. Promoting greater participation in this field requires theoretical as well as practical frameworks that empower key stakeholders engaged in collaboration for ESD. Theoretically, ESD presents key interlinkages with pedagogical approaches to understanding and imparting ESD practices and values. We see the interlinkage between institutional ethos and thematic interest with regard to implementing ESD initiatives. 2005 to 2014 was dubbed by the United Nations as the decade of Education for Sustainable Development. ESD literature analysed through the lens of the SDG 4 emphasizes the importance of collaboration for effective implementation of ESD goals and policies. The overall goal of the UN Decade of Education for Sustainable Development (DESD) was to integrate the principles, values and practices of sustainable development into all aspects of education. This educational effort encourages changes in the behaviour of stakeholders in order to create a more sustainable environmental integrity, economic viability and just society for present and future generations.

UNESCO's final DESD Report released in 2014 presents how approaches to partnerships evolved during the DESD. Greater inclusivity and stakeholder diversity underpinned UNESCO's ESD strategy. This encouraged the participation of the private sector, youth and media groups. To bridge the gap between policy and practice, UNESCO emphasizes the importance of a clear research agenda, reliable monitoring and evaluation mechanisms and a strong multi-stakeholder dialogue. The development of the UN Interagency Committee for DESD (IAC), the DESD Reference Group and the Monitoring and Evaluation Expert Group (MEEG) were all developed to foster continued engagement between stakeholders at the global level (governing institutions, member states), the national level (NGOs, CSOs, NFPs), the local level (indigenous groups, faith-based community organizations), etc. (Buckler 2014, p. 172).

There are three broad gaps highlighted across the range of literature reviewed. The first gap presented in the research is the definitional cohesion of ESD. This review engaged with a series of definitions that presents key limitations for interdisciplinary understanding of ESD, as well as effective implementation of ESD. The UNESCO roadmap for implementing the Global Action Programme for ESD and the SDGs provides an explicit definition outlining the key sectoral and temporal expectations for SD. This normative standard provides a cohesive definition that in turn could accelerate sustainable solutions at the local level in order to increase the participation of ESD stakeholders at multiple levels. However, in order to strengthen multi-stakeholder networks at the local level, it is necessary to improve the quality of local platforms for learning and cooperation.

The second gap evident in literature is temporal expectations relating to short-term and long-term ESD implementation strategies. Critical analysis of Franco et al. (2018), Vaughter et al. (2015), Annan-Diab and Molinari (2017) and Borges et al. (2017) presents key short-term outcomes of ESD implementation. In all articles, there was no succinct reflection on the key expectations for the outcomes of these studies. The general nature of temporal expectations reflected in approaches to ESD policy advancement and implementation provides a gap in knowledge for predicting the viability of certain approaches to ESD.

This leads to the third gap which highlights the applicability of a normative approach to a diverse range of cultural backgrounds. The majority of the studies reviewed were undertaken in a western educational context. This presents a gap to understanding the viability of capacity building schemas reflecting ESD principles in higher education institutions (HEIs) in different cultural circumstances. This gap was raised by Ryan et al. (2010) in their assessment of Sustainability of Higher

Education within the Asia and Pacific and Franco et al. (2018). The authors drew attention to the cultural diversity inherent in different educational contexts in Asia and the Pacific, whilst Franco et al. (2018) provided a global overview. Furthermore, the studies were not only reflective of a gap in understanding the diversity in cultural conceptions of ESD but also the impact of socio-economic conditions on capacity building efforts for ESD (Ryan et al. 2010).

5.2.1 Factors Compromising Collaborative Governance for ESD

Literature shows that there are three key rising factors that affect collaboration for ESD: the rise of technology, globalization and changing demands and rising socio-economic inequality. This section will address how ESD strategies can remain agile to the demands of these rising phenomena. The discussion section will present additional limiting and fostering factors identified by participants engaged in ESD Networks globally.

5.2.1.1 Technology

The rise of technology in education presents key transformations in methodology, partnerships and demands for investors and knowledge sets. There are two sides to increasing technology advancement which impacts ESD. Firstly are new technological innovations. These new jobs will require a broader set of skills than those we are currently teaching in our classrooms. Technology is also changing the way we engage with content. With the advent of big data and predictive analytics, information is accessible—and more personalized to our needs and interests. Secondly is the use of technology to increase educational access. Global access to education is a major caveat to ESD implementation. However, technology has aided in increased access to education. For example, through key organizational partnerships with local businesses and school, the non-for-profit 'Pencils of Promise' has provided access to education to more than 220, 000 children (Wittles 2015). This presents an opportunity for governing bodies, private institutions and businesses to collaborate at the global, regional and local levels to provide greater access to ESD.

5.2.1.2 Globalization and Changing Demands

Globalization and technological innovation have transformed the required skill sets to succeed in today's world of work. Extended retirement ages and career lifespans have a great impact on how employers approach skill development in the workplace, more specifically, with 8.5% of people worldwide (617 million) aged 65 and over. A percentage is projected to jump to nearly 17% of the world's population by 2050

(1.6 billion) (World Bank 2017). As the population ages, career lifespans are increasing as well. Amongst global workers today, 72% plan to keep working after retirement, and 58% expect to enter a new line of work to have more flexibility. Thus, the ageing population needs continuous learning and reskilling opportunities which technology can provide. Millennials will comprise more than one of three adult Americans by 2020 and 75% of the workforce by 2025. This presents key challenges for employers to manage demands for employment. Furthermore, it also increases the standards of employers for skill sets that are agile to the changing world of work (UN DESA 2017). Managing demands at both ends requires continued engagement and dialogue by public and private sector actors with the working population. Due to the challenges of accelerating environmental change, resource scarcity, increasing inequality and injustice as well as rapid technological change, greater challenges and opportunities for access and means of education are developing.

5.2.1.3 Rising Socio-economic Inequality

Intercountry inequality remains a persistent issue. There is a gap of 32 percentage points between children completing primary school in low- versus high-income countries and 52 percentage points for secondary school. Furthermore, increasing environmental degradation and political volatility and conflict have escalated levels of resource scarcity and human displacement. This not only impedes access to education and basic resources but also exacerbates existing inequalities for women and girls. As indicated by the World Bank, education gaps for women and girls directly affect economic growth (Lusk-Stoven and Woden 2017). Thus, the interlinkage of economic and education policies and strategies is impaired by the lack of access to education by girls and women at the global level. Rising socio-economic inequalities exacerbate pre-existing conceptions of gender roles that impede the realization of ESD targets.

5.3 Methodology

This study utilized a qualitative strategy, involving a variety of qualitative methods and techniques, in order to reduce methodological limitations (Singleton and Straits 2010). Triangulation was adopted to increase the reliability of the data collected (Yin 2009). Specific techniques employed in this study include surveys, individual interviews, literature review, focus groups, field observations and policy analysis. The research primarily emerged through literature and policy reviews. An analysis of secondary sources on ESD and governance for ESD was initially conducted. Furthermore, policy analysis was also undertaken in this review and encompassed policy reports from international organizations and sustainability policy statements across participant higher education institutions and their stakeholders.

Researchers were also interested in reliable primary sources, particularly the perceptions of various networks engaged in ESD in order to form a composite picture of priority areas and governance for ESD gaps. Twenty-eight (28) ESD Networks participated in a global survey translated into three languages, namely, English, Spanish and Portuguese, to increase our understanding of the context of CSD in non-English-speaking countries.

5.4 Discussion

This section provides a multilevel governance analysis (global, national and institutional) on ESD and key stakeholders' perceptions on limiting and fostering factors in collaboration for ESD. It aims to better understand whether ESD policy developed at the global level can be imagined at the national, institutional and local levels and highlights the factors involved in collaboration for ESD. Conclusions are drawn from the literature and policy review and highlight key inconsistencies and gaps that resulted from incompatible perspectives and motivations for realizing the SDG targets at all levels of governance. This section is also based upon research findings from data collected through a global survey with stakeholder networks engaged in ESD networks globally. This hopefully helps us understand how to develop rhetoric that can advance from policy to impact.

5.4.1 Global Level of Governance

The interdisciplinarity of ESD implementation is the need to establish partnerships with the broader community with institutional leaders promoting sustainability plans and strategies. Furthermore, adherence to global agreements should play a major factor in this advancement of policy. Stafford-Smith et al. highlight the interdisciplinarity of ESD implementation. Their research shows that key policy imperatives and limitations with sectorial and stakeholder diversity are evident in ESD practice. Thus, as mentioned in our previous scholarship analysis, the UNESCO's roadmap, the Millenium Development Goals (MDGs) and now the Sustainable Development Goals (SDGs) provide a normative standard for policy advancement and implementation. This normative standard provides a cohesive definition that in turn aids in the acceleration of sustainable solutions at the local and community levels in order to increase the participation of ESD stakeholders on multiple levels. Stafford-Smith also suggested that in order to strengthen multi-stakeholder networks at the local level, it is necessary to improve the quality of local platforms for learning and cooperation. This could aid in the integration of ESD programmes and perspective into the planning and decision-making processes of the community.

Another issue found in the literature has to do with implementation of ESD projects. Findings show that the major challenge is to move forward with regard to policy commitments and demonstrating projects to be fully implemented across the curriculum, teaching and operations, whether in formal systems or in non-formal learning and public awareness raising platforms. Globalization has implications for implementation and is argued to be a key factor in homogenizing approaches to education that omit cultural and environmental identity. This highlights the challenge of globalizing standards and implementation for education to address SDGs. Homogenizing education through global policy agreements necessitates greater partnership and collaboration with actors at the regional and local levels. This also requires a better understanding of limiting and fostering factors that prevent collaboration for ESD. Another key gap in implementation has to do with the motivation and goals of current governments to meet the targets and standards of ESD.

5.4.2 National Level of Governance

ESD has galvanized pedagogical innovation and education policy, including curricula changes that now promote learning for sustainable development at the national level. The national curriculum often serves as the most significant piece of educational policy and can provide the surest means to secure the implementation of ESD. In many countries, the initial entry point of inclusion for ESD has been National Plans for Sustainable Development. However, these mandates for ESD do not always translate quickly into strong integration into educational policy. For example, out of 70 reporting countries in the UNESCO survey, 66% indicated having an ESD strategy or plan, and 50% identified the inclusion of ESD in relevant policy; but when reporting on their major achievements during the decade, only 28 countries (40%) indicated actual integration of ESD into the curriculum or standard teaching objectives as one of their achievements. Similarly, in a detailed review of progress made on ESD in seven countries in Southeast and East Asia, only three countries reported clear inclusion of ESD into recent curricular revisions even though all countries had related policies including ESD. It is worth noting, though, that due to the normal cycles and timing of policy and curricular revisions, it can take several years before a country is able to achieve this type of change.

One of the obstacles to change has been the reluctance or the inability to integrate social and environmental concerns into policymaking and practice. Politicians have been slow to take up the challenge, both from lack of understanding and a piecemeal approach to policy and from a lack of political will. Discussions in one of the high-level groups during the UNESCO (United Nations Educational, Scientific and Cultural Organisation) ESD mid-decade Bonn Conference underlined this issue when delegates identified politicians and policymakers as a key target for ESD.

5.4.3 Institutional Level of Governance

As presented in the literature review, there are three key ideologies that underpin education: Neo-classical, Liberal-Progressive and Socially Critical. These ideologies impact the methodology and strategy of public and private higher education institutions (HEIs). There are two key areas of analysis for understanding ESD strategy at the institutional level: access and investment. In order to understand if ESD is being implemented, access is important. At the local level, this multi-stakeholder analysis presents key points of disconnect amongst global, governmental and local actors. This impairs the capacity building abilities of private institutions based on the political and economic motivations of governments. Thus, with the hindered interdependency of HEIs and organizations on government funding, there are temporal limitations on the ease with which ESD can be implemented in the short term in order to have more sustainable long-term outcomes. The interlinkage of education and economic policies presents a key institutional challenge to establish a framework that can sustainably foster ESD. This can be best managed and sustained through greater partnerships. As presented in the section below, partnerships for collaboration in ESD present a key opportunity to sustain strategies and means of implementation, yet limiting and fostering factors in collaboration for ESD need to be paid greater attention.

5.4.4 Stakeholder Networks Governing ESD

Evidence from the research indicates there are several aspects that have a substantial impact on stakeholder networks and their interactions. These factors can either foster or limit collaboration for ESD. The principal limiting factors include but are not limited to contextual factors, governance issues, lack of resources (both tangible and intangible) and academic constrains. On the other hand, there are enabling or fostering factors that have the potential to strengthen stakeholder networks and foster collaboration, such as knowledge and expertise, skills across stakeholder networks and research approaches and resources. These various factors are explained further below.

5.4.4.1 Limiting Factors

Contextual dynamics have impacted on higher education institutions and their stakeholders. Exacerbated by a long history of confrontation, some networks have witnessed political violence in the contexts where they operate. This, unfortunately, has influenced the ways they collaborate or do not collaborate with external stakeholders. The ESD Network in Guatemala argues that the political context in which they are immersed prevents them from actively engaging in sustainability initiatives as they feel. This has not only aggravated the level of discontent in research locations but has also resulted in the destruction of potential opportunities to foster impact sustainability research and education initiatives, since stakeholders reap very little benefit from collaborative endeavours in these circumstances. According to the ESD Networks in Nigeria and Mexico, these contextual factors have been exacerbated due to 'high level of illiteracy among the local population at the local level' preventing communities from becoming active participants in available sustainability initiatives as well as a gender-biased culture 'a genuine problem when the primary researchers or project leaders are women'.

These limiting contextual factors have also been detrimental for the governance environment and stakeholders' interactions. Some participants in ESD Networks from Germany, Colombia, Mexico and Guatemala describe the governance environment at HEIs as hierarchical, 'closed', 'bureaucratic' and charged by external misconceptions about being very theoretical. In addition, other stakeholders agree there is a lack of institutional support, as the government does not provide a favourable environment for sustainability research and education. According to the ESD Network in Brazil, 'it is not a government priority and therefore there is lack of funding'. Yet, evidence indicates that both the role of the government and a collaborative approach to sustainability neither provide an adequate development of sustainability initiatives nor provide for broader interactions amongst networks. According to participants from the ESD Networks in Brazil, Chile and Portugal, 'individual interests' and 'competition between different actors' have prevailed upon the 'collaborative capacity' of stakeholder networks and their willingness to share. According to the ESD Networks in the United States, Ecuador, Colombia and Nigeria, this has resulted in increased workloads, difficulty in communicating and demonstrating the mutual benefit (win-win) of sustainability initiatives, issues when establishing institutional relationships and trust as well as unequal allocation of resources.

Stakeholder networks are active supporters of sustainability initiatives; however, at times collaboration becomes difficult due to lack of resources: according to the ESD Networks in Thailand and Indonesia, in some cases financial support is 'unconsecutive' or 'unstable for longer time period'. Insufficient funding is becoming a major issue in ESD Networks in Canada, Sweden and the United States. Limited financial resources also compromise the expanding capacity of existing networks – ESD Network in the United States adds. Likewise, ESD Network in Mexico argues that dispersed funding and not funding at all are issues that have an impact on the network itself. ESD Networks in the United States, India, Canada, Portugal and Chile agree that lack of financial resources also compromises the availability of other resources, such as time, information and institutional support.

Research approaches to sustainability science with a potential impact at the community level are two factors that require further attention. Regarding community engagement, some participants acknowledge the fact that local communities find it difficult to engage with the overall research process – at least this is the case of the ESD Networks in Indonesia. This can be due to the fact that 'researchers are in and out, lacking in more permanence in the communities', the ESD Network in Mexico adds. In addition, ESD Networks in Nigeria, Uganda and Mexico agree that lack of awareness of the potential impact of research projects at the community level as well as the reproduction of traditional training and research away from community needs exacerbates the problem. ESD Networks in Mexico, Japan, Greece and Italy agree that further community engagement needs to be supported by interdisciplinary, innovative and rigorous research designs with cohesive KPIs, aims and strategies.

5.4.4.2 Fostering Factors

There are various factors that can unleash the potential of stakeholder networks globally. Participants reported on four main factors that can help stakeholders collaborate and engage more effectively, namely, knowledge and expertise, skills across stakeholder networks and research approaches and resources.

Evidence from ESD Networks in Greece and Brazil indicate that stakeholder networks have broad knowledge and expertise on sustainability and the sustainable development goals (SDGs). According to ESD Network in Portugal and Uganda, such knowledge is rooted on firsthand experience, projects connecting society and the environment as well as climate change and social sustainability issues at the local level. Research found that leveraging such knowledge and expertise across networks can boost stakeholders' impact on research and education for sustainable development. ESD Networks from Nigeria, Sweden, Thailand, Germany, Ecuador and India reported other fostering factors such as leadership, collaboration and openness to collaborate, goodwill, negotiation skills and stakeholder's diversity. A successful case on ongoing stakeholder engagement and collaboration can be found in the ESD Network in Mexico and its stakeholders:

> …(the Department) … has more than 30 years working in conservation and natural resource management projects, as well as in social development and particularly in the development of environmental education projects. It is the Department that has the largest number of collaboration agreements in the region with various institutions and has the coordination of the 'network'.

If allocated effectively, both skills and knowledge can foster impact research on sustainability. According to participants, research is a factor that can also enable existing collaborative initiatives towards sustainability. ESD Networks in Colombia, Albania, the United States, Guatemala and Japan agree that research projects should be participatory, community-focused, indigenous-orientated and context-based.

Regarding resources, only a few ESD Networks in developed locations, namely, Italy and Sweden, have reported on availability of financial and physical resources. Findings show that in some cases, ESD Networks in developed countries reported better performance due to more effective allocation of resources compared to those based in developing countries:

> We have a large variety of groups, across sectors, that collaborate via many networks… 1) We have hosted a variety of different events and collaborated with multiple partners, to serve diverse audiences, across sectors, integrating the skills and resources of a variety of partners, with numbers exceeding our targets. 2) Different working groups have offered their time and expertise to plan and execute the events.

Whilst developing locations lack tangible resources, evidence from ESD Networks in Indonesia and Nigeria indicate there is availability of human resources with a strong interest in volunteering and becoming active participants in research and education initiatives on sustainability, yet lack of resources, at times, prevents them from further engaging in existing initiatives.

5.5 Impact Sustainability: Final Remarks

Education for Sustainable Development in general presents a history where equal access, just implementation and high level of investment have proven difficult. This section highlights some final remarks and recommendations on collaborative governance and collaboration for ESD. Education itself is complex, and ESD exacerbates the definitional, political, financial and environmental caveats for action. This chapter argued that partnerships and greater collaboration are imperative for realizing ESD initiatives that aim to achieve the SDG 4 targets. Through a literature review, this chapter analysed the key theoretical and pedagogical debates surrounding ESD and provided an overview of current strategies at all levels of governance, namely, global, regional and institutional levels. The chapter also identified stakeholders' perceptions on collaboration processes for sustainable development. This provided a deeper understanding of the key challenges and opportunities that stakeholders face when developing strategies agile enough to growing demands of a changing society. Thus, one can see that Education for Sustainable Development requires not only a strong dialogue to overcome definitional limitations but also necessitates greater collaboration and stakeholder investment in order to overcome key barriers such as political change, rapid economic growth, lack of infrastructure and a changing workforce.

Two key recommendations have emerged from this chapter. The first recommendation calls for greater dialogue with local actors when developing and engaging in ESD. More specifically, the transition from the global agreement to national policy to stakeholder networks impact necessitates greater attention to merging normative and holistic approaches to ESD practice and understanding at the local level. It is advisable to take into account the limiting and fostering factors that compromise collaboration for ESD: as per limiting factors, context, governance issues, lack of resources (both tangible and intangible) and academic constrains, and fostering factors such as knowledge and expertise, skills across stakeholder networks and research approaches and resources.

The second recommendation focuses on political affiliations and a better understanding of ESD arrangements at all levels of governance and how they impact the culture of ESD partnerships and collaborations for implementation. A key challenge highlighted in this chapter is the inconsistency of understanding across a wide variety of sectors. However, deeper knowledge about how culture impacts how different stakeholder networks view ESD, in general, would aid in a deeper understanding of the broader policy environment for each network. This, in turn, can facilitate further dialogue with regard to ESD implementation within different cultural contexts at different levels.

References

Annan-Diab F, Molinari C (2017) Interdisciplinarity: practical approach to advancing education for sustainability and for the sustainable development goals. Int J Manage Educ 15(2):73–83

Borges J, Cezarino L, Ferreira T, Muniz Sala O, Unglaub D, Ferreira Caldana A (2017) Student organizations and communities of practice: actions for the 2030 agenda for sustainable development. Int J Manage Educ 15(2):172–182

Buckler C (2014) Shaping the future we want: UN decade of education for sustainable development; final report, 1st edn. UNESCO, Paris, pp 168–172

Franco I (2014) Building sustainable communities: enhancing human Capital in Resource Regions. PhD Dissertation. The University of Queensland, Brisbane

Franco I, Saito O, Vaughter P, Whereat J, Kanie N, Takemoto K (2018) Higher education for sustainable development: actioning the global goals in policy, curriculum and practice. Sustain Sci:1–22

Healey P (2006) Transforming governance: challenges of institutional adaptation and a new politics of space 1. Eur Plan Stud 14(3):299–320

Lusk-Stoven O, Woden Q (2017) Girls' education overview. World Bank. https://www.worldbank.org/en/topic/girlseducation

Minnery J (2007) Stars and their supporting cast: state, market and community as actors in urban governance. Urban Policy Res 25(3):325–345

Ryan A, Tilbury T, Corcoran P, Abe O, Nomura K (2010) Sustainability in higher education in the Asia-Pacific: developments, challenges, and prospects. Int J Sustain High Educ 11(2):106–119

Singleton JR, Straits BC (2010) Approaches to social research, 5th edn. Oxford University Press, New York

United Nations Department of Economic and Social Affairs (2017) The sustainable development goals report. United Nations, Departments of Economic and Social Affairs, New York, p 2017

Vaughter P, Wright T, Herbert Y (2015) 50 shades of green: an examination of sustainability policy on Canadian campuses. Can J High Educ 45(4):81

Wittles O (2015) A platform that's changing the world. https://pencilsofpromise.org/platform-thats-changing-world/

World Bank (2017) Population ages 65 and above (% of total) | Data. (2019). https://data.worldbank.org/indicator/SP.POP.65UP.TO.ZS?end=2017&start=2017&view=bar

Yin RK (2009) Case study: research design and methods, vol 5, 4th edn. SAGE, Los Angeles

Chapter 6
SDG 5 Gender Equality

Not Just a Women's Issue: Sustainable Leadership in Male Dominated Industries – The Case of the Extractive Industry

Isabel B. Franco, Paulina Salinas Meruane, and Ellen Derbyshire

Abstract Male-dominated and gender-segregated fields, such as the extractive industry, present key limitations for sustainable leadership opportunities and career growth for women. By identifying these existing barriers and addressing necessary actions to be taken, research findings show that success in this area largely depends on the collaboration of multiple stakeholders, that is, governments, corporations, higher education institutions and civil society organizations. This study will provide a qualitative assessment of current leadership and organizational discourse in order to build knowledge and understanding of the limiting factors and barriers that prevent women from embarking on a sustainable leadership pathway in the early stages of their career. These limiting factors are grouped into three categories, namely, sociocultural, corporate and governance factors. Some specific challenges identified by participants include maternal, family, cultural conceptions of gender roles and norms as well as workplace diversity. These variables all contribute to a gender and culturally normative ecosystem that present competitive barriers for career development for women.

I. B. Franco (✉)
Institute for the Advanced Study of Sustainability, United Nations University Shibuya-ku, Tokyo, Japan

Australian Institute for Business and Economics, The University of Queensland, Brisbane, Australia
e-mail: connect@drisabelfranco.com

P. S. Meruane
School of Journalism, Universidad Catolica del Norte, Antofagasta, Chile
e-mail: psalinas@ucn.cl

E. Derbyshire
Faculty of Business, Economics and Law, Business School, The University of Queensland, Brisbane, QLD, Australia
e-mail: ellen.derbyshire@uq.net.au

© Springer Nature Singapore Pte Ltd. 2020
I. B. Franco et al. (eds.), *Actioning the Global Goals for Local Impact*, Science for Sustainable Societies, https://doi.org/10.1007/978-981-32-9927-6_6

Keywords Leadership · SDG 5 gender equality · Higher education · Sustainable development · Extractive industry · Career development

6.1 Introduction

Equal opportunities for greater engagement and participation in leadership roles are not just the responsibilities of women. The limiting factors for engagement are underpinned by a broader set of androcentric, sociocultural, governmental and organizational norms within male-dominated industries, such as extractives. Thus, the responsibility lies in a collaborative effort between the private sector, government, higher education institutions and civil society organizations to promote equal opportunities towards sustainable leadership of women in the extractive industry. Although recent trends indicate an increase in participation of women in leadership roles in male-dominated fields, these opportunities remain highly conditional and limited by systemic conceptions of gender roles (Ibañez 2010; National Council of Innovation and Competitiveness 2014; Council of Mining Competencies 2015; Navarro et al. 2016). The significance of this issue results from the challenges imposed on the ability of women to gain leadership opportunities from training in higher education through to their career transition. Issues such as lack of recognition, limited skill development for career growth, educational barriers, work and life balance, reconciliation of maternal responsibilities and career development are all underpinned by a predominantly androcentric policy and government approaches, compromising the achievement of the sustainable development goal 5, gender equality (hereinafter SDG 5). Thus, a framework that promotes the interests and growth of men only hinders opportunities for career development for women into leadership roles and reduces industry productivity.

The persistence of these challenges is mainly due to lack of education at the corporate level regarding ways of promoting sustainable leadership opportunities for women. Research findings presented in this chapter show that the private sector and its stakeholders require a nuanced understanding of existing challenges and opportunities to take actions that have a positive impact on women in male-dominated industries. At the global level, the international community has developed compliance mechanisms, international standards and regulations aimed to promote sustainable leadership of women in male-dominated fields. However, this is not only a women's issue as it requires multi-stakeholder collaboration. This chapter provides a qualitative analysis of a series of interviews, focus groups, literature review and policy analysis. The study presents key androcentric sociocultural, corporate and governance factors that hinder the sustainable leadership of women in male-dominated fields, such as the extractive industry. This analysis presents competing perspectives, presented in academic literature and policy discourse, to facilitate a conversation on the value of a more inclusive approach to career development, work dynamics and productivity for women in the extractive industry (Franco 2014).

6.2 Literature Review

The sustainable leadership of women in the extractive industry has become a topic of relevance, particularly in resource regions. However, despite the resource boom, the challenges of a segregated industry have limited women's options for accessing leadership positions, at all levels, from the operational to the management levels. There are several factors that limit the participation of women in contexts in which leadership is perceived as more commonly a male activity (Bartel and Dutton 2001; Eagly and Karau 2002). From their student life to their transition into the industry and their rise within it, women face significant challenges. This has implications not only for women but also the productivity of the industry, which will be implicated by the lack of management strategies to address equality and sustainable leadership for women in the workplace. The increase in costs associated with the rotation of personnel, the time of the personnel dedicated to recruiting and training new workers, the loss of knowledge and the social relations within the company are just some of the factors that compromise the productivity of the industry due to a lack of equality (ICMM 2003; IFC 2009, 2013).

There is a consensus in the literature that suggests the need for sustainable leadership of women over time. However, to what extent the exercise of that leadership is realized and the factors that limit the sustainable leadership of women are issues that require greater attention (Swanson et al. 1996; Eagly and Carli 2007; Cardoso and Marques 2008; Kark and Eagly 2010; Donoso et al. 2011; Ibarra et al. 2013; Hoobler et al. 2016; Navarro et al. 2016; Terrill 2016; Meister et al. 2017). Swanson et al. (1996) identified some hindering factors such as sexual discrimination, lack of confidence, gender cultural roles, tensions between children and professional demands, racial discrimination, difficulties in decision-making and lack of models or mentors, among others. Cardoso and Marques (2008) add socio-economic, ethnic or racial and gender factors. Donoso et al. (2011) in an investigation with Spanish university students conclude that negative evaluation results in education processes are factors that condition the professional career of women. The authors explain that women show fear of evaluation due to criticism from others and the perception that they are forgiven less for mistakes, especially if they occupy a position of leadership and power. Also, Navarro et al. (2016) highlight other challenges including motherhood, multiple roles, gender stereotypes, promotion, working conditions, macho culture, job assignments, harassment and disrespect, recruitment and selection, compensation and social networking. The authors agree these factors constitute barriers to the incorporation of professional women in the industry.

In the transition from student to professional, women seem to have problems with adapting to the new environment. This is primarily related to the physiological challenges imposed through a system that is inherently masculine (Bell and Sinclair 2016). In Chile, there are workers who have chosen to perform hysterectomies in order to eliminate menstruation, and others have stopped breastfeeding and minimized their visits to the bathroom during the workday, while others control and masculinize their attitude, behaviours and emotions. These changes impose maternal limitations, which impact the personal life of women. This aspect highlights a distinct compromise that women consider to not be apparent in the male experience in the extractive industry. Also, regarding the transition to motherhood, women lack incentives to return to the company after having obtained maternity leave. This is one of the factors that most heavily affects the sustainable leadership of women in extractive industries, even in industrialized countries such as Australia. Women after motherhood usually only return part-time or simply do not return at all (Terrill 2016).

Once adapted to the role of mothers, other factors emerge which influence their leadership in the workplace. The reconciliation of work and family life is a key struggle for women, which is deeply associated with the guilt and anguish of being away from home and unable to exercise care for their children. According to Salinas and Al Dajani, the greatest sacrifice for women is to get away from their families and not be able to attend proxy meetings or school events and not being present on important dates such as birthdays or anniversaries. This emotional pressure is exacerbated by work pressure and discrimination in the workplace. All these factors that impact from the university life of women to positions of leadership in the industry are not currently given proper consideration and therefore have a significant impact outside and within the industry.

6.3 Methodology

The social reality of women in leadership within the extractive industry remains volatile and organizationally subjective. Based on a qualitative methodology, this chapter presents key discourse analysis of perceptions of female leadership at the university, professional and sectoral levels. The research presented in this chapter compares the perceptions of women students, professionals and relevant stakeholders in the extractive industry in Chile and Colombia (Salinas and Cárdenas 2009; Denzin and Lincoln 2012). The research design is descriptive for comparative purposes between the two countries in Latin America.

The study conducted at the university level revealed that there are a series of factors that compromise women's leadership at the organizational level. The women interviewed noted that the dominance of masculine interests appeared early in the classroom setting and remained evident in the workforce. This is more predominant in the interrelationships between students and educators and employers and employees. Primarily, this was apparent through negative reinforcements and comments made by supervisors, which increased the difficulty of carrying out their professional practices and projects, among others (Salinas and Romaní 2017). Therefore, the focus of this article lies in the perceptions of female students, professionals and other relevant actors in the extractive industry. The challenges that women face and the assets that they require in order to develop sustained leadership skills and career development opportunities are also presented in this chapter. As a result, this study will identify key strategies and recommendations to overcome existing barriers, which prevent women from engaging in a sustainable leadership path and achieve SDG 5.

6.3.1 Case Study Analysis

This study was conducted in Latin America, with a special focus on Chile and Colombia. However, research findings can be applied to other locations in Latin America and elsewhere. Chile, a nation with large copper reserves in the south central region of the country, evidenced the advent of virtuous mining, which required human talent for the long-term sustainability of operations, with women playing the protagonists key to ensuring the success of mining projects in the long run. However, the country is still experiencing challenges in regard to the participation and leadership of women in the industry. Evidence shows that the 1996 decree banning the entry of women to large copper deposits was repealed. Despite indications of progress in this area, it is evident that the issue of female leadership in a segregated industry such as the extractive industry is premature (National Council of Innovation and Competitiveness 2014). According to available information, the participation of women is 23% in disciplines relevant to the mining industry, namely, industrial civil engineering, civil metallurgy, civil mining, geology, civil engineering and construction prevention risk. This proportion drastically decreases in technical careers in which women only reach 4.3% of enrolment (Higher Education Information Service (SIES)) 2014). Research in the Colombian case, for example, shows alarming figures: the program with least enrolment by women is mechanical engineering (9%), followed by electrical engineering (10%), electronics and telecommunications (14%) and systems and telematics (26%) (Ministry of Education of Colombia n.d.).

Regarding employment in the Chilean case, evidence shows that women's participation in the workforce in the extractive industry represents 89% women supervisors vs 69% men; 73% of women maintainers vs 49% of men; and 57% of women operators vs 17% of men (Ramos 2017). Some initiatives implemented by the government to foster women's participation are as follows: Chilean Standard NCh3262-2012 that issues the "System of Gender Equality and Reconciliation of Work, Family and Personal Life" (CODELCO 2017), the program "Good Labour Practices with Gender Equality", and "Equal Seal", which seeks to install good work practices in organizations, improving the incorporation of women in the industry as well as fostering their career development (SERNAM 2015). Likewise, the "Labour Reform" incorporates some clauses with a gender perspective and encourages unions to modify their regulatory frameworks to guarantee the inclusion of women in their boards (SERNAM 2017). In the Colombian case, women represent only 14.9% in the extractive industry, including small, medium and large mining (ECLAC 2017). The National System for Human Capital Development provides a general policy that aims to foster human talent for the industry and provides guidelines for the participation of women (Franco 2014).

6.3.2 Data Collection

For the purposes of data collection, 27 open interviews were conducted. Additionally, two focus groups including women and key stakeholders in the extractive industry in Colombia and Chile were also analysed. Focus groups addressed topics such as challenges and opportunities for the participation of women in the extractive industry, from higher education to employment. Interviews were conducted between May and October 2016. The focus groups were conducted in Colombia and Chile, in the first semester of 2017, while the data analysis and development of the manuscript was carried out in the second semester of 2017 and the beginning of 2018.

6.3.3 Data Analysis

Discourse analysis enabled data analysis. This methodological technique was conducted through a detailed reading line by line of each of the interviews. Subsequently, a simplified version coding grounded theory (Strauss and Corbin 2002; Trinidad et al. 2006) facilitated the development of categorical discussion either open or lateral. This bi-level analysis established several key categories, which added a greater depth of analysis to the discourse by paying specific attention to both the superficial-structural aspects and those with the deepest roots, even hidden from the subjects (Santander 2011). The findings are presented in conjunction with relevant literature and field observations.

6.4 Discussion

For purposes of this study, the factors that compromise sustainable leadership of women are a set of challenges that hinder their development at both the university level and at the labour markets (Swanson et al. 1996; Cardoso and Marques 2008; Donoso et al. 2011). These challenges are understood as "events or conditions, either within the person or in their environment that hinder the progress of the race" (Swanson and Woitke 1997: 446), that is, it is a wide range of obstacles that interfere in the professional development of a person. In the analysis of the research results, three types of factors were identified, namely, sociocultural, corporate and governmental. Also, within these three categories, we identified eight specific factors and possible actions to counteract the adverse effects of these barriers on women and that prevent leadership at the university and the workplace: *gender equality, gender diversity, sexual discrimination, lack of trust, maternity, working conditions, macho culture* and *the reconciliation of one's personal and professional life.*

6.4.1 Sociocultural Factors

Evidence of symptoms of stress and discrimination are all changes experienced by women due to the difficulties in adapting during higher education and making work and family compatible. Women are asked to adapt to the industry without mourn. The demands of a predominantly male industry have exacerbated the lack of trust and given rise to a macho culture, both of which were identified in this study as the most predominant sociocultural barriers. Participants in focus groups also expressed possible actions to mitigate the adverse impact of the previously identified factors.

6.4.2 Lack of Trust

Regarding *lack of trust*, there were coincidences between the speeches of men and women. Both affirm that it is a trait that is expressed in different ways since entry into higher education, making it difficult for students to enter and stay in the industry. Some of the discursive categories found are as follows: "they are more fragile", "they are afraid of the work environment", "they are more introverted", "they tend to keep quiet", "they are insecure because of their personality" and "they have less capacity to hold back emotionally". These factors exacerbate the perception of their male peers who perceive them as unbelievable actors in their profession (Bartel and Dutton 2001).

> … in critical thinking exercises they tend to remain silent... they are more introverted, ….men are more critical... it seems to me. (Participant, individual interview, Chile)

> There is still fear among women. (Participants, group interview, Colombia)

> ... there is a generalised thought that women are not ready to enter this world, because they are very fragile... men tend to think that women have less capacity to think and much less ability to hold back emotionally when there is a crisis or an emergency. These are the biggest barriers that need to be overcome. (Participant, individual interview, Chile)

6.4.3 The Macho Culture

The *macho culture* is expressed in the discourses in a transversal way; interviewed women highlight the practices that exacerbate this cultural barrier in the classrooms: "classmates make fun of women's questions"; "men think women have less ability to think." "The boys have to accept women and leave the macho attitude". On the other hand, men's speeches focus on the reactions of male students, with a certain complacency, although they also highlight the strength of the traditionally male culture, which is accentuated in the industry. Likewise, men and women agree on the impact of the macho culture and the insufficient number of women to transform the sector: "masculine vision predominates in the industry", "the macho culture is very predominant", "the female incorporation in the industry responds more to political favours than to a true intention of the companies to insert them in the workforce" and "everything that has been done so far is insufficient", they say. Some of these sociocultural aspects had already been documented by Swanson et al. (1996):

> The policies of the industry respond to a macho culture ... women are hired due to political favours and not by market demand. (Participant, group interview, Colombia)

> ... there are more political favours than other things. Because we have seen how (the company) needs to meet certain goals, then they say "we need to have more women managers, or super-women or women in the operation." It is not like an internal issue for the company, at least here in Chile. (Participant, individual interview, Chile)

6.4.4 Corporate Factors

The factors that compromise the achievement of SDG 5 have an effect on women and on the overall corporate productivity. Research findings show that the government does not recognize the value of female labour, which prevents them from applying to executive positions. Corporations do not adequately reward women's organizational skills, multitasking and commitment to the community. Corporate factors have been identified as mainly paradigm problems of equality, diversity and sexual discrimination. Participants in focus groups also expressed possible actions tending to mitigate the adverse impact of these factors.

6.4.5 Gender Equality

The participants agree that the training should be the same for male and female students. This factor is considered as a homogeneous strategy, where the students must adapt to the "Male Mining Worker" model. The limitations that women experience not only have an impact on their behaviour but also on their physical identity (Bell and Sinclair 2016). In this regard they specify: "training should be conducted without gender distinction", "mining deals equally with men and women", "as a policy, the treatment is the same in the race", "care is taken not to make distinctions", " the woman have to adapt to a system made for men "," she must adapt to the mining world" and " she must show that she possesses the same capabilities as the man". They even emphasize that no differences should be made, since discriminatory. However, discourse analysis in the Colombian case showed that the "machista" culture is equally responsible for creating key paradigms within perceptions of gender in universities and workplaces.

> We guide ourselves through the curriculum, so we do not see if they are men or women, they are a professional future, we must teach them in the same way, we explain the same to them, what they will do, what they should do, as they should do. (Participant, individual interview, Chile).

We are in a macho culture ... there is a long way to go. (Participants, group interview, Colombia)

Look at the training is really for everyone, there are no branches just for women nothing, everything is even, the training is the same, the idea is to come out of here with the same tools and enter into equal conditions and the one that remains in the place where they work is by their own merit. (Participant, individual interview, Chile)

6.4.6 Gender Diversity

While in the Colombian case, the participants argue that the problem of *gender diversity* is an issue of supply, given the scarcity of jobs for women in the sector, in Chile there is diversity discursive in relation to this topic. The interviewees admit that students should be foreseen in the development of personal and social skills. Therefore, they must "strengthen their personality", "acquire psychological and personal skills"and "implement leadership subjects". It is also required that they "develop the attitude", "the will", "develop soft skills", "a psychological work is needed", "the university should give skills to women", "there should be a subject that promotes the entry of women into mining" and "women must have tolerance to frustration", among others.

> In the case of the students of the last years we try to make several soft skills talks, personal skills so that they can perform better in any type of situation and put them in different situations, including how to negotiate salaries [...] We here give you the tools and we hope they take advantage of all their theoretical domain and make it work [...]. (Participant, individual interview, Chile).

> Job offers for female profiles are scarce in the sector. (Participants, group interview, Colombia).

> ….I remember that there was a course that helped students develop these characteristics (soft skills). I think those are the characteristics that women have to develop. (Participant, individual interview, Chile)

The female students made a discursive turn and questioned adaptation as a mechanism for the inclusion of women in the field, emphasizing *gender diversity* and highlighting the particularities that women contribute to the industry: "on the contrary, it is men who must learn", "they are much more careful" and "they are much more productive".

> Instead of women adapting to language or customs, we have to bring men to the customs and language that women have. Make them more sensitive, more understanding, more delicate, so that they can empathize. (Participant, individual interview, Chile)

> Women are very detailed and that is what they value. In some more advanced cultures it has been determined that women are more productive, because in some way we have much more responsibility, we are more committed, we have the ability to analyse things in a more different, then that adds value to the business. (Participant, individual interview, Chile)

6.4.7 Sexual Discrimination

In terms of *sexual discrimination*, the discourses of men show a structural argument; where this factor is adduced to systemic causes, some of the discursive categories require: "there are few women in careers", "women are unaware of their rights", " there is no will to hire women" and "there is a lack of women specialists". Eagly and Carli (2007) document some of these limitations and argue that one of the most obvious barriers is the fact that men expect their leaders to be women. However, in the case of the women interviewed, predominantly critical discourses point to the prevailing ideology of gender discrimination in both higher education and the workplace. These discursive categories reflect the disadvantages faced by students and professionals, through conditions of entry to the races, the female imagination and occupational segregation of industry, among others. Among the discursive categories that exemplify this factor are "there are gender barriers", "there is a detriment in the representation of women", "they are always offered less remuneration" and "they think they are not ready to enter this world (extractive industry) ", among others.

> There should be a way to break the paradigm and change the mentality between men and women. (Participants, group interview, Colombia)

> ... there is no real intention to incorporate them strongly, if they do not ask for it. (Participant, individual interview, Chile)

6.4.8 Governance Factors

The research findings show that the policy actions are key topics that concern most of the participants. In terms of government relations, the research findings indicate that the lack of collaboration between different stakeholders, government, business, universities and civil society prevents the sustainable leadership of women in the industry. Also, the absence of a strong government role has become unclear within policies on gender such as a balance of family and work, policies for maternity and child care, lack of new technologies and adequate infrastructure so that women can hold positions that are currently considered to be masculine. The main governance factors identified by the participants are working conditions, maternity leave, reconciliation and multiple roles.

6.4.9 The Working Conditions

The *working conditions* are related to the currently inadequate infrastructure of companies to successfully facilitate women incorporation in the workplace. The working conditions are related to the currently inadequate infrastructure of companies to successfully facilitate women incorporation in the workplace. But also to the greater opportunities for men to occupy positions in the companies, given that they have extra working time, allowing them to more easily meet the requirements of the head offices. An issue that for women is more restrictive, especially in the case of having small children (Navarro et al. 2016). The interviewees expressed this factor in: "there are no toilets, changing rooms, equipment problems". The women went further, stressing that in mining "everything is made for men: the structure, the hierarchical base, equipment, bathrooms, etc." and "the positions are held by men", among others.

> Older men have that prejudice with us [...] they are used to working with men, (they say) "stay beyond the shift because I need you" and there they stay because they have no problems, they know they have their women in the houses that are going to be taking care of the son and they do not make problems, but in the case of the women if they say "Hey, stay" one has already organized his schedule [...], then if they occupy more of my time to one they complicate it, then under that point of view they say: "She does not serve me because she is super squared in her schedule". (Participant, individual interview, Chile)

> There is also a lack of investment in technology to expand the participation of women in operational work. (Participant, group interview, Colombia)

6.4.10 Maternity

Motherhood was transversely identified by respondents. Men and women agree that it is an obstacle that is clearly described by its impact on employment screening of students and professionals in the field (Navarro et al. 2016): "Motherhood is the hardest thing", "they see their maternity as a factor that compromises their careers", "it is a barrier to growth at work" and "you have the vision that a woman who works does not have to have children".

> ... the problem is motherhood, several of them have compromised their careers in a certain way... this should not be the case .... (Participant, individual interview, Chile)

> Public policies are discriminatory. (Participants, group interview, Colombia)

> And in general they try to avoid hiring women because of the pregnancy issue, it is one of the biggest problems we have with our female member... the issue of motherhood. (Participant, individual interview, Chile)

6.4.11 Multiple Roles

Finally, "multiple roles" is a factor closely related to motherhood. This factor is emphasized in the wide range of tasks that are assigned to women, highlighting the poor compatibility between caring for children and the shift work system. This also has to do with the variety of problems that students face: "sometimes family and economic problems are bigger than academic issues".

> Working with women gets a little complicated… working at very long distances from the city, the house or in the shift system often does not make them compatible when they have children. (Participant, individual interview, Chile)

> Women have skills that are valued in the industry, such as organization and multi-task. (Participants, Gupal, Colombia)

Although the eight factors identified by the participants that compromise the sustainable leadership of women in the extractive industry are clearly prevalent, they are diverse in nature. The discursive categories analysed in each one of them show that there are more challenges than opportunities for women to experience fair and sustainable leadership in a highly segregated sectors, such as the extractive industry.

6.5 Impact Sustainability: Final Remarks

The challenges that women face in the extractive industry are determined by both internal and external factors, namely, sociocultural, corporate and governmental. The results show that this series of factors has a significant impact on the physical and emotional health of women, students and professionals, also compromising the productivity of companies and revealing the absence of governance frameworks to effectively promote the sustainable leadership of women within the industry.

Evidence of symptoms of stress, discrimination, feelings of guilt and changes in their feminine identity are all changes experienced by women due to the difficulties in making family life compatible with work and when facing an environment dominated by a masculine hegemony. Actions working towards the empowerment of women who face key barriers due to their own sense of self-discrimination should be included in corporate agendas. This needs to be mixed with public campaigns to promote self-confidence, support networks and forums of dialogue between men and women to create empathy and eliminate feelings of guilt, fear, inferiority and stereotypes that dominate the perception of their role in society. These limitations imposed by women, themselves and society indicate the sociocultural adversity faced by women in regard to sustainable leadership.

Businesses also face losses in productivity due to high turnover rates as a result of poor management of human capital. They also face potential fines for non-compliance with labour standards, as well as the deterioration of community relations. Through their corporate social responsibility agendas and strategies for sustainability, companies are able to transform. This, in collaboration with government, works to develop more holistic approaches that apply a sustainable lens towards education throughout childhood and higher education institutions. This will result in the installation of a sustainability mindset becoming naturalized within society. The strategies of sustainability should be more inclusive and accessible for the promotion of initiatives by civil society for sustainable leadership opportunities for women and the achievement of SDG 5.

Other actions which companies can execute to counteract these obstacles are, among others, the design and compliance of corporate policies that allow women to apply for executive positions with more attractive salaries in which their academic and professional skills are properly recognized and promoted. Such skills include but not limit to multitasking, demonstrating a high level of commitment to the company and communities where the industry operates which according to the research findings, in many cases, is usually greater than that of their male colleagues.

Regarding factors of governance, in spite of the principles and policies developed by international organizations to counter the adverse industry standards for balance between family and professional life of women, the study showed that these provisions have not been holistically incorporated in businesses or in government policies. Both case studies addressed the political climate of sustainability governance, as well as current mechanisms for compliance established by the government and used in industry. The study showed that a strong government presence in the coordination with the company and higher education institutions is one of the key requirements for institutional commitment which will be reflected in corporate policies and dialogue processes at all levels of governance: local, national and international. The institutional commitment from different levels of governance demonstrates the potential to create change, agreements, incentives and more inclusive evaluation measures to develop indicators for flexible and non-sexist working opportunities, as well as increased access to shifts that promote work-life balance policies throughout maternity and child care. Furthermore, policies that contest harassment and educational programs to break the paradigm of male providers and female housewives are also required. Finally, policies for the inclusion of new technologies and infrastructure improvements such that women can perform traditionally masculine positions and regulations to promote the selection of women through merit and not solely through a political process would also be highly beneficial.

It is thus necessary to highlight it is not just a women's issue but stakeholders' responsibility, namely, business, government and civil society, to foster women's sustainable leadership and contribute to the achievement of SDG 5 in the extractive industry. Therefore, the institutional commitment of all stakeholders is key to realizing the possibility for women to exercise gender equality.

References

Bartel C, Dutton J (2001) Ambiguous organizational memberships: constructing organizational identities. In: Social identity processes in organizational contexts. Psychology Press, New York, pp 115–130

Bell E, Sinclair A (2016) Bodies, sexualities and women leaders in popular culture: from spectacle to metapicture. Gender Manag 31(5/6):322–338

Cardoso P, Marques F (2008) Perception of career barriers: the importance of gender and ethnic variables. Int J EducVocat Guid 8:49–61

Codelco (2017). https://www.codelco.com/prontus_codelco/site/edic/base/port/proyectos.html

Council of Mining Competencies (2015) Recovered from http://www.ccm.cl/mujeres-en-la-mineria/

Denzin N, Lincoln Y (Coords) (2012) The field of qualitative research. In: Qualitative research manual, vol 1. Gedisa Editorial, Barcelona

Donoso T, Figuera P, Rodríguez M (2011) Barriers to gender in the professional development of university women. J Educ 355:187–212

Eagly AH, Carli LL (2007) Women and the labyrinth of leadership. Harv Bus Rev 85:9

Eagly AH, Karau SJ (2002) Role congruity theory of prejudice towards female leaders. Psychol Rev 109(3):573

ECLAC (2017) Technical-professional education and science and technology: keys for the economic autonomy of women. http://www.prigepp.org/emails/2017/07_04/landing/index.html

Franco I (2014) Building sustainable communities: enhancing human Capital in Resource Regions. PhD Dissertation. The University of Queensland, Brisbane

Higher Education Information Service (SIES) (2014) Main indicators of higher education in a gender perspective. http://www.mifuturo.cl/index.php/estudios/estudios-recentes

Hoobler JM, Masterson CR, Nkomo SM, Michel EJ (2016) The business case for women leaders: meta-analysis, research critique, and path forward. J Manag. https://doi.org/10.1177/0149206316628643

Ibañez M (2010) On the other side of occupational segregation by sex. Men in female occupations and women in male occupations. Int J Sociol 68(1):145–164

Ibarra H, Ely R, Kolb D (2013) Women rising: the unseen barriers. Harv Bus Rev 91(9):60–66

ICMM (International Council on Mining and Metals) (2003) 10 Principles. Retrieved March 23, 2017, from http://www.icmm.com/our-work/sustainable-development-framework/10-principles

International Finance Corporation (IFC) (2009) Women in mining: a guide to integrating women into the workforce. International Finance Corporation, World Bank Group, Washington, DC

International Finance Corporation (IFC) (2013) Investing in women's employment: good for business, good for development. International Finance Corporation Group, Washington, DC

Kark R, Eagly AH (2010) Gender and leadership: negotiating the labyrinth. In: Handbook of gender research in psychology. Springer, New York, pp 443–468

Meister A, Sinclair A, Jehn KA (2017) Identities under scrutiny: how women leaders navigate feeling misidentified at work. Leadersh Q 28(5):672–690

Ministry of National Education (n.d.) Women in higher education. Ministry of National Education, Bogota

National Council of Innovation and Competitiveness (2014). Mining: a future platform for Chile. Report to the president of the republic, Michelle Bachelet. Mining and Development Commission of Chile. [online]. Available in: http://programaaltaley.cl/wp-content/uploads/2015/10/Mineria-Una_Plataforma_de_Futuro_para_Chile_web.pdf

Navarro E, Román M, Infante M (2016) International review of studies of career barriers under the gender perspective in the construction industry. Innovate 26(61):103–117

Ramos J (2017) Challenge and opportunity of productivity in Chile and female labor participation (mining). National Productivity Commission. Strategic Conferences Exponor Chile 2017. Retrieved from http://www.exponor.cl/conferencia_estrategica.html

Salinas P, Cárdenas J (2009) Social research methods, Ecuador. Intiyan, Editions Ciespal 52

Salinas P, Romaní G (2017) Labor projection of female students in mining careers in Chilean higher education. Form Univ. [Online]. 2017, vol. 10, no 3, pp 31–48. ISSN 0718-5006. https://doi.org/10.4067/S0718-50062017000300005

Santander P (2011) Why and how to do discourse analysis Moebio tape 41:207–224. www.moebio.uchile.cl/41/santander.html

SERNAM (2015) Good labor practices with gender equality gender applied to people management. Claudia Echeverría Turres – women and labor area seminar on gender equity and good labor practices, Ministry of Foreign Affairs, September. http://www.minrel.gob.cl/minrel/site/artic/20090831/asocfile/20090831111948/seminario_rree_sernam_18_08_15.pdf

SERNAM (2017) More rights for women. Best Chile most happy and worthy. Ministry of Women and Gender Equity. March of the Women of the Government of Chile

Strauss A, Corbin J (2002) Bases of qualitative research. Faculty of Nursing of the University of Antioquia. Contus editions, Medellín

Swanson J, Woitke R (1997) Theory into practice in career assessment for women: assessment and interventions regarding perceived career barriers. J Career Assess 5:443–462

Swanson J, Daniels K, Tokar D (1996) Assessing perceptions of career related barriers: the career barriers inventory. J Career Assess 4(2):219–244

Terrill J (2016) Women in the Australian mining industry: careers and families. The University of Queensland, Brisbane

Trinidad A, Carrero V, Soriano R (2006) Grounded theory: the construction of theory through interpretational analysis. Center for Sociological Research, Madrid

Chapter 7
SDG 6 Clean Water and Sanitation

Sustainable Use of Energy and Water Resources in the Mining Sector: A Comparative Case Study of Open-Pit and Alluvial Mining Technology

Natalia A. Cano Londoño, Jessi Osorio Velasco, Felipe Castañeda García, and Isabel B. Franco

Abstract Environmental impacts associated with the use of water and energy resources are among the most significant problems for the mining industry, requiring the implementation of new solutions in line with Sustainable Development Goal 6 – Clean Water and Sanitation. Currently, the challenge is converting mineral wealth into development opportunities while responding to the needs of future generations. This is specifically regarding the investment of nonrenewable resources in the implementation of strategies to promote the efficient use of both renewable and nonrenewable energy sources. This chapter aims to evaluate the cradle-to-gate consumption of renewable (water) and nonrenewable energy sources in both open-pit and alluvial mining systems. Additionally, life cycle assessment (LCA) has been performed to both estimate and analyze their impact on water resources. This is extended by the presentation of opposing reductive strategies including the optimization of process efficiency and use of circular economies. This research is significant within the Colombian context as water usage is often a determining factor in the attainment of key environmental and social licenses. Furthermore, the

---

N. A. Cano Londoño (✉) · J. O. Velasco
School of Mines, Universidad Nacional de Colombia, Medellín, Colombia
e-mail: nacanol@unal.edu.co; jeosoriove@unal.edu.co

F. C. García
Grupo Mineros, Medellin, Colombia
e-mail: luis.castaneda@mineros.com.co

I. B. Franco
Institute for the Advanced Study of Sustainability, United Nations University Shibuya-ku, Tokyo, Japan

Australian Institute for Business and Economics, The University of Queensland, Brisbane, Australia
e-mail: connect@drisabelfranco.com

© Springer Nature Singapore Pte Ltd. 2020
I. B. Franco et al. (eds.), *Actioning the Global Goals for Local Impact*, Science for Sustainable Societies, https://doi.org/10.1007/978-981-32-9927-6_7

results of this investigation clearly show how water usage and the magnitude of its related impacts differ between opposing forms of extraction.

Keywords SDG 6 Clean Water and Sanitation · Sustainability · Water use · Energy use · Efficiency · Gold mining

7.1 Introduction

Despite the great efforts made by society to dematerialize, that is, to reduce the amount of energy and materials required for economic function, through strategic reuse and recycling of materials (Ruiz-Mercado et al. 2017), global projections indicate the primary metal production is set to expand into the future. This can be seen as a direct result of societal impacts such as population growth (Norgate and Haque 2010; Awuah-Offei 2016). However, the mining sector is under increased scrutiny to reduce its energy consumption as well as all related environmental, social, and economic impacts generated as a result of its operation (Norgate and Haque 2010).

Although mineral extraction is seen as an essential source of resources, it is also deemed to have extraordinarily detrimental economic, social, and environmental impacts on both the global and local scale (Vintró et al. 2014). While the mining sector supplies vital raw materials and energy to a large number of industries, its activities are still commonly considered to be a threat to the environment, especially due to their related effects on the air, water, and soil quality. These detrimental effects also include greenhouse gas emissions, destruction of ecosystems, damage to protected areas, and pollution (Vintró et al. 2014; Bustamante et al. 2017). Furthermore, these impacts are expected to increase exponentially as globally the ore grade (metal content) has been plummeting for some time (Hubbert 1956; Valero Delgado 2013).

Mining companies consequently face unprecedented social pressure to assume strong commitments in responsible management of their environmental and social environments (Botín and Vergara 2015), particularly in regard to the achievement of the Sustainable Development Goal 6 – Clean Water and Sanitation (hereinafter SDG 6). Sustainability is becoming more and more commonly used to describe this paradigm which supports the configuration of the environmental, social, and economic future of humanity (Kharrazi et al. 2014).

There is a social conception that mining cannot be deemed sustainable, as its operations have a finite lifespan, and the dependence of humanity on nonrenewable resources cannot progress indefinitely (Sterman 2012). Additionally, as the mismanagement of these nonrenewable resources will eventually cause them to expire, mining is seen as having destructive environmental and social impacts (Young and Septoff 2002). These are some of the main controversial matters that opponents use to abort mining projects. Furthermore, the primary obstacle of mining projects today often does not lie in the environmental license nor in the relevant governmental operating license; it lies in obtaining a *social license*, a situation that triggers social, economic, and environmental problems (Franco 2014).

The mining industry, as will be discussed throughout this chapter, can contribute to sustainable development, in the sense that if properly managed, it can provide long-term opportunities for economic growth and social development with "acceptable" environmental impacts (International Council on Mining and Metals (ICMM 2012a). In particular, mining produces minerals, metals, and energy, which have been the central driver of development since before the industrial age (Auty and Warhurst 1993; Trigger 2005; ICMM 2012a, b).

The United Nations Sustainable Development Goals (SDGs) provide an ambitious set of targets for improving environmental sustainability, economic development, social cohesion, and human development by 2030 (United Nations 2015). The functions of these goals are eliminating poverty, combating inequalities within and between individual countries, building righteous societies, and combating gender inequality (Pactwa et al. 2018). The primary input in the gold extraction process is water. Consumed water often comes from a variety of sources; the quality of these sources is regularly the cause of controversial debate within the industry. Instilling confidence of mining's contribution to the responsible and sustainable use of this water is instrumental in the development of the industry. Therefore, this paper focuses on SDG 6 "Clean Water and Sanitation" (Drelich 2012).

This chapter will critically analyze the consumption of both renewable and nonrenewable resources while using life cycle assessment (LCA) to estimate the industry's detrimental water impact. The methodology included in this chapter consists of three key stages: (1) analyzing energy and mass balance of both open-pit and alluvial mining systems, (2) comparing data within the literature to the processing of other minerals worldwide, (3) and estimating the detrimental water impact through LCA and drawing comparisons between data found within the literature. The environmental impacts included within these steps were classified through the ReCiPe method based on 1 kg of gold as a functional unit under the Ecoindicador 99 methodology, the Ecoinvent 3.1 database, and the Umberto NTX Universal Software. This chapter will then present conclusions and strategies to achieve more efficient resource consumption within the mining sector.

7.2 Literature Review

7.2.1 Description of Gold Mining System Technology in Colombia: Alluvial Mining and Open-Pit Mining

One of the major challenges in mining often emerges in the extraction process. Challenges differ depending on the type of deposit, namely, alluvial, philonian, or spread deposits. Geological environments rely on various different extraction systems. These also respond to specific local requirements in accordance with the production and economic capacities and can then be used to limit the environmental impacts experienced in the production process (Orche 1998).

This chapter presents both open-pit and alluvial mining as case studies to assess the sustainability of gold mining. In recent years, open-pit mining has been seen as the preferred method, while alluvial is seen as slightly less conventional (Norgate and Haque 2012). The case study mine sites are located in Antioquia, Colombia. The open-pit mine is located in South-East Antioquia, and the alluvial mine-site is in North-East Antioquia.

Generally, an extractive process can be understood as inclusive of both the prospecting and exploration stage and the separation of the mineral of interest. For both systems the operational process included the cut and stripping stage, the extraction stage, and finally the processing and transformation of a gold ingot. Data was supplied annually over a 6-year period for alluvial mining and an 11-year period for open-pit mining.

7.2.1.1 Open-Pit Mining

Figure 7.1 below depicts the open-pit mining extractive process. Land is prepared through removal of the vegetation cover and organic soil in the clearing and stripping stage, and residual biomass is then stored for future land restoration. Mineral excavation is then carried out by conventional extraction methods, which may include drilling, blasting, loading, and hauling.

Ore beneficiation is carried out through physical-chemical processes. It begins with the size reduction of the excavated mineral by means of primary crushing and both primary and secondary milling. Irrigation water is used in this step to minimize the detrimental impacts of total particulate matter (PST) emitted. This stage produces two different process lines; the first flow moves straight onto the flotation process. The flotation process helps to concentrate the sulfide minerals containing gold (96.3% of gold and 79.5% of silver are recovered in this stage, respectively). Then, the mineral resulting from flotation goes through an intensive leaching process (34.7% and 10% of gold and silver are recovered in this process, respectively). Finally, tailings are stored in a tailing pond. Then it is directly transported to the gold recoverable by gravity (GRG) process.

Gold and other metals extracted from both the leaching process (cyanidation) and gravimetric separation are adsorbed on activated carbon in a carbon-in-pulp circuit (CIP). They are then released into an elution column under certain conditions of pressure and temperature. This gold-rich solution continues the electrowinning process where a selective precipitation is made through electrolysis. Once the electrowinning of gold and silver is obtained, it is sent to the casting furnace. Assuming no losses in the smelting and casting process, approximately 1904 ton/year of gold and 2155 ton/year of silver are produced, which amounts to around 952 gold ingots and 1077 silver ingots, with a 900 millesimal fineness.

Tailings are conformed by flotation tails. Evidence shows that 96.5% of the total industrial wastewater is generated in the beneficiation process, with leaching and carbon adsorption tails corresponding to the remaining 3.5%. These last two are treated by a detoxification system, in which the solution of circuit is oxidized by

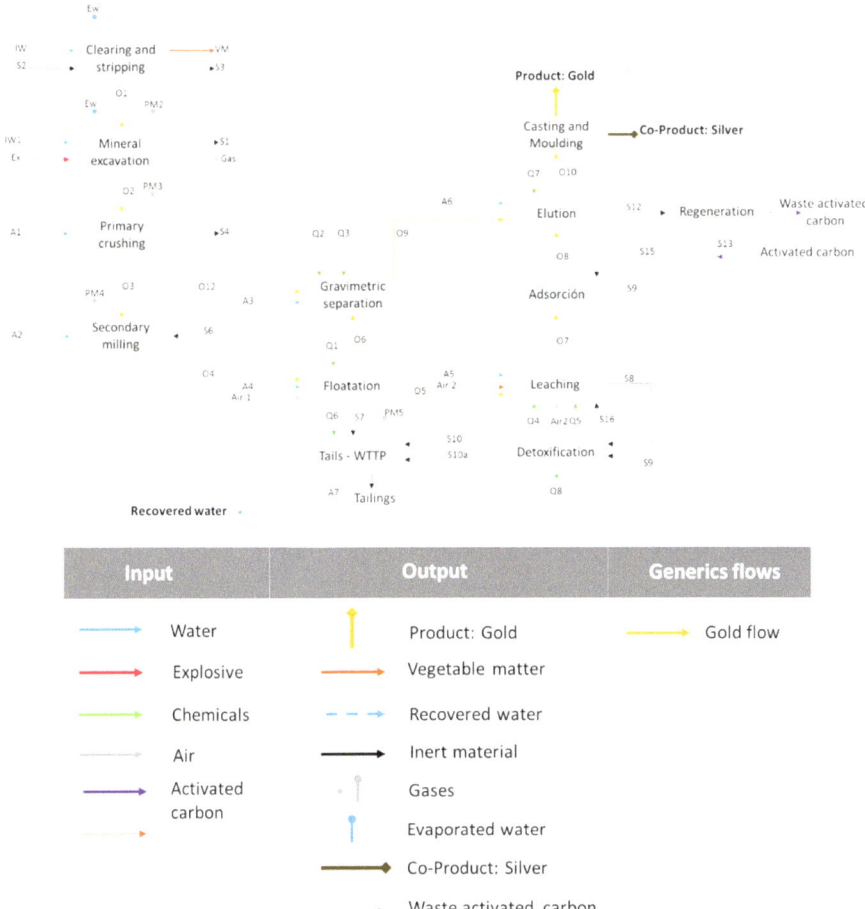

Fig. 7.1 Description of open-pit mining process

applying hydrogen peroxide (H2O2) before being stored in the tailings pool, in which a dewatering process is undertaken and recirculation is carried out. This process consumes 98% of the extraction and beneficiation process' water.

Sterile carbon resulting from the elution process (gold uncharged carbon) is sent to the carbon-reversing furnace for reactivation and reuse in the CIP process. Finally, 83.98% of the water in the entire process is recirculated.

7.2.1.2 Alluvial Mining

Figure 7.2 below presents a detailed description of the alluvial mining process. It is important that the selected sediment deposits are conducive to the exploitation of alluvial gold. This selection is carried out in the exploration process. After this

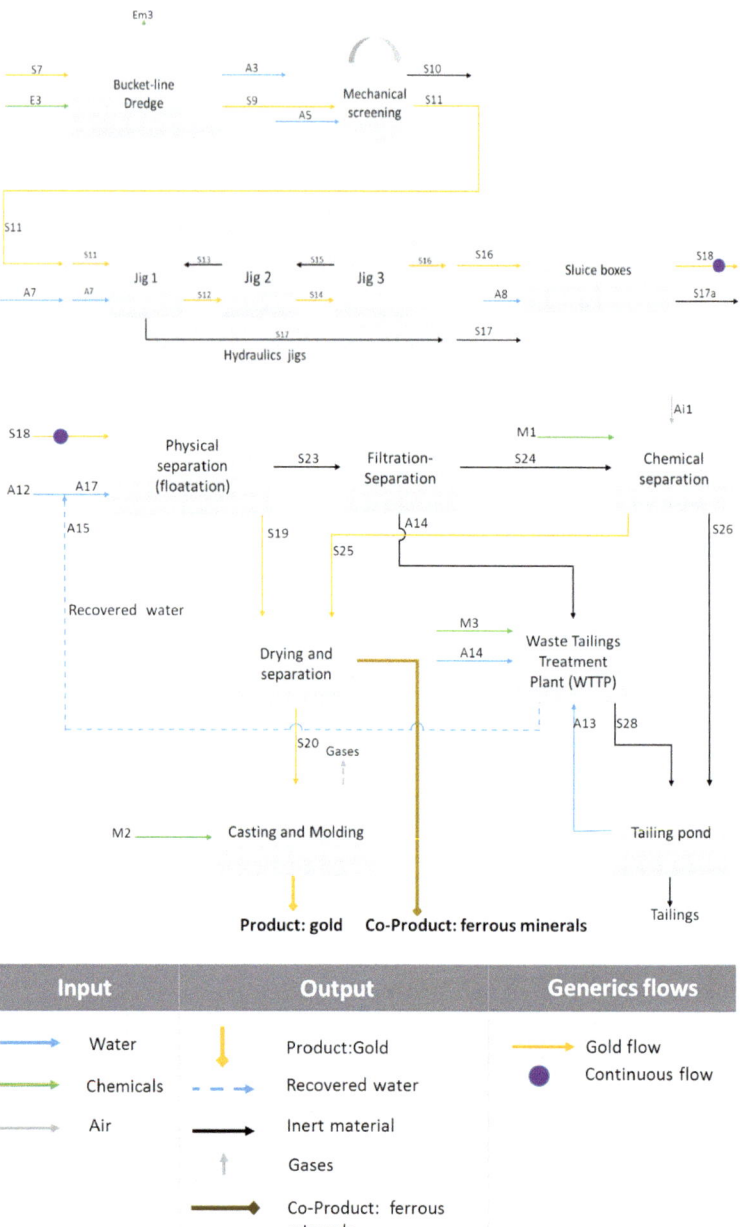

Fig. 7.2 Description of open-pit mining process

selection, preparation and access to the exploitation zone continues through the clearing and stripping by suction dredgers. This process converts vegetation cover to bare soil by cutting and removing the superficial horizons of the soil (UNDP 2016).

Mineral beneficiation activities occur simultaneously with the beginning of operations. Mineral excavation (consisting of the excavation of sand, gravel, clay, and the mineral of interest) is carried out by dipper dredger (dredging line step), which extracts the ore from the alluvial deposits. This is followed by an analysis of the gold physical beneficiation by size classification (mechanical screening) and gravimetric concentration using hydraulic jigs and sluice boxes. As a result of the first stages of this process, a waste line (typically a sterile material such as gravel or sand) is produced. Additionally, a second wet process flow, which is rich in gold and mixed with sands, ferrous metals, and other impurities, continues down the process line in order to increase the concentration and purification of gold (flotation stage). The 11% of ore (dry basis) enters along a continuous stage in both the beneficiation line and the separation stage, where 99% of the process stream moisture is chemically removed for further concentration. This process is carried out with the objective of recovering 4% of the gold not recovered in the flotation stage. These gold-rich flows (wet basis), which are produced in the flotation and chemical separation process, continue in the drying line, and simultaneously the gold is separated from the ferrous minerals which correspond to about 3% of the gold-rich flow line. The gold obtained from these concentrates is then melted and casted. Assuming no losses in the smelting and casting process; 3103 ton/year is melted, which amounts to approximately 155 ingots with a 900 millesimal fineness.

Tailings generated in both the filtration-separation and chemical separation stages are submitted to a Waste Tailings Treatment Plant (WTTP), where 99% of the water used in the beneficiary process is recovered and reused in this same process, together with the water obtained from dewatering the tailings pond. Ferrous metal is then stored for future economical uses, as a coproduct of the process.

7.3 Methodology

7.3.1 Life Cycle Assessment

Life cycle assessment (LCA) is an environmental approach that considers both natural resource consumption and pollutant emissions throughout a product or activities' entire life cycle (Blengini et al. 2012). The methodology is standardized by ISO 14040 (ISO 1998) and made up of four stages: goal and scope definition, inventory analysis, impact assessment, and interpretation. For the first stage, the research's motivation is to determine the selection of impact categories and characterization methods while defining the system boundaries and the functional unit. The second step, inventory analysis, involves collating all relevant data including resource inputs, products, and emissions. Impacts are then quantified through characterization methods which depend on the relevant impacts of interest and the chosen assessment method. Finally, the results are analyzed to aid in the relevant decision-making process.

Furthermore, ReCiPe methodology (uses hierarchies, to include long-term effects) has been chosen to include midpoint indicators, including freshwater ecotoxicity, freshwater eutrophication, marine ecotoxicity, marine eutrophication, and water depletion, although these indicators are obviously only related to the assessments relevant to water-based ecosystems.

7.4 Discussion and Results

7.4.1 Energy and Renewable Input Resources

Environmental impacts associated with energy consumption and water use are among the most important issues for the mining industry (Pimentel et al. 2016). Consumption of energy, nonrenewable and renewable resources (water) is clearly emphasized for each stage listed above, with the challenge being to improve the efficiency of both mining systems. It should also be pointed out that greater environmental impacts are expected from the stages with the highest consumption of nonrenewable and renewable resources. As such it is clear that the global production-consumption cycles of minerals, water, and energy are inextricably connected.

7.4.1.1 Energy Consumption in Mining Activity

Northey et al. (2013) suggest characterizing energy consumption by type (electrical, diesel input for machines, gas) and electricity source (produced on-site or off-site) to find specific opportunities for improving resource consumption and sustainability reporting (Northey et al. 2013).

Values for direct and indirect energy consumption provide an approximation of the electricity, gas, and energy acquired by mines or generated in situ. Figure 7.3 shows energy consumption, energy losses (38.93% of the total energy input), and total useful energy (61.07% of the total energy input) for open-pit mining systems. Additionally, 63.56% of the total energy consumed comes from electricity, 35.91% from diesel, and 0.53% from gas. Highest electricity, gas, and diesel energy consumed in alluvial mining are exhausted in the "grinding mill" stage (67.14%), other services use 100%, and "mineral excavation" stage uses 99.41% of total consumption for the entire process. The highest loss of energy happens in the "services and mineral excavation" stage with a first law efficiency equal to 35.0%.

Comparatively, in alluvial mining technology, energy losses (10.93% of the total energy input) and total useful energy (89.07% of the total energy input) are presented in Fig. 7.4. Furthermore, 99.55% of the total energy consumption comes from elec-

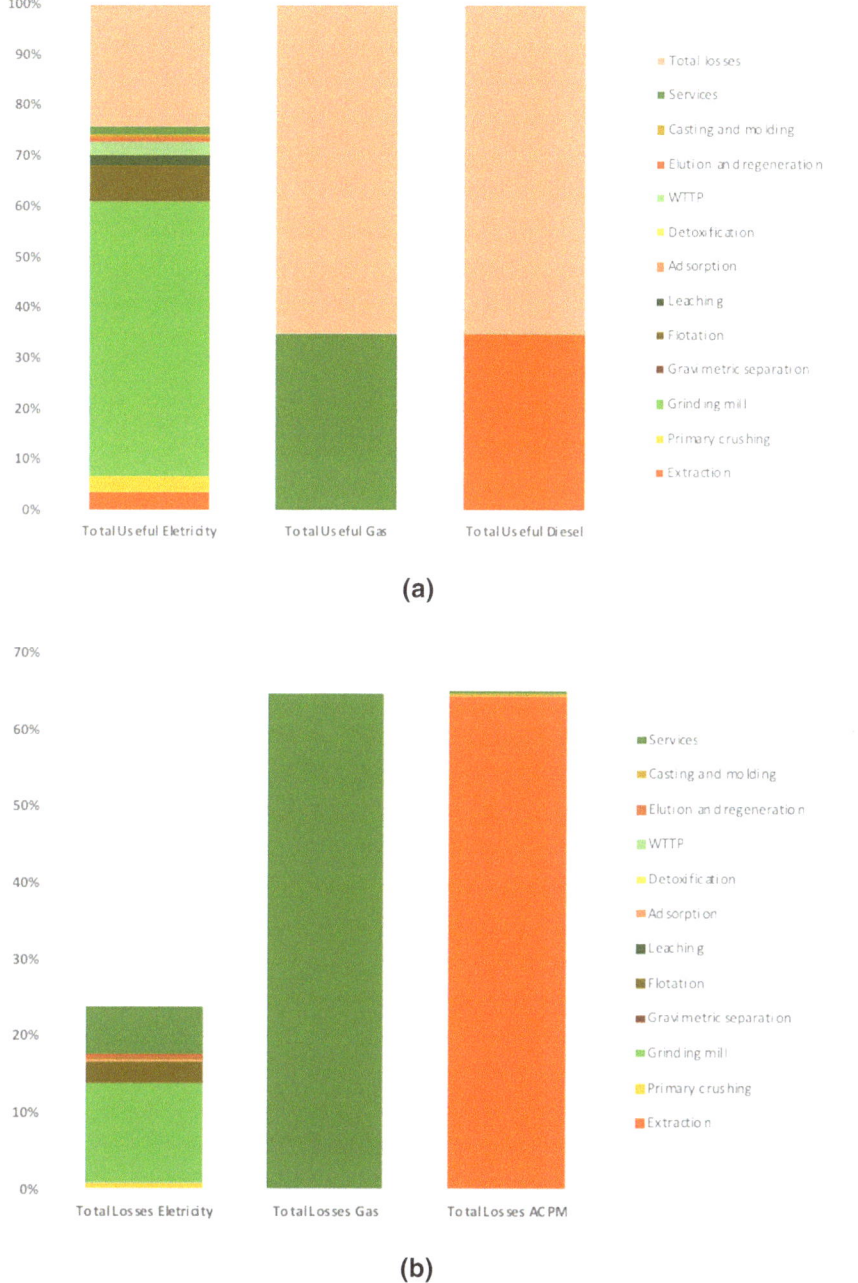

Fig. 7.3 Energy consumption and loss for each stage of the process in open-pit mining technology. (**a**) Energy consumption for each stage open-pit mining. (**b**) Energy loss for each stage of the open-pit mine. (Note: Efficiency of first law (Cullen and Allwood 2010; Pellegrino et al. 2004; Romero Rueda et al. 2012). Extraction ($\eta = 38\%$), primary crushing ($\eta = 81\%$), grinding mill ($\eta = 81\%$), gravimetric separation ($\eta = 90\%$), flotation ($\eta = 72\%$), leaching ($\eta = 90\%$), adsorption ($\eta = 55\%$), detoxification ($\eta = 72\%$), WTTP ($\eta = 90\%$), elution and regeneration ($\eta = 72\%$), casting and molding ($\eta = 82\%$), other services ($\eta = 21\%$))

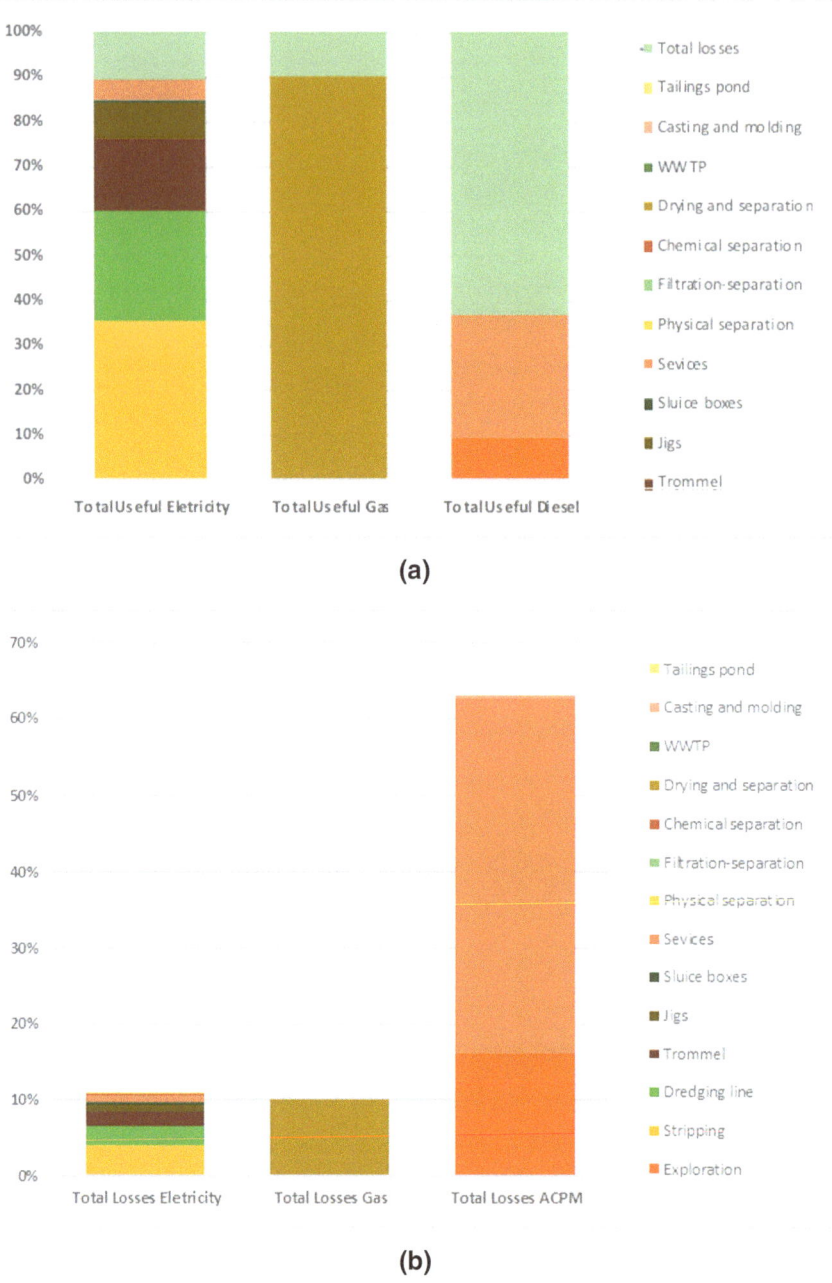

Fig. 7.4 Energy consumption for each stage of the process in alluvial mining technology. (**a**) Energy consumption for each stage of the process. (**b**) Energy loss for each stage of the process. (Note: First law efficiency (Cullen and Allwood 2010; Pellegrino et al. 2004; Romero Rueda et al. 2012). Exploration ($\eta = 37\%$), stripping ($\eta = 90\%$), dredging line ($\eta = 90\%$), mechanical screening ($\eta = 90\%$), hydraulic jigs ($\eta = 90\%$), sluice boxes ($\eta = 72\%$), services ($\eta = 78\%$), physical separation ($\eta = 90\%$), filtration-separation ($\eta = 72\%$), chemical separation ($\eta = 72\%$), drying and separation ($\eta = 90\%$), WTTP ($\eta = 72\%$), casting and molding ($\eta = 37\%$), tailings pond ($\eta = 72\%$). Energy consumption is not specific)

tricity, 0.44% from diesel, and > 0.01% from gas. The highest electricity, gas, and diesel energy consumed in alluvial mining are exhausted in the "clearing and stripping" stage 39.43%, "drying and separation" stage 100%, and "mineral excavation" stage 74.20% of total consumption of the entire process. The highest loss of energy is presented in the "clearing and stripping" stage with a first law efficiency equal to 90%. However, this mining system has installed their own run-of-river hydroelectric plant, in which 1.46E+18 kJ is generated annually, contrary to open-pit mining where all consumed electricity is generated off-site.

7.4.1.2 Renewable Resources in Mining Activity (Water Consumption)

Regarding nonrenewable (removed material mining progress) and renewable resources (water), Figs. 7.5 and 7.6 demonstrate the consumption of all stages of the mineral extraction process. For both methods, the increase of gold fraction in contrast with the decrease of sterile mineral fraction is obvious. In addition, water consumption is evident in each stage of the process.

Both systems use a great proportion of water in the "ore benefit and metallurgical extraction" phase. However, the highest water inputs are in the "grinding mill" stage (3.59, E+07 ton) and the "mechanical screening" stage (7.46, E+07 ton) for open-pit and alluvial mining, respectively.

It is noteworthy that out of the 83.98% of water consumed in the mining process, 24.65% is returned to the same catchment area. Finally, after balancing the relevant water inputs and outputs included in the process, water consumption amounts to 4.79 E+02 ton/year for open-pit mining and 2.36 E+04 ton/year for alluvial mining. This corresponds to 16.02% and 74.90% of the total water inputs, respectively. However, both water recovery and the intensity of usage are dependent on cooler climates or arid regions (Northey et al. 2013), as higher temperatures in arid regions

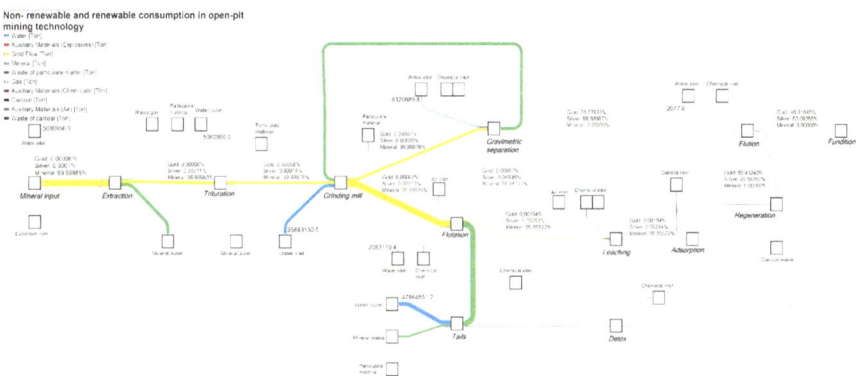

Fig. 7.5 Nonrenewable (inert material removed) and renewable (water) consumption in open-pit mining technology from cradle to gate. Nonrenewable (inert material removed) and renewable consumption is not specific

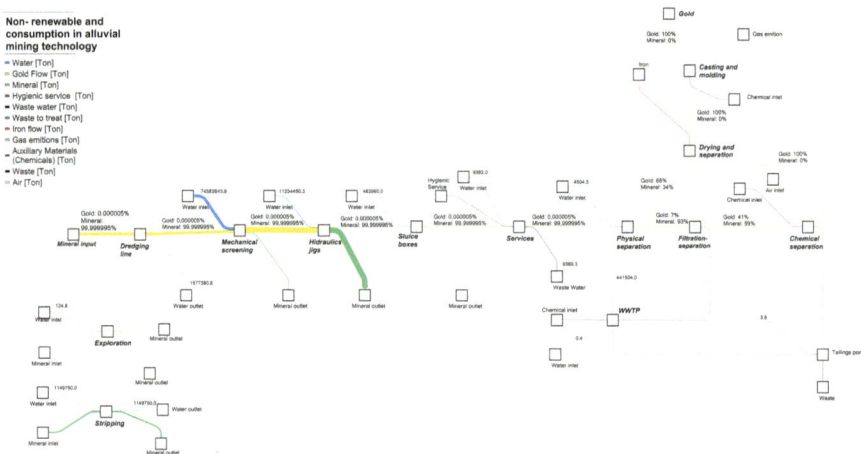

Fig. 7.6 Nonrenewable (inert material removed) and renewable (water) consumption in alluvial mining technology from cradle to gate. Nonrenewable (inert material removed) and renewable consumption is not specific

result in greater water loss throughout the site, via evaporation. As such this reduces the amount of water recoverable through tailing dewatering (Castillo et al. 2001). Furthermore, there is greater demand for water in dust suppression as soils in these regions have lower moisture contents (Gambatese and James 2001).

Mining processes associated with other minerals including copper, iron, and bauxite exceed the referenced consumption shown in Table 7.1; however, it is clear that this comparison is broad and requires a more thorough study to be valid.

7.4.2 Water Impact Categories in Life Cycle Assessment

Despite the intensive water usage present in both mining technologies, this study has not considered the full water footprint. Its analysis is based on a LCA. However, it is only considered to be a full LCA if all of the following metrics are analyzed (Pfister and Ridoutt 2013; Boulay et al. 2014; ISO 2014; Boulay et al. 2015): water availability indicator, quality indicators of water degradation (eutrophication, ecotoxicity, acidification, among others), and endpoint modeling (human health, ecosystems, and resources) (ISO 2014). Despite this, only quality indicators of water degradation and specific water sources will be assessed due to the following reasons:

- According to Padowski et al. (2016), Colombia presents low values when assessing global water security, underwater availability, and accessibility to water services. Other studies approach freshwater withdrawals (Lawrence et al. 2002; UNDP 2016).

7 SDG 6 Clean Water and Sanitation

Table 7.1 Energy and water consumption in other studies

	Open-pit	Alluvial	Papua New Guinea[a]	Peru[b]	ROW[c]	Haque and Norgate (2014)	Norgate and Haque (2012)	Mudd (2007)
Energy kJ/kg Au	1.68E+14	8.20E+13	4.72E+07	6.39E+07	1.54E+08	3.47E+08	4.65E+08	1.49E+11
Water kg/kg Au	2.51E+09	1.43E+08	1.33E+06	6.89E+07	4.28E+05	2.88E+02	2.59E+02	6.35E+05

Collection of a range of data of the Australian mining industry (Mudd 2010)

Note: Copper (6.10E+10 kJ/kg), iron (1.53E+02 kJ/kg), bauxite (5.49E+01 kJ/kg) (Norgate and Haque 2010; Norgate et al. 2007)

[a]Papua New Guinea, gold-silver mine operation with refinery. From the transport of raw materials to the mine, until the refining of gold and silver (Gobain et al. 2016)

[b]Peru Yanacocha Mine, cradle-to-gate from raw material until refining gold (Gobain et al. 2016), gold-silver mine operation with refinery, PE, (Author: Matthias Tuchschmid (obsolete) active)

[c]Row (rest of the world), open-pit gold-silver mine operation with refinery. This multi-output process "mining and refining, gold-silver deposit" delivers the two coproducts: gold and silver. This data set includes the combined mining and refining of gold and silver in open-pit mines in Peru

- According to the water footprint framework formalized in ISO 14046 standard for LCA. No consensus-based approach exists for applying this standard. Furthermore, results are not always comparable when different scarcity or stress indicators are used for the characterization of impacts (Boulay et al. 2017). Additionally, regionalized assessment is still a challenge with current databases and software, referring to a "global" region without specific geographic information (Boulay et al. 2017).
- A range of methodological and data limitations hamper the efforts to conduct water footprint studies of mining, such as private mining companies not releasing data to the public (Northey et al. 2016). General life cycle assessment studies, which include "water use," are mainly focused only on water consumption and neglect the environmental impacts of this consumption.

Table 7.1 compares the life cycle assessment water indicators of both open-pit and alluvial mining combined with mining processes taken from the Ecoinvent 3.1 database. It shows that for all water impact categories, alluvial mining technology presented the lowest value in relation to open-pit mining and the standard results from Perú, Papua New Guinea, and ROW (from the Ecoinvent 3.1 database). This is of course excluding water depletion where it obtained the highest values for its excessive water usage.

The unusual behavior presented by alluvial mining in comparison with other mining systems could be explained by the stark difference in its extractive technique. Conventional extractive processes such as open-pit mining and those presented by Ecoinvent 3.1 all seem to have similar environmental impacts as they all use a similar conventional ore extractive method. However, it is worth noting that values for all impact categories showed in Table 7.2 are higher for the open-pit technology than in the Ecoinvent processes.

Figure 7.7 shows the contribution of each process to open-pit mining's water environmental impact categories. From the figure, 98.9% of the total impacts to the water resource (9.63×10^4) are generated by sulfide tailings. This is a direct result of its huge need of area and its contents of toxic substances. Tailings generate almost 99% of total impacts in the following impact categories: freshwater ecotoxicity, freshwater eutrophication, marine ecotoxicity, and marine eutrophication.

In the case of alluvial mining technology (see Fig. 7.8), 99.9% of the total water impacts (2.82×10^4) come from water extracted from the river to build the digging pond and use in the physical separation of gold by gravimetric concentration, which mainly affected water depletion. The rest of the impact categories are largely affected by the run-of-river electricity production.

In regard to mining technologies' impact on water resources, freshwater ecotoxicity and marine ecotoxicity were clearly the most affected impact categories in comparison to freshwater and marine eutrophication indicators.

It is also clear that according to the laboratory values for the tailings composition, specific substances cause the high toxicity values from the open-pit mining process. The phosphorus content of the tailings slug contributes to nearly 75% of open-pit mining's impact on freshwater ecotoxicity and close to 99% of freshwater

Table 7.2 Water impact categories. A comparison of gold processes: alluvial, open-pit mining, and mining systems from Ecoinvent 3.1. database

Damage categories	Impact categories	Alluvial mining technology	Normalized values	Open-pit mining technology	Normalized values	Ecoinvent 3.1. database Peru	Normalized values	Ecoinvent 3.1. database Papua New Guinea	Normalized values	Ecoinvent 3.1. database ROW	Normalized values
Ecosystem (water)	Freshwater ecotoxicity (kg 1,4-DB eq/year)	1.29E+01	3.01E+00	7.39E+04	1.72E+04	1.41E+04	3.28E+03	6.33E+03	1.47E+03	1.24E+04	2.89E+03
	Freshwater eutrophication (kg P-eq/m^3)	4.20E-02	1.45E-01	4.94E+02	1.71E+03	3.80E+02	1.31E+03	1.70E+02	5.87E+02	3.24E+02	1.12E+03
	Marine ecotoxicity (kg 1,4-DB eq/year)	1.14E+01	4.63E+00	2.11E+04	8.57E+03	1.25E+04	5.08E+03	9.07E+03	3.68E+03	1.18E+04	4.79E+03
	Marine eutrophication (kg N-eq/year)	3.94E-01	5.37E-02	8.55E+01	1.16E+01	1.06E+02	1.44E+01	1.20E+02	1.64E+01	1.26E+02	1.72E+01
	Water depletion (m^3/year)[1]	2.82E+04		4.91E+02		1.89E+01		1.23E+03		1.34E+02	

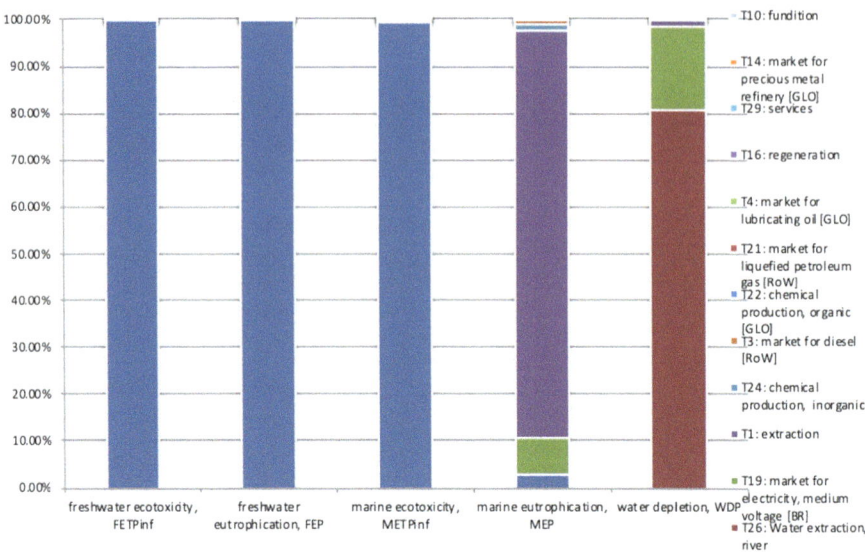

Fig. 7.7 Water impact categories in open-pit mining technology by processes

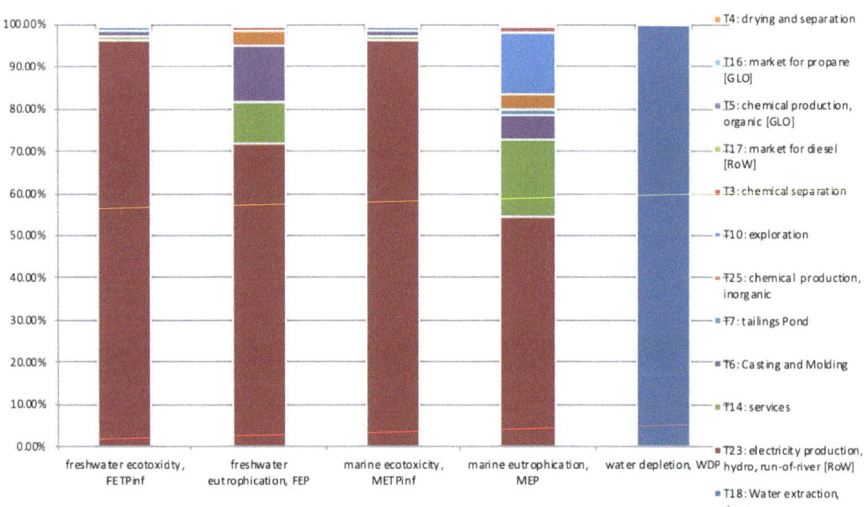

Fig. 7.8 Water impact categories in alluvial mining technology by processes

eutrophication. Furthermore, copper contributes nearly 36% of the marine ecotoxicity impact and nitrogen close to 72% of marine eutrophication.

In the alluvial mine, 83% of freshwater toxicity stems from copper, 75% of freshwater eutrophication from phosphate, 81% of marine ecotoxicity from copper, and finally 53% of marine eutrophication from oxides of nitrogen. These compounds enter the tailings pond from use in the chemical separation process.

7.5 Impact Sustainability: Final Remarks

Renewable and nonrenewable resource consumption can be reduced through strategic management of water, energy, and relevant mining materials, which can contribute to the achievement of SDG 6. Evidence shows that 83.98% of water consumption is reused in open-pit mining, while 24.65% of water used in alluvial mining can be returned to the same catchment area. It would also be possible to increase the energy efficiency presented in each stage of the process through improving the adopted technology, specifically focusing on those stages with both the highest consumptions ("grinding mill" and "extraction" stages in open-pit and the "stripping and dredging" stage in alluvial mining systems) and the highest losses ("extraction and services" stage in open-pit mining and the "exploration and services" stage in alluvial mining). Finally, in terms of the excavated inert material in alluvial mining, a minimum soil tenor equal to 100 mg gold/m^3 must be ensured for the extractive activity to be technically, economically, energetically, and environmentally viable. The same is true for open-pit mining with a minimum soil tenor equal to 470 mg gold/ton. Furthermore, the material from the alluvial process returns to the artificial pool, while in open-pit it is temporarily stored for beneficiation processes after the mining system's end-of-life (EoL) .

Open-pit mining technology presents higher values in relation to each water resource impact category except water depletion, where it is significantly higher in alluvial mining technology. This is due to intensive use of water resources with a value equal to 2.82×10^4 m3/year. The total water consumed in open-pit and alluvial mining technologies is equal to 5.70×10^7 ton/year and 9.79×10^7 ton/year, respectively. However, in the alluvial extractive process, the amount of water returned to the river basin is equal to 2.41×10^7 ton/year; and water recirculated into the process amounts to 4.79×10^7 ton/year and 4.42×10^5 ton/year in the open and alluvial systems, respectively, which reduces the water consumption values to 9.83×10^6 ton/year and 7.33×10^7 ton/year, respectively, as shown in Figs. 7.5 and 7.6.

Improvements in mining projects' sustainability lie in the efficient use of the resources coupled with the optimization of process efficiency. This leads to the reduction of emissions and generated waste, therefore reducing pollution released in the environment. With these strategies, virgin resources are diminished, which implies a lower extraction rate. Furthermore, as waste is inherent in the process, it cannot be avoided. As such, the project must work to improve its material recycling and reuse strategies.

When designing the waste reuse processes, it is essential to imagine a loop economy, a regenerative system in which resource consumption, waste, emission, and energy loss are minimized by the slowing, closing, and narrowing of material and energy loops (Stahel and Reday 1976). This works to encourage reuse, remanufacturing, refurbishing, and recycling when possible. However, primary metals will be still required throughout the transition toward a more sustainable solution, as it is unreasonable to assume that the recycling of end-of-life (EoL) products will entirely replace primary extraction in the near or distant future.

In order to manage the extraction of nonrenewable resources in a sustainable manner, it is necessary to determine various factors such as the criticality of the material for society, its economic importance, the stability of its deliverance, its substitutability, and finally its recycling potential. By assessing these parameters, the following strategies can be adopted to improve sustainable resource consumption, namely, substitution of a resource for another less scarce resource, increasing materials' efficiency, or increasing the recyclability of resources (Henckens et al. 2016).

References

Auty R, Warhurst A (1993) Sustainable development in mineral exporting economies. Resour Policy 19(1):14–29. https://doi.org/10.1016/0301-4207(93)90049-s

Awuah-Offei K (2016) Energy efficiency in mining: a review with emphasis on the role of operators in loading and hauling operations. J Clean Prod 117:89–97. https://doi.org/10.1016/j.jclepro.2016.01.035

Blengini G, Garbarino E, Šolar S, Shields D, Hámor T, Vinai R, Agioutantis Z (2012) Life cycle assessment guidelines for the sustainable production and recycling of aggregates: the sustainable aggregates resource management project (SARMa). J Clean Prod 27:177–181. https://doi.org/10.1016/j.jclepro.2012.01.020

Botín J, Vergara M (2015) A cost management model for economic sustainability and continuous improvement of mining operations. Resour Policy 46:212–218. https://doi.org/10.1016/j.resourpol.2015.10.004

Boulay A, Motoshita M, Pfister S, Bulle C, Muñoz I, Franceschini H, Margni M (2014) Analysis of water use impact assessment methods (part A): evaluation of modeling choices based on a quantitative comparison of scarcity and human health indicators. Int J Life Cycle Assess 20(1):139–160. https://doi.org/10.1007/s11367-014-0814-2

Boulay A, Bayart J, Bulle C, Franceschini H, Motoshita M, Muñoz I et al (2015) Analysis of water use impact assessment methods (part B): applicability for water footprinting and decision making with a laundry case study. Int J Life Cycle Assess 20(6):865–879. https://doi.org/10.1007/s11367-015-0868-9

Boulay AM, Bare J, Benini L, Berger M, Lathuillière MJ, Manzardo A et al (2017) The WULCA consensus characterization model for water scarcity footprints: assessing impacts of water consumption based on available water remaining (AWARE). Int J Life Cycle Assess:1–11. https://doi.org/10.1007/s11367-017-1333-8

Bustamante N, Danoucaras N, McIntyre N, Díaz-Martínez JC, Restrepo-Baena OJ (2017) Review of improving the water management for the informal gold mining in Colombia. Revista Tecnica de La Facultad de Ingenieria Universidad Del Zulia 40(3):181–192. https://doi.org/10.17533/udea.redin.n79a16

Castillo R, Sanchez JM, Araya V (2001) Large mines and the community: socioeconomic and environmental effects in Latin America, Canada, and Spain. World Bank Publications, Washington, DC, pp 87–142

Cullen J, Allwood J (2010) Theoretical efficiency limits for energy conversion devices. Energy 35(5):2059–2069. https://doi.org/10.1016/j.energy.2010.01.024

Drelich J (2012) Water in mineral processing, proceedings of the first international symposium. Society for Mining, Metallurgy, and Explotation, Colorado

Franco I (2014) Building sustainable communities: enhancing human capital in resource regions. PhD dissertation. The University of Queensland, Brisbane, Australia

Gambatese J, James D (2001) Dust suppression using truck-mounted water spray system. J Constr Eng Manag 127(1):53–59. https://doi.org/10.1061/(asce)0733-9364(2001)127:1(53)

Gobain S, Sa I, Frank A (2016) Dataset information (UPR), 01(2014), pp 1–15

Haque N, Norgate T (2014) The greenhouse gas footprint of in-situ leaching of uranium, gold and copper in Australia. J Clean Prod 84:382–390. https://doi.org/10.1016/j.jclepro.2013.09.033

Henckens M, van Ierland E, Driessen P, Worrell E (2016) Mineral resources: geological scarcity, market price trends, and future generations. Resour Policy 49:102–111. https://doi.org/10.1016/j.resourpol.2016.04.012

Hubbert MK (1956) Nuclear energy and the fossil fuel. In: Drilling and production practice. American Petroleum Institute, Chichester

ICMM (2012a) Our work: sustainable development framework. International Council on Mining and Metals, London

ICMM (2012b) Mining's contribution to sustainable development, in TechRep. International Council on Mining and Metals, London

ISO, 14040 (1998) Environmental management-life cycle assessment-principles and framework. International Organisation for Standardisation, Geneva

ISO (2014) ISO 14046:2014 International Organisation for Standardization (ISO). Environmental Management e Water Footprint e Principles, Requirements and guidelines, Geneva, Switzerland

Kharrazi A, Kraines S, Hoang L, Yarime M (2014) Advancing quantification methods of sustainability: a critical examination emergy, exergy, ecological footprint, and ecological information-based approaches. Ecol Indic 37:81–89. https://doi.org/10.1016/j.ecolind.2013.10.003

Lawrence PR, Meigh J, Sullivan C (2002) The water poverty index: an international comparison. Department of Economics, Keele University, Keele

Mudd G (2007) Global trends in gold mining: towards quantifying environmental and resource sustainability. Resour Policy 32(1–2):42–56. https://doi.org/10.1016/j.resourpol.2007.05.002

Mudd G (2010) The environmental sustainability of mining in Australia: key mega-trends and looming constraints. Resour Policy 35(2):98–115. https://doi.org/10.1016/j.resourpol.2009.12.001

Norgate T, Haque N (2010) Energy and greenhouse gas impacts of mining and mineral processing operations. J Clean Prod 18(3):266–274. https://doi.org/10.1016/j.jclepro.2009.09.020

Norgate T, Haque N (2012) Using life cycle assessment to evaluate some environmental impacts of gold production. J Clean Prod 29-30:53–63. https://doi.org/10.1016/j.jclepro.2012.01.042

Norgate T, Jahanshahi S, Rankin W (2007) Assessing the environmental impact of metal production processes. J Clean Prod 15(8–9):838–848. https://doi.org/10.1016/j.jclepro.2006.06.018

Northey S, Haque N, Mudd G (2013) Using sustainability reporting to assess the environmental footprint of copper mining. J Clean Prod 40:118–128. https://doi.org/10.1016/j.jclepro.2012.09.027

Northey S, Mudd G, Saarivuori E, Wessman-Jääskeläinen H, Haque N (2016) Water footprinting and mining: where are the limokitations and opportunities? J Clean Prod 135:1098–1116. https://doi.org/10.1016/j.jclepro.2016.07.024

Orche E (1998) Minería americana del oro en las épocas precolombina y del imperio. Bol Geol Min 109:599–616

Pactwa K, Woźniak J, Strempski A (2018) Sustainable mining – challenge of Polish mines. Resour Policy. https://doi.org/10.1016/j.resourpol.2018.09.009

Padowski JC, Gorelick SM, Thompson BH, Kummu M, Ward PJ, De Moel H (2016) Measuring global water security towards sustainable development goals. Environ Res Lett 11:124015

Pellegrino JL, Margolis N, Justiniano M, Miller M, Thedki A (2004) Energy use, loss, and opportunities analysis for US manufacturing and mining. Energetics Inc., Columbia

Pfister S, Ridoutt B (2013) Water footprint: pitfalls on common ground. Environ Sci Technol 48(1):4. https://doi.org/10.1021/es405340a

Pimentel B, Gonzalez E, Barbosa G (2016) Decision-support models for sustainable mining networks: fundamentals and challenges. J Clean Prod 112:2145–2157. https://doi.org/10.1016/j.jclepro.2015.09.023

Romero Rueda I, de Armas Teyra M, Pérez Mena BM, Guerrero Rojas Y (2012) Evaluación energética de motores asincrónicos ante armónicos y desbalance de voltaje en una empresa minera. Minería y Geología 28:49–61

Ruiz-Mercado G, Gonzalez M, Smith R, Meyer D (2017) A conceptual chemical process for the recycling of Ce, Eu, and Y from LED flat panel displays. Resour Conserv Recycl 126:42–49. https://doi.org/10.1016/j.resconrec.2017.07.009

Stahel WR, Reday G (1976) The potential for substituting manpower for energy, report to the Commission of the European Communities

Sterman JD (2012) Sustaining sustainability: creating a systems science in a fragmented academy and polarized world. In: Sustainability science. Springer, New York, pp 21–58

Trigger D (2005) Mining projects in remote Australia: sites for the articulation and contesting of economic and cultural futures, Culture, economy and governance in Aboriginal Australia. University of Sydney Press, Sydney, pp 41–62

UNDP (2016) United Nations Development Programme. Human Development Reports

United Nations (2015) Transforming our world: The 2030 agenda for sustainable development. Resolution adopted by the General Assembly

Valero Delgado A (2013) Mineral resource depletion assessment. Eco-efficient construction and building materials: life cycle assessment (LCA), eco-labelling and case studies, pp 13–37. https://doi.org/10.1533/9780857097729.1.13

Vintró C, Sanmiquel L, Freijo M (2014) Environmental sustainability in the mining sector: evidence from Catalan companies. J Clean Prod 84:155–163. https://doi.org/10.1016/j.jclepro.2013.12.069

Young J, Septoff A (2002) Digging for change: towards a responsible minerals future, an NGO and community perspective. Mineral Policy Centre, Global Mining Campaign, Washington, DC

Chapter 8
SDG 7 Affordable and Clean Energy

eWisely: Exceptional Women in Sustainability Have Energy to Boost – Contribution of the Energy Sector to the Achievement of the SDGs

Isabel B. Franco, Caitlin Power, and Josh Whereat

Abstract The world has experienced a rapid demand of energy sources, both fossil fuels and renewables. Nevertheless, this scenario has given rise to complexities in resource regions, compromising how women cope with the impacts of unsustainable use of energy sources. Women adjacent to energy resource commonly experience loss of assets, compromising overall sustainability. Comparing two case studies, Japan and Colombia, the research presented in this chapter argues that key stakeholders in the energy sector, both renewables and fossil fuels, need to further engage in the enhancement of women's assets and capacities toward the achievement of the Sustainable Development Goal 7 Affordable and Clean Energy. Building the capacity of exceptional women in sustainability can foster overall sustainable development in both cases and contribute to the achievement of SDG 7 and the Agenda 2030. Based on a qualitative methodology, this study also presents a capacity-building approach toward the achievement of SDG 7 in alignment with the other Sustainable Development Goals (SDGs).

Keywords SDG 7 Affordable and Clean Energy · Women · Sustainability · Latin America · Asia-Pacific

---

I. B. Franco (✉)
Institute for the Advanced Study of Sustainability, United Nations University Shibuya-ku, Tokyo, Japan

Australian Institute for Business and Economics, The University of Queensland, Brisbane, Australia
e-mail: connect@drisabelfranco.com

C. Power · J. Whereat
Faculty of Humanities and Social Science, The University of Queensland, Brisbane, QLD, Australia
e-mail: caitlin.power@uqconnect.edu.au

8.1 Introduction

There is an increased concern in the literature about unsustainable energy practices and potential effects on vulnerable community groups, particularly on women and girls. This study explores how women can boost their assets and capacities to cope with the effects of unsustainable energy consumption, such as use of pollutants and waste, use of fossil fuels, inadequate compensation for the loss of livelihood options, and lack of corporate accountability for adverse unsustainable energy practices. In the practicality, the international community and local stakeholders (higher education institutions, the private sector, governments, and civil society organizations) have joint efforts to build women's capacity by delivering technical assistance programs under the umbrella of the Sustainable Development Goals (SDGs), particularly SDG 7 Affordable and Clean Energy.

SDG 7 Affordable and Clean Energy ensures access to affordable, reliable, and sustainable energy and is crucial in achieving many of the SDGs – from poverty eradication via advancements in health, education, water supply, and industrialization to mitigating climate change. Research shows that existing capacity-building approaches to energy do not strategically target women and fail in incorporating priority capacity-building areas valuable for women. However, these efforts represent a potential for fostering sustainable energy consumption patterns and therefore overall sustainability.

By investing in capacity-building and education structures that boost women's capacities in line with a vision for greater access to affordable, reliable, and sustainable energy for women further facilitates sustainable livelihood development and access to the rewards of the fulfillment of the global goals. This manuscript argues that boosting women's capacities to cope with the effects of increasing unsustainable energy consumption can foster overall sustainability in the long run. Based on a comparative case study methodology, the research reported in this manuscript identifies priority capacity-building areas valuable for women and proposes a capacity-building approach for sustainability in selected case studies. Some of the women participants in this study are based in locations where eWisely (Exceptional Women in Sustainability) operates. eWisely is the largest connector of women in sustainability with nearly 8000 leaders and followers in +60 locations. This research has also been supported by Research eWisely Lab.

Existing capacity-building initiatives for sustainability appear to have achieved a very low level of impact, as these actions are often disconnected to the local context or are irrelevant for women and girls. Yet there are many unanswered questions in this area, and the research reported in this article aims to increase our understanding in this field. What the main capacity-building priority areas are and how valuable they are for women are some of the questions that will be addressed in this article. Identifying and improving core capacity-building areas seems from this study to be the most effective way to enhance the ability of women to cope with pressing sustainability challenges over time. Based on stakeholders' perceptions in two case studies, this article pays attention to the nature and importance of these capacity-building priority areas.

This chapter argues that capacity-building impactful areas are those valuable for women and the contexts they are immersed. The focus of this research is to identify and investigate the actual level of impact of existing capacity-building initiatives, and recommend priority areas, so that stakeholders in the case study locations can play a stronger role in helping women boost their assets toward sustainable energy consumption practices. This article also explores existing capacity-building approaches to assist women in protecting their assets and capacities and reverse the effects of unsustainable energy consumption. Scholarship debates show that capacity-building has been a subject of analysis by scholars from various disciplines like education, economics, and more recently sustainable development. Coined as a long-term process to strengthen individuals' and organizations' skills to solve problems and achieve objectives, the notion capacity-building for women has been well covered in the literature; however, there are few scholars who deal with its impact to reverse unsustainable energy consumption at the local level. Although the subject of capacity-building is appealing from a rhetorical standpoint, at the community level, these initiatives appear to have achieved a very low level of impact, particularly for women. However, why this is the case has not been explored extensively. This research gap needs further exploration, specifically in developing contexts, where capacity-building for women is essential in the face of escalating sustainability issues. Therefore, it is necessary to investigate the actual level of achievement of these initiatives, to identify barriers to their impactful implementation in local development agendas and to recommend ways of overcoming these barriers, so that stakeholders can play a stronger role in assisting women build resilient assets and reverse existing unsustainable practices at the local level. Following a qualitative methodological approach, this article compares two cases, Japan, Asia, and Colombia, Americas.

8.2 Literature Review

8.2.1 Women and Sustainable Energy

Unsustainable energy practices are becoming a pressing issue in the global sustainable development agenda. Therefore, the importance of investigating this subject through gender lens is to further examine the effects on women and girls. Boosting women's capacity and assets to cope with unsustainable energy consumption patterns can create a positive change in the contexts in which they are immersed (Denton 2002). Sustainable development, one of the key buzzwords which flooded academic literature and policy documents in the 1990s, has been reinvigorated into a new phase of influence under the United Nations (UN) Sustainable Development Goals, 17 global aims to be achieved by 2030, the successor of the Millennium Development Goals. "Sustainable development" is a term defined in the 1987 Brundtland Report as "development that meets the needs of the present without compromising the ability of future generations to meet their own needs" (Brundtland Commission 1987: 41). The importance of this often-quoted statement has had a growing influence as the threats and effects of energy issues become more evident in the present and worrisome for the future generations.

There is an increased concern in the literature that unsustainable energy practices have a major impact on vulnerable community groups, particularly on women and girls. A major issue of sustainable development policies and literature is the lack of active women's participation and involvement and therefore their lack of capacity to cope with unsustainable energy practices. While women's rights have progressed throughout the past century, there are still many gaps in the literature and active participation in many forms across the globe. Amartya Sen (1999) posited the need for increasing women's agency as "women are increasingly seen as active agents of change: the dynamic promoters of social transformation." Sen (1999: 191) also argues that "the survival disadvantage of women compared with men in developing countries seems to go down sharply- and may even get eliminated- as progress is made in women's agency." This statement highlights not only the importance of women's agency but also the vulnerability of the contexts in which women are immersed.

The complexities of the contexts in which women are embedded have escalated in the past years due to pressing socio-environmental unsustainable practices at the local level (Cecelski 2000; Denton 2002; Dankelman 2010; Alston 2014). Cecelski (2000), for example, argues that from a social standpoint, "women's economic contribution is often unpaid, unrecognized and undervalued" resulting in a lack of investment in technological advancements to relief their burden which limits their agency. On the environmental domain, it is argued that "the threats posed by global warming have failed to impress on policy-makers the importance of placing women at the heart of their vision of sustainable development" (Denton 2002). Unfortunately, the role of women in society and the household places them at a great disadvantage to men particularly when disasters occur, a situation that worsens during recovery stages. This is due to the fact that women are often left with sociocultural norms and caregiving responsibilities which reduce their mobility in seeking shelter from disaster risks. They also lack access to energy sources, clean water, safe sanitation, and health supplies which adds to their burden (Dankelman 2010). Therefore, in order for climate change policy to be effective, it must take into account the interests of all stakeholders, signifying that women and those particularly in developing countries must have a voice in policy decisions that are more likely to affect them (Denton 2002).

Boosting women's capacity to tackle adverse unsustainable energy practices through supply of clean, regular energy will alleviate poverty of many people across the world. "Difficult, time-consuming work of collecting and managing traditional fuels (which) is widely viewed as women's responsibility, is a factor in women's disproportionate lack of access to education and income, and inability to escape from poverty" (Kaygusuz 2011: 936). In this regard, Alston (2014) argues that there is an urgent need for gender mainstreaming in policy documents particularly those with regard to climate change as the failure to do so "risks cementing gender inequalities in post-disaster and reconstruction efforts because of the inherently inequitable power relations, resource allocations and underpinning assumptions on which responses to climate disasters are based." Overlooking women's agency and their potential to cope pressing issues can become detrimental to future policy decision on climate change and overall energy sustainability (Alston 2014).

8.2.2 Women, Sustainable Energy, and Capacity-Building: Making the Links

To reverse the impacts of unsustainable energy practices such as indiscriminate use of natural resources, use of fossil fuels and pollutants, waste production, inadequate compensation for the loss of livelihood options, and lack of corporate accountability for adverse environmental impacts, international organizations have promoted a set of technical assistance programs and community capacity-building initiatives in developing countries (UNDP 1997). These global actions have been the subject of analysis by scholars from various disciplines such as education, economics, and more recently sustainable development. Global agencies such as the United Nations (UNDP 2011) and the scholarly literature (Loza 2004) posit capacity-building as a core area to be addressed in both developed and developing contexts. Capacity-building initiatives for women have become a recent subject of debate. It is argued that women, particularly in developing countries, lack education, and therefore stakeholders should boost their capacities as well as those of the broader community's (Gylfason 2001).

While the idea of women's capacity-building as a long-term process is appealing from a theoretical standpoint, there are major challenges in its real-life application. Capacity-building initiatives lack continuity due to the absence of a long-term commitment, resources, coordination, and collaboration among stakeholders (Franco 2014), relationships and dynamics of the community, management of power and resource imbalances, and development of a community identity (Clifford and Petrescu 2012). However, the major challenge is that institutional stakeholders are likely to believe that they know what the impactful areas of women's capacity-building are. Yet, an investigation of the situation shows this is not the case. This scenario creates confusion and resentment at the local level as institutional perceptions tend to overlook context-based barriers and impactful areas for capacity-building in sustainability (Wiek et al. 2012). Instead, the scholarship argues that community-based initiatives seem to be more effective when facing sustainability challenges. However, the scholarly literature indicates there are gaps in this area that need to be explored seriously (Nakata and Viswanathan 2012). On the one hand, the literature shows that top-down capacity-building, usually implemented by the international community to achieve sustainable development goals through education, media, and information-intensive campaigns, can increase awareness but is not impactful enough to reverse unsustainable energy practices. It is argued that capacity-building, particularly for women, is perceived by locals as a product to be sold, is usually poorly designed, and underestimates the barriers women face to cope with unsustainable energy practices. Moreover, these initiatives fail in paying attention to the human side of capacity-building and therefore incorporating women's perceptions about priority areas and barriers to engage in sustainable patterns (Kempton et al. 1985, 1992; Redman 2013; Franco 2014). On the other hand, policy-makers lack capacity to design impactful capacity-building initiatives for women (Mckenzie-Mohr 2000).

8.3 Methodology

This manuscript is based on ongoing research that explores the linkages among women capacity-building and sustainability. This study is qualitative in nature and was initially undertaken in Colombia. The Japanese case was later explored, and a global survey was conducted to identify priority capacity-building areas on sustainability for women and girls in selected locations. Survey results have been triangulated through individual and group interview. Fifteen semi-structured interviews were conducted with various stakeholders actively involved in capacity-building for sustainability in Colombia, while ten stakeholders were approached in the Japanese case. Literature and policy review were also conducted to explore exiting global trends and capacity-building priority areas for sustainability. Focus groups were also conducted and involved the participation of multiple stakeholders, namely, higher education institutions, the private sector, governments, and civil society.

Two case study areas have been selected, Risaralda, Colombia, in the Americas, and Okayama, Japan, in Asia due to the complexities in existing capacity-building approaches in both cases. Women in sustainability as well as representatives from higher education institutions and government representatives in both cases facilitated data collection and have been actively involved in the ongoing research.

8.4 Discussion and Results

This section aims to discuss priority areas of capacity-building to help women boost their assets, cope with unsustainable energy practices, and realize their full potential. This section also discusses and underlines that women's abilities can be enhanced if priority areas are identified and adequate capacity-building is developed and delivered in the selected case study areas. Research findings indicate that priority or impactful areas for capacity-building are those aligned with women's aspirations (Amartya Sen 1999) in collaboration with public and private stakeholders and with the Sustainable Development Goals (SDGs). Capacity-building areas for sustainability should follow the global agenda for sustainable development without neglecting the local contexts in which vulnerable groups, particularly women, are embedded. Neglecting local expectations hinders women's ability to foster overall sustainability and questions the effectiveness of existing capacity-building approaches (Puk and Behm 2003).

8.4.1 Risaralda, Colombia

In Risaralda, Colombia, priority areas for sustainable development depend largely on the contexts in which women are immersed. Risaralda is a region located in the Colombian Andes in South America. This geographical area holds extensive reserves of gold, silver, and coal. With the escalation of fossil fuel projects, stakeholders in Risaralda have attempted to maximize social benefits for vulnerable groups, particularly for women. While at the local level stakeholders join efforts to develop a capacity-building approach that tackles key community issues (Franco 2014), at the regional level, high priority areas for capacity-building in the Americas seem to have been identified by stakeholders (Franco et al. 2018). Yet, priority areas for women in alignment with the SDGs require further attention.

Collaboration for capacity-building for sustainability is characterized by active community engagement. Despite the escalation of fossil fuel projects, communities, particularly women's capacities, have been boosted to cope with existing sustainability challenges. The success in the implementation of priority capacity-building projects depends on both state and non-state actors. Colombia is a resource country, experiencing the scalation of large fossil fuel projects. Very often, communities depend largely on extractive industries, and their development aspirations are usually aligned with resource extraction. The extractive industry in Risaralda has somehow provided communities, particularly women, with capacity-building initiatives. However, these actions are limited compared to the adverse impacts caused by the extraction of minerals and metals.

Disruptive industries such as mining, oil, and gas will eventually affect the livelihoods of women in farming, jewelry design, artisanal mining, large-scale mining, and other economic activities. Women farmers, for example, are choosing artisanal mining over agriculture (Franco and Kunkel 2017). Research shows that although women's expectations have been considered in the implementation of capacity-building agendas, stakeholders involved, namely, governments, the private sector, and education institutions, need to further assist women and surrounding communities in enhancing their capacities to cope with the indiscriminate use of energy resources. This will in turn have a positive impact on women and overall community sustainability. Women consultation in relation to their needs and expectations is highly valued by female leaders, resulting in immediate benefits for them and the broader community (Franco 2014). Some female coffee and jewelry producers and women entrepreneurs have already expressed the importance of implementing capacity-building better aligned with their development aspirations:

> We have been trained in jewelry design ... the company has also provided us with some financial assistance to attend international fairs so that we can promote and sell our products. (Community Members, Interview)

The role of non-state actors such as extractive industries in boosting women's capacity to cope with unsustainable energy consumption has been a driver to enhance resilient women's and community's assets. Ensuring sustainable energy practices does not restrict to the role of the government but also involves other parties such as the extractive industry, governments, civil society, and education institutions (Davies 2005; Franco 2014). Joint efforts of involved parties are further required to help women boost their capacity and develop asset-based adaptation strategies to cope with unsustainable energy practices:

> Both, women and men have a strong sense of community. They own agriculture-based community associations that stimulate the local economy, foster employment and leadership. For example, they own associations for blackberry and coffee production and commercialization. (Corporate Representative, Interview)

Another significant finding in the Colombian case study is the active participation of the women in the formulation of sustainable development agendas particularly in resource-rich regions. Women have benefited from capacity-building initiatives helping them to cope with livelihood transformations due to the impact of fossil fuel projects:

> The company helped the municipality to open a plant for waste collection which has the potential for generating income for women and the local community. (Community Leader, Jewellery CBO, Interview)

Such initiatives include development activities in agribusiness, dressmaking, jewelry, coffee production, and agriculture (Franco 2014). These initiatives are the result of effective government-corporation partnerships and in response to requests from women. Upfront investment in priority areas will assist women in realizing their full potential in the case study area. Based on stakeholders' and community's perceptions, these areas are income, employment, education and training, work experience and apprenticeships, and infrastructure development (Franco 2014; Franco and Kunkel 2017).

SDG	Impact capacity-building
SDG 1	Generating and managing income
SDG 8 Decent Work and Economic Growth	Employment opportunities Work experience and apprenticeships
SDG 9 Industry, Innovation, and Infrastructure	Infrastructure for development

The case study showed that stakeholders should boost these areas in order to help women become more resilient and cope with the impacts of unstainable use of energy resources in the Colombian case.

8.4.2 Okayama, Japan

Okayama is a prefecture located in the southern part of Japan, Asia. Its economy comprises major industries such as petrochemicals, coal chemicals, and transport equipment, chemical, steel, and general machinery/tool. Energy consumption patterns have increased the complexities in multi-stakeholder collaboration for capacity-building and education for sustainable development (Abe 2017).

Research findings show that capacity-building should focus on connecting nature to people and sharing common ground to protect the environment. So far, the environment is not given the right emphasis; hence many initiatives have been undertaken so that human society and nature can thrive together. Capacity-building for sustainability has been backed by strong government policies in Asian nations such China, Japan, and the Philippines (Ryan et al. 2010). However, it is argued that despite strong government support, there is a lack of collaboration for capacity-building for sustainability if compared to other regions (Naeem and Peach 2011).

Multiple stakeholders in the Japanese case are collaborating in building community capacity in sustainability. Led by the local government, the existing approach targets from children at early stages in the education system to adults through non-formal education. Based on participants' perceptions, priority areas for capacity-building are as follows:

SDG	Impact capacity-building
SDG 6 Clean Water and Sanitation	Water quality
SDG 15 Life on Land	Biodiversity
SDG 11 Sustainable Cities and Communities	Traditional knowledge
SDG 4 Quality Education	Teacher education on sustainability
SDG 12 Responsible Consumption and Production	Corporate social responsibility

Although the broader community is systematically being integrated in the existing capacity-building approach to sustainability, the major challenge encountered by stakeholders is the inclusion of women in decision-making in the identification of capacity-building priority areas:

> It is an interesting question because when we talk about the ration of participants of community-based activities the majority are women but when it becomes of decision making, most are men. (Higher Education Representative, Japan)

> Young mothers with children are involved in community capacity-building' because they have more time. (Civil Society Representative, Japan)

Vague responses were also provided by participants when addressing this issue: 'we are including these issues broadly' – stated by one of the participants (Higher Education Representative, Japan). However, lack of women participation in decision-making processes around capacity-building seems a general issue and does not only pertains to the subject of this research:

> When women graduate from school they work but after giving birth their participation in the workforce declines. When they turn 40 and their children are grown up they get a part time job. (Higher Education Representative, Japan)

Overlooking women's voices in the development of capacity-building approaches to sustainability does only escalate unsustainable patterns (Kempton et al. 1985, 1992; Redman 2013; Franco 2014) but also reflects the lack of capacity to design impactful capacity-building initiatives for women (Mckenzie-Mohr 2000). This eventually will prevent stakeholders from fostering overall sustainability in the long run, at least in the Japanese case.

8.4.3 A Capacity-Building Approach for Sustainable Energy

Research indicates that both cases acknowledge the participation of women in capacity-building approaches to sustainable energy. However, the Japanese case differs from the Colombian case study in fostering women's decision-making to cope with sustainability challenges around energy and related issues. Interestingly, women in Colombia are more empowered and have been able to integrate their development aspirations in the design of existing capacity-building approaches. While women are strategically targeted in existing capacity-building initiatives in Colombia, in the Japanese case, the inclusion of women in capacity-building initiatives is mainly due to time convenience or to meeting government requirements. Interestingly, in both case study areas, multi-stakeholder collaboration and alignment with SDGs are perceived as determinant factors for the success of existing capacity-building approaches.

Research also shows that priority areas are those most valuable for women and the broader community. This inductive derivation indicates their critical importance to enhance women's capacities and protect their assets. Despite other forms of capacity-building sometimes being mentioned, these were constantly identified as central. A capacity-building approach to sustainable energy targeting some or most of these areas will assist women and the broader community in coping with sustainability challenges in energy and related pressing issues. Figure 8.1 shows a preliminary proposal of a capacity-building approach for sustainable energy in alignment with the SDGs. Interestingly, SDG 4 Quality Education seems to be a commonality in both cases with education and training for women and teacher education as impact capacity-building areas relevant for women.

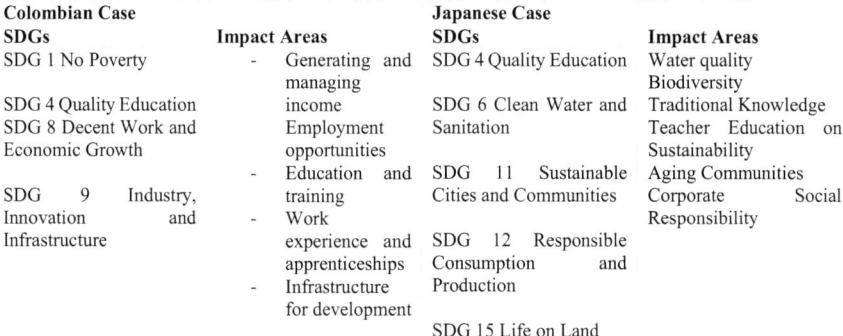

Fig. 8.1 Capacity-building approach to sustainable energy

8.5 Impact Sustainability: Final Remarks

The escalation of unsustainable practices in the use of energy sources is likely to result more detrimental for vulnerable community groups, particularly for women and girls. Successful integration of priority areas for capacity-building can boost women's capacities to cope with the unstainable use of energy resources. This chapter recommended a strategic approach to boost women's abilities to sustain their assets despite the escalation of unsustainable practices in the use of energy resources. It also highlights existing issues when integrating a gender lens perspective to examine sustainability issues.

This chapter also indicated that priority capacity-building areas which are the most valuable for women are those that help them achieve their own sustainable development aspirations. Although collaboration of state and non-state actors is pivotal to boost women's capacity, a bottom-up approach to capacity-building for sustainability is more likely to foster overall sustainability than corporate and/or government top-down approaches. Community-oriented and women-driven agendas can help women become more resilient to cope with unsustainable practices in energy consumption or associated matters. However, such approaches need to be also aligned with the global agenda on sustainable development. It is not solely local actors' responsibilities to make capacity-building approaches valuable for women as the international community also need to share responsibilities with local stakeholders and develop technical assistance programs accordingly.

In addition, consultation and active participation in decision-making for the development of capacity-building approaches to sustainability need to be guaranteed in both cases. However, nuanced attention needs to be paid to the Japanese case where gender inclusion issues are complex. In both cases consultation should not be merely the token provision of information but should be such that women are involved in decision-making about matters such as priorities for the allocation of funds and asset transformation. Consultation processes should also include women across the broader community to help them boost their assets and cope with unsustainable energy practices in the case study locations.

References

Abe H (2017) First RCE thematic conference: towards achieving the SDGs, Okayama, 5 December, UNU-IAS, Tokyo

Alston M (2014) Gender mainstreaming and climate change. Women's Stud Int Forum 47(B):287–294

Brundtland Commission (1987) Report of the World Commission on Environment and Development: Our Common Future. UN Documents. Available at http://www.un-documents.net/our-common-future.pdf

Cecelski E (2000) The role of women in sustainable energy development. National Renewable Energy Laboratory, Golden

Clifford D, Petrescu C (2012) The keys to university–community engagement sustainability. Nonprofit Manag Leadersh 23(1):77–91
Dankelman I (2010) Climate change, human security and gender. In: Dankelman I (ed) Gender and climate change: an introduction. Routledge, London
Denton F (2002) Climate change vulnerability, impacts, and adaptation: why does gender matter? Gend Dev 10(2):10–20
Davies JS (2005) Local governance and the dialectics of hierarchy, market and network. Policy Studies 26(3–4):311–335
Franco IB (2014) Building sustainable communities: enhancing human capital in resource regions–Colombian case. PhD thesis, Sustainable Minerals Institute, The University of Queensland
Franco IB, Kunkel T (2017) Extractives and sustainable community development: a comparative study of Women's livelihood assets in the Americas. Int'l J Soc Sci Stud 5:54. https://doi.org/10.11114/ijsss.v5i11.2724
Franco I, Saito O, Vaughter P, Whereat J, Kanie N, Takemoto K (2018) Higher education for sustainable development: actioning the global goals in policy, curriculum and practice. Sustain Sci:1–22
Gylfason T (2001) Natural resources, education, and economic development. Eur Econ Rev 45(4–6):847–859
Kaygusuz K (2011) Energy services and energy poverty for sustainable rural development. Renew Sust Energ Rev 15(1):936–947
Kempton W, Harris CK, Keith JG, Weihl JS (1985) Chapter 6: Do consumers know "what works" in energy conservation? Marriage Fam Rev 9(1–2):115–133
Kempton W, Darley JM, Stern PC (1992) Psychological research for the new energy problems: strategies and opportunities. Am Psychol 47(10):1213
Loza J (2004) Business–community partnerships: the case for community organization capacity building. J Bus Ethics 53(3):297–311
Mckenzie-Mohr D (2000) New ways to promote proenvironmental behavior: promoting sustainable behavior: an introduction to community-based social marketing. J Soc Issues 56(3):543–554
Naeem MA, Peach NW (2011) Promotion of sustainability in postgraduate education in the Asia Pacific region. Int J Sustain High Educ 12(3):280–290
Nakata C, Viswanathan M (2012) From impactful research to sustainable innovations for subsistence marketplaces. J Bus Res 65(12):1655–1657
Puk T, Behm D (2003) The diluted curriculum: the role of government in developing ecological literacy as the first imperative in Ontario secondary schools. Can J Environ Educ 8(1):217–232
Redman E (2013) Advancing educational pedagogy for sustainability: developing and implementing programs to transform behaviors. Int J Environ Sci Educ 8(1):1–34
Ryan A, Tilbury D, Blaze Corcoran P, Abe O, Nomura K (2010) Sustainability in higher education in the Asia-Pacific: developments, challenges, and prospects. Int J Sustain High Educ 11(2):106–119
Sen A (1999) Development as freedom. Oxford University Press, Oxford
UNDP (United Nations Development Program) (1997) Capacity development: technical advisory paper 2. Retrieved March 26, 2014 from, http://mirror.undp.org/magnet/cdrb/Techpap2.htm
UNDP (United Nations Development Program) (2011) Capacity development. Retrieved June 23, 2011, from http://www.beta.undp.org/undp/en/home/ourwork/capacitybuilding/approach.html
Wiek A, Farioli F, Fukushi K, Yarime M (2012) Sustainability science: bridging the gap between science and society. Sustain Sci 7(1):1–4

Chapter 9
SDG 8 Decent Work and Economic Growth

A Decent Day's Pay for a Decent Day's Work – Lessons to Be Learnt from Fair Trade Small Producers' Experiences in Global Markets

Ana Cristina Ribeiro-Duthie

Abstract This chapter focuses on the Sustainable Development Goal 8: Decent Work and Economic Growth. The theoretical framework from evolutionary economics is reviewed and used to facilitate the understanding of developing countries' economic aspects. The relations to sustainable development actions are addressed through case studies centred on fair trade initiatives – a market solution that links developed and developing economies through production and trade relations based on good market practices. In a model that seeks to address inequalities generated by conventional trade, the fair trade standard assures stable prices and income together with decent work conditions for small producers from developing economies. The methodological approach of this qualitative study was a literature review associated with a comparative analysis of case studies published about fair trade certified small-scale producers. Small rice farmers from Thailand and small quinoa farmers from Bolivia are presented to illustrate the potential to produce social benefits within a sustainable development framework proposed by the fair trade model. Drawing from both case studies, the chapter ends with recommendations for multilateral trade agreements and policies design regarding agriculture development.

Keywords Economic growth · Decent work conditions · Evolutionary economics · Sustainable development goals · Fair trade · Social change

---

A. C. Ribeiro-Duthie (✉)
Politics and International Relations Program, School of Social Sciences,
University of Tasmania, Hobart, Australia
e-mail: anacristina.ribeiroduthie@utas.edu.au

9.1 Introduction

Conventional analysis on economic development has thus far focused on the conditions and drivers that the developing economies are unable to fulfil. As such, analysis regarding those parameters within the context of least developed or developing nations will often be incomplete and sometimes counterproductive. Although some least developed and developing economies seem to break the laws of classical and neoclassical economics lessons to improve development in their own way – success using such an approach requires a broader analysis of socio-economic factors. Recognising this, one could argue that evolutionary economics is useful in the sense that it accommodates learning and knowledge of individual actors participating actively in real life economics. However, limitations for a holistic evolutionary analysis of the least developed and developing economies' realities remain, despite the contribution of evolutionary theorists in explaining the role of entrepreneurial innovative solutions for existing bottlenecks for development in these countries. Some of the solutions are from within the market despite not relying solely on market mechanisms to succeed as a liberalist perspective would sustain.

In the end, there is no complete answer regarding which economic model would be best suited to analyse least developed, underdeveloped or developing economies. These terms were proposed by the UNDP (United Nations Development Programme) when mapping countries according to their score in the Human Development Index. Recently the term "underdeveloped" was substituted at the UNDP classification for "least developed countries". However, "developing" has also been used to differentiate from "developed". As conventional analysis of the determinants of economic development appears to fail to embrace the complexity of those countries' realities, it follows that conventional solutions for developmental inclusiveness may indeed also fail. While a theoretical approach is necessary, urgent action is required to overcome deep developmental constraints. The sustainable development goals (SDGs) work as a call for change to these pressing issues.

Inspired by the words of the managing director of the International Monetary Fund (IMF), Christine Lagarde, at the 2017 World Economic Forum, any international institution may contribute to address the economies "left behind". In her view, globalisation has been incredibly effective and is not going to be stopped; however it is not working for all (Lagarde 2017) – which needs to be both recognised and addressed. In Lagarde's view, "globalisation is embedded in the way capital moves around, in the way in which a lot of people around the world work, transact, move, (…) and are used to getting certain products cheaply" (Lagarde 2017). However, a big part of the world, approximately "3.6 billion people that we are not talking about enough (…) who constitute 50% of the global economy" (Lagarde 2017) must be included through effective actions so they too can benefit from globalisation. According to Lagarde, the "improvement of (their) living standards depends massively on international trade (…) on participating in the global economy" (Lagarde 2017). As the IMF managing director states: "we cannot go ahead without them". Lagarde highlights that "globalisation has to work for all and to do that, *changes*" and efforts have to be made (Lagarde 2017 – my emphasis). This can be heard as a call for socio-economic responsibility and sustainable development.

The SDGs constitute a valuable guide to addressing those left behind, and the 17 SDGs capture several aspects that can be considered essential for global well-being. In line with those concerns, it can be seen that the fair trade movement has already started connecting many of the dots. In this case, perhaps the "effort" suggested by the fair trade movement is that "getting products cheaply" cannot be maintained at the expense of the 3.6 billion people who require the same decent work conditions, stable source of income, food security and living standards improvements as everybody else in society. Considering this, the standards for production and consumption need to be raised to attend social responsibility requirements and not only to assure cheapest prices. Sustainability certification schemes such as Fairtrade propose ethical standards for production and trade to progress the economy through assurance of decent work conditions while respecting human rights and the environment. Fairtrade is the trademark of Fairtrade International, also known as FLO (Fairtrade Labelling Organisation), the first fair trade certifier that established the standards in 1988. This model has been reproduced by several fair trade organisations thereafter, following closely or not the standards, principles and values of fairness in trade. Fair trade is used in this study to refer to the movement in general, not alluding to any specific organisation. Fair trade "draws explicitly on civic values and qualifications in promoting labor rights" (Riisgaard 2015, p. 124). Work, health and safety standards are required for fair trade certification within an ambitious agenda that is consistent with the SDG 8. As such, this chapter addresses economic growth and decent work and its links to sustainable development actions through case studies centred around the fair trade initiative – a market solution which links developed and developing economies through production and trade based on market good practices. Fair trade can be understood as an emerging "response to the negative effects of globalization" (Murray and Raynolds 2007, p. 4), not as a mere "anti-globalization" but as a "new globalization" (Murray and Raynolds 2007, p. 6).

In its 30 years of existence, fair trade has called attention to the old economic assumption of lowest prices being a priority, which has ensured poverty for many working people given the slavery-like work conditions practised in the search for profit maximisation. The practice of lowering costs at any price, with no respect for human rights or labour policies, is addressed by the Fairtrade standards with stringency and transparency, following the ILO conventions for good working conditions, described below. Among the initiatives to demonstrate the global emergence of the SDG 8, this study focuses on fair trade food production which constitutes an inclusive decent work opportunity for rural and remote areas within the Global South, and a potential driver of economic growth.

Hence, to assure the orientation that "the 50% of the global economy" (Lagarde 2017) must be included as more than a mere rhetoric, perhaps efforts also need to be made regarding changing the assumptions that classical economic models rely upon. This means to assure the 50% of the global economy left behind – by policies and decision makers – progress, efforts must also be made by those who are future-focused. This move can be surprisingly hard, as some people are attached to beliefs and values even if those beliefs seem to be contradicted by facts. Social change requires a dramatic change of behaviour, of course at the individual level but also at an organisational level. Our approach is then to follow and readdress Lagarde's recommendation that we should "adopt the right narratives, the right language" in order "to make globalisation work for all, not only for some happy few" (Lagarde 2017). This chapter aims to contribute with analysis and recommendations for policy design related to agriculture development identified through fair trade initiatives, given its potential for economic growth and decent work conditions assurance.

9.2 Literature Review

9.2.1 The Fair Trade Model and the SDG 8

Fair trade is a social movement striving to change conventional trade and promote economic growth by exposing small producers in developing countries in global markets. How is this proposed social change enabled? By establishing a fair price sufficient to provide a stable income and to protect small producers from price fluctuations in exchange for a production based on stringent ethical standards in line with the fair trade certification label. Fair trade aims to promote the following values and principles: creating opportunities for small producers in social and economically disadvantaged areas; establishing and maintaining long-term commercial relations with respect to democracy, transparency and accountability; paying a fair price to producers; paying a social bonus for community development; pre-funding to assure production sustainability; rejecting child labour; rejecting forced labour; practising non-discrimination (gender, association); promoting capacity building; and protecting the environment (FLO 2016–2017; WFTO 2016).

Welcomed by small producers and farmers associations (Carlisle 2016; Barratt Brown 2007; Shiesari and Gruninger 2014), this alternative trade model received positive reviews from several academic studies (Sharma et al. 2018; Englund and Berndes 2015; Becchetti et al. 2012; Udomkit and Winnett 2002; Rice 2001) but also criticism (Garret et al. 2016; Thavat 2011; Carimentrand and Ballet, 2010). The fair trade movement was launched in 1988 as a solidary movement to aid coffee farmers in Mexico. It is possible to observe the progress of a sustainability agenda in the fair trade organisations' corporate social responsibility reports across the years. The consistency of the fair trade aims with the sustainable development goals was demonstrated in a recent work from Ribeiro-Duthie et al. (2018). This evolution can be understood as developing in line to the maturity of the concept of sustainable development since its inception on the world stage by the Brundtland Commission in 1987. This means that fair trade organisations have shown effective commitment to targets expressed in the sustainable development goals (SDG) since they were released by the United Nations (UN) in 2015. Even before that, fair trade organisations demonstrated their alignment with the Millennium Development Goals established by UN and released in 2008. Prior to this, sustainability values were present in case studies and social responsibility reports made available by the Fairtrade Foundation (FF) since 2001.

The Fairtrade (and fair trade) model, through establishing a floor price above the market price, works to address inequalities in the market and to protect small producers from prices volatility risks. This minimum price standard assures a stable income to small farmers, which allows them to plan and invest in their production. Whenever the market price goes above the *fair* price, the market price is guaranteed for fair trade producers. A certified fair trade producer will have the guarantee of a buyer and assurance of a stable income from his production. This trade initiative can make difference for small producers' working conditions, livelihoods, their families, the community where they grow food, the local economy and international trade. At every level of the fair trade supply chain, a degree of social change is shared. Half of the global population subjected to hunger is constituted by small farmers (FAO 2012), as such the change promoted by the fair trade model, despite viewed as small by some analysis of the movement, can have impact for small producers and wealth generation. Fair trade constitutes less than 5% of the global trade, while the main fair trade organisation produces revenues above USD$ 8 billion a year (as per Table 9.1). This example shows that sustainable development innovative solutions are welcome to economic growth. In accordance to Lagarde (2017), every international institution can play a role to include those marginalised from the global economic transactions. Small-scale farmers must not be left behind.

Fair trade related to food production is worth stressing due to constraints of agricultural practices combined with realities of small producers. These include price variations caused by supply/demand mechanisms which weakens small farmers; small farm size which grows a smaller crop volume and exposes small producers to loss when prices are lowered due to competition from large-scale producers who may benefit from economy of scale; small land in remote areas facing high transportation costs with effects on distribution and final prices; lack of access to market; and lack of information technology and literacy skills making small producers vulnerable in market exposure and price negotiations. This list illustrates challenges for agriculture development and the several risks small farmers are exposed to. The case studies selected embody a social change in process and can exemplify the delivery of SDG 8 addressed by the fair trade standards on working conditions. Guidelines are clearly expressed in the standards of Fairtrade for small-scale producer organisations. According to Fairtrade International standards, requirements

Table 9.1 Demonstrating Fairtrade International revenues per year in billions of euros. Data collected and compiled from FLO reports 2003–2004 to 2016–2017

2004	2005	2006	2007	2008	2009	2010	2011	2012	2013	2014	2015	2016
$832	$1.1	$1.6	$2.3	$2.9	$3.4	$4.3	$4.9	$4.8	$5.5	$5.9	$7.3	$7.8

for certification include assurance of a detailed set of criteria which are partially reproduced below from the latest Fairtrade version (Fairtrade 2019):

- Freedom from discrimination based on "race, colour, sex, religion, political opinion, national extraction or social origin".
- "No tests for pregnancy, HIV/AIDS or genetic disorders".
- "No tolerance of gender based violence".
- "Freedom from forced or compulsory labour". In case such practice is identified among the member organisations, procedures should be implemented to avoid that adults are "employed in abusive, exploitative and unacceptable work conditions as defined by ILO Conventions 29 and 105".
- No offer of housing conditional to employment is tolerated. "Spouses have the right to work elsewhere".
- "No children under 15 years old employed". In case such practice is identified, remediation procedures should be placed to avoid the continuous abuse.
- "No hazardous work for children under 18 years old".
- Payments are made under regular intervals and documented.
- Salaries for workers are set at official minimum wages for similar occupations, following national or regional average wages. In addition, "you and your members gradually increase salaries above the regional average and the official minimum wage".
- "Ensure that all permanent workers have a legally binding contract and are aware of their rights and duties, responsibilities, salaries, and work schedules as part of the legal labour contract".
- Ensure hiring and working conditions to subcontracted workers according to this standard, and whenever possible assign them as permanent workers.
- Occupational health and safety based on ILO Convention 155 is assured to all workers.

Fairtrade International counts upon 1.5 million small producers in 74 countries. It is worth noting that Fairtrade is not the sole fair trade organisation in the market, but it is a primary reference point for the movement, which provides statistical data on a regular and consistent manner, and this aspect facilitates research and assessment.

Interestingly, the fair trade movement has been confronted regarding working conditions exactly due to the requirement of decent and high standard work conditions for small producers. In this case, the criticism is that the smallest producers cannot afford to meet the (high) standards, or to invest in the necessary decent work requirements, nor to change their informal work arrangements to achieve labour rights standards (Smith 2014; Taotawin 2011; Thavat 2011). A concern that is understandable, although those arguments can be disputed. Fair trade's decent work criteria can be seen by some as too strict and even unattainable, as a pattern lifted from developed economies and transported directly to underdeveloped or developing countries. However, it is of note that work, health and safety (WHS) involve a significant cost for any responsible business and should not be misrepresented. The assurance of WHS should be a prime standard for anyone anywhere, and unfortunately it can be used as a gateway to lower production costs in conventional practices. This type of practice is common in least developed/developing countries given the intense job competition, unbalanced power relations and asymmetric information. In this sense, the evolutionary perspective seems able to explain the insight of fair trade initiatives in terms of social entrepreneurship, whose limitations can be addressed to advance the necessary social change agenda.

9.2.2 Evolutionary Economics and Developing Economies Constraints to Be Overcome

Given the challenging characteristics of developing or underdeveloped countries, the approach of evolutionary economics when it integrates change appears to meet a basic non-excludable condition for understanding the dynamics of the underdeveloped/developing economies. There is a trend of simply addressing non-developed countries as developing countries – this use is followed in this study whenever possible. However, the term underdeveloped was of common use when some articles used as reference here were published; in this case the term "underdeveloped" is kept, with no derogatory stress to one or another. When considering economic models, classical and neoclassical economic analysis are not sufficient in explaining underdeveloped countries' economic challenges as they focus on market equilibrium and optimal conditions for operating markets.

Meeting Pareto Optimal conditions is far from feasible for underdeveloped economies, whose characteristics are remote to optimal equilibrium and skip objective economic laws. Foster and Metcalfe (2012) contribute to outlining underdeveloped economies features such as realities of strict markets; bounded rationality; institutions poorly developed; uncertainty; and economic knowledge incomplete. Following this analytical pathway could diversify the economic approach to such realities. Underdeveloped/developing economies characteristics are endogenously related to uncertainty and instability of markets. In such contexts, possible achievements of fair trade initiatives – which bring together developing and developed economies in a viable trade model – can be understood.

As "evolutionary theories place the diversity of behaviour, rather than the uniformity of behaviour" (Foster and Metcalfe 2012, p. 422), it seemed appropriate to stress those categories as a framework in this analysis. Uncertainty – a constitutive part of those economies – has the potential to bring a positive effect as such characteristic is a great field to innovation, according to Schumpeter (Schumpeter 1939; Dopfer 2011). In the same extent, predictability may not apply to those underdeveloped or developing countries, exactly because the history of their market behaviour does not always dispose of a stable or predictable pattern to allow forecasting. This feature may largely undermine investors' interest and consequently credit availability, leaving to these countries a marginal role to advance their productive processes and to participate in international trade.

Nevertheless, under an evolutionary regard, there is hope. According to Witt (1992), a crucial characteristic of evolutionary economics is associated with change, as the same author explains: the emphasis is on "becoming rather than being in the economy" (Witt 1992, p. 405). Following Witt's insight, the inspirational factor on the term "evolution" can be then linked to "the capacity of an economy, or some part of it, to generate change from within" (Witt 1992, p. 406). Therefore, an evolutionary perspective seems more suitable in case of growth within a context that skips from classical economic conditions to run models given the admission of variances from inside the economy. Underdeveloped/developing economies are operating, despite no Pareto Optimal conditions, and in some extraordinary cases, outcomes would surpass predictions. Thus, the historical evolution of developing economies should be regarded in their specificities.

According to a Schumpeterian perspective, while invention relies more on technological or scientific development (Research and Development), which is not always available in underdeveloped or developing economies, innovation can be in place with the use and application of an invention in a different manner, facilitated by the presence of financial credit. This aspect may explain how some countries meet the conditions to creatively innovate whereas others do not. Under the Schumpeterian point of view from Dopfer (2011), the main aspect related to the neoclassical model failure is this dynamic of individual behaviour not being properly considered. In the classical doctrine "the activities of individuals had no role to play" at an objective "machinery" (Dopfer 2011, p. 105).

Dopfer (2011) compares the individual role in the neoclassical model to a passive behaviour, while in his Schumpeterian inspired model, the individual is actively engaged in changing the reality (Dopfer 2011, p. 105), which includes economy. This dynamic of individuals overcoming objective laws (Dopfer 2011, p. 106) seems applicable to underdeveloped and developing economies in their peculiar manner of improving their economies even though not fully equipped to do so. In these instances, Schumpeter's proposal of radical innovation may be suitable to contemplate underdeveloped and developing economies. Evolutionary economic perspective certainly can be complimentary and bring new insights for admitting plasticity and uncertainty of markets as able to promote innovation.

With the Neo-Schumpeterian evolutionary economists, not only behaviour with its potential to change over time is included in the model, but also technological, organisational and institutional change are recognised as "the core drivers of economic growth" (Foster and Metcalfe 2012, p. 425). From analysing underdeveloped or developing countries, in many cases, there is no basic technological development; organisations did not achieve a homogeneous pattern for operating when compared to developed economies; and not all institutions accomplish with basic needs of individuals. Despite those unfulfilled requirements, sometimes entrepreneurs may surpass those challenging realities creating (and innovating). According to Schumpeter (1942), innovation occurs as a disruptive process, when energetic entrepreneurs break the circle destroying an existing course of the economy by introducing new combinations. The entrepreneur "revolutionizes the economic structure from within" in a constant process of "creative destruction" (Schumpeter

1942, p. 83), and this is an essential role for economic development (Schumpeter 1952/1997).

In this sense, despite economies of underdeveloped and developing countries are not standardised models to which classical economic laws could be applied to generate predictable patterns of solutions, their realities of uncertainty could benefit from an evolutionary analysis as well as to contribute for building economic knowledge. Based on Malerba et al. (1999), one could say that singular realities may constitute land to grow knowledge even if based on empirical evidence. This leads to his assumption that "theorizing" is about some particular contexts (Malerba et al. 1999, p. 37). It seems that within the evolutionary perspective, there is room to focus on diverse realities and on what this shift may add to the understanding of economic dynamics. Malerba et al. (1999) contribute showing through an evolutionary history-friendly approach that "what economists know about the economic world comes in the form of relatively detailed knowledge about particular contexts" (Malerba et al. 1999, p. 37). Cimoli and Katz (2003) when analysing Latin America, for instance, observed that "undergoing fundamental changes in the model of production organization" were "gradually enforcing the creation of a new, more competitive (…) pattern of development" (Cimoli and Katz 2003, p. 389). However, they also observed that "productivity growth has not been rapid enough, nor has the pattern of production specialization been transformed in the direction of activities with higher value added" (Cimoli and Katz 2003, p. 389) in those countries. Thus, the process of creative destruction cannot assure to this group of underdeveloped and developing economies a macro level of stability. Of course, there remain market asymmetries and bounded rationalities, as well as lack of institutional and organisational support to drive their process of development – which are termed meso rules, or institutional structures (Dopfer 2011). It seems that the uncertainty of economic environment in those countries cannot assure innovation if basic needs provided by meso structures are not a guarantee. The entrepreneur must spend his energy fighting for standards of living and working. Therefore, beyond theory, empirical evidence reinforces the need of reviewing some classical concepts and recognising the limitations of evolutionary economic approach as well.

The example of fair trade may illustrate how the alternative model addresses innovation but also depends upon meso structures, which the trade model recognises in importance by assuring them through the social premium. The fair trade model provides stable income to marginalised producers from underdeveloped/developing countries at one hand. On another hand, through the "social premium", it creates conditions to finance "meso structures" for the producer's community socio-economic and environmental development. However, the fair trade movement is a singular contribution within the course of the entire economy. It seems there is potential that the movement's impact could be expanded with support of government policies and multistakeholder partnerships, broadening its social change benefits and sustainable development targets and vision.

9.3 Is There Anything Definitive About Fair Trade?

Despite limits for a complete analysis of developing economies' realities, there are contributions which facilitate the comprehension of some mechanisms in role in such economies. Economic development has been dominated by two main competing schools of thought – Marxism and Liberalism – and both do not fully contemplate underdeveloped economies' specificities. On the one hand, Marxism is considered to reduce developing economies problems to politics constraints, whereas Liberalism reduces the problems to market constraints.

The main critique to Marxism originates from the structuralism approach (Choat 2010), which proposes that Marxism does not go beyond its political criticism and the strict belief in the role of state to distribute income. As a result, it does not recognise the market reality and its potential dynamics as having a distributional effect. On another hand, an analogous criticism can be applied to the Liberalism where a government role is undermined in favour of the "invisible hand" that would solely allow the market flow and grow with the competition.

The higher price instead of the lowest price for products and the high standards in which they are required to be produced and distributed within the fair trade model is an undeniable challenge to classical economic assumptions – which has succeeded as the increasing rate of fair trade revenues demonstrates, feeding the legacy of the ethical consumerism model. Table 9.1 demonstrates the increasing revenues of the first entrant in the market, Fairtrade International. If there is anything definitive about fair trade, it seems to be the response of consumers to sustainability standards, reflected in the increasing rate of Fairtrade sales.

Although the wealth and income generation and allegedly changing realities of small producers and communities, this socio-economic-environmental change is also criticised. Fair trade faces criticism for using market tools and language to challenge the same market: the movement's challenge to the conventional trade model is accused of limitation because it relies upon commoditisation, and thus it is not politically activist enough. Fair trade is accused of excessive stringency for the small producers whose precarious conditions – financial, institutional and organisational – are out of the scope of the requirements for ethical trade. Fair trade is criticised because the proposed fair price higher than market prices for products holding the fair trade seal creates a surplus that would not return entirely to small producers, but it is shared along the supply chain with an amount going, for example, to retailers. Fair trade has been criticised for the space given to large corporations as Starbucks and Nestle, which enjoy of the fairness halo effect of holding the fair trade labelling but actually purchasing a very small amount of fair trade coffee (Barratt Brown 2007). It can be observed that the movement is judged both from a Liberal and a Marxist perspective. This reinforces the relevance of an evolutionary perspective proposed in this analysis. Fair trade is proposed to be simply seen as an example from within the market that uses the market language and tools to challenge the same market and produce social change. For doing so, its contribution can be recognised as a transitional model, able to generate "change from within" and perhaps capable of creative destruction to deliver sustainable development. Still a fundamental question remains: how revolutionary is the fair trade movement? Has it succeeded to include the reality of underdeveloped/developing countries, or just by imposing the high standards of developed economies? I let the forthcoming case studies speak by themselves, with their specificities contributing to the debate on fair trade potential contributions for sustainable development based on empirical knowledge.

9.4 Methodology

The methodological approach of this qualitative study was a literature review associated with a desktop-based review of case studies for a comparative analysis about fair trade certified small-scale producers. Small rice farmers from Thailand and small quinoa farmers from Bolivia are analysed regarding the SDG 8 in light with the potential of the fair trade initiative to produce positive outcomes, from which recommendations are drawn.

9.5 Case Studies Findings

9.5.1 Thai Rice Case

Thai rice farmers are brought into evidence to illustrate the fair trade initiative potential benefits. Findings were depicted from studies by Becchetti et al. (2012). These small rice producers are from the northern, north-eastern, and central regions of Thailand. They count approximately 1.100 producers who have joined Green Net, a non-governmental organisation founded in 1993 to stand up for sustainability practices in agriculture. Green Net was created given the concern of producers and consumers with the use of pesticides and chemicals in the country and their impacts over public health and environment (Green Net). The change to livelihoods of fair trade certified small-scale rice producers was tested by Becchetti et al. (2012), what is a rare finding related to staplefoods, being more common the studies about fair trade top sales commodities such as coffee, cocoa and fruits.

Rice farmers in Thailand have been the target of a national government policy to turn conventional rice farmers into organic ones regarding environmental protection and population's health. In 2004 the organic rice farms comprised 0.049% of all rice farms in the country (Becchetti et al. 2012), and Thailand ranks among the world's top rice producers. Fair trade has worked in the country through cooperatives and NGOs as Green Net, joining efforts to increase sustainable development farming practices. Green Net was granted the Fairtrade (from Fairtrade Labelling Organisation) certification in 2002.

The initial Fairtrade premium was used to buy mills and computers to assist producers' cooperative. Green Net operates paying the small producers in advance, and those producers organised into groups buy the paddy and store them. Once rice is grown, they are milled and delivered to Green Net for packaging and trade, as Green Net receives export orders and systematises sales. Thai rice is traded with countries as Germany, France, Italy and the UK.

According to Becchetti et al. (2012) analysis, fair trade is an opportunity to increase the market access for small producers and to reduce the vulnerability of producers who depend on "monopolistic transportation intermediaries" (Becchetti et al. 2012, p. 119). Drawing from reviews, those authors rejected the existence of stereotypical exclusive relationships between fair trade affiliated producers and the fair trade channel, as some farmers continued to sell part of their production at the local market and to intermediaries. Becchetti, Conzo and Gianfreda's tested hypothesis demonstrated that fair trade affiliation added economic value more than on the control groups, showing an increase in income of fair trade affiliated producers. The latter were able to sell a higher percentage of their production, keeping family size and the amount for self-consumption equal in both groups compared. This meant that 1.100 producers secured a higher income while working with assurance of a set of good working conditions. Becchetti et al. (2012) remarked the importance of promoting a culture of impact studies to explore further and assess this alternative in trade.

9.5.2 Bolivian Quinoa Case

The present example demonstrates that globalisation can have side effects, and a model to include small farmers can also increase inequalities in the presence of economic growth. While several authors recommend impact evaluation of fair trade, Carimentrand and Ballet (2010) stress that numerous studies about the impact of certification schemes and fair trade exist. According to them, most studies indicate the positive effects of fair trade on "artisans, agricultural producers, and employees" (Carimentrand and Ballet 2010, p. 3); however their study reports on a counterexample: the negative effects of quinoa peaking production to attend the fair trade export market.

Quinoa production from Peru, Bolivia and Ecuador summed 17,747 metric tonnes in 1970. In 2005 their production together had reached 53,443 metric tonnes. It is reported that Bolivian quinoa production has increased from the 1990 onwards, according to Carimentrand and Ballet (2010). Thus, Bolivian quinoa production departed from 344 tonnes in 1990, to deliver 1423 tonnes in 2000, and 7641 in 2006 (Carimentrand and Ballet 2010). In the case studied, quinoa is grown in eight administrative provinces from the Bolivian Altiplano, and each province is divided into municipalities that include different communities. The middle of the year 2000 is marked by the arrival of agro-industrial business hiring producers to attend large commercial contracts given the increasing demand for quinoa in the global market.

Carimentrand and Ballet (2010) argue about the risks of mainstreaming commodities' markets as it happened in the case of the royal quinoa grown in Bolivia. The commodity boom in the European market led to the introduction of mechanised cultivation to attend the sharply increased demand.

The imbalanced capacity of producers to afford mechanisation increased the disparity in production. Advantages of mechanisation had effects on land ownership, and the plain areas suitable to mechanised cultivation started being disputed. This caused social disruption with fights for land among producers and communities. As a result, disparities among producers have soared. The added value of the fair trade produced quinoa was transferred to agro-industrial companies who started hiring producers to guarantee large-scale commercial contracts. Therefore, the direct transference of benefits to the small quinoa producers, who have assured the ethical consumption market to the fair trade model, was diverted.

In the associations and cooperatives, models to protect small producers' interests by adding up strengths, there was deviation of collective interests to individual interests (Carimentrand and Ballet 2010). In ANAPQUI (The National Association of Quinoa Producers) and APROQUIRY (The Regional Association of Quinoa Producers), the proportion of small producers' participation was reduced, according to examples provided by Carimentrand and Ballet (2010). In this case, one can see that the associations of producers did not address collective interests, and the wild fight for the external market created inequality due to mismanagement and priority to commercial contracts. In short, there was a booming economic growth with revenues from increasing sales but no evidence of sustainable development assured to small producers.

Authors report that the manual practices coexist with mechanised cultivation in the Bolivian Altiplano; as a result the disparities among different provinces and communities in the country remain (Carimentrand and Ballet 2010). This is where externalities of globalisation for mainstreaming a product in the absence of enough planning and sustainable development partnerships – such as government and institutions – can lead, if stringent guidelines towards sustainability are disregarded or misrepresented.

9.6 Discussion

From both cases one can observe a prime lesson to learn: the global market access requires planning and surveillance under sustainable development guidelines to avoid externalities of over production. Further analysis indicates the advantages of assisting and protecting small farmers with price stabilisation, anticipation of payments, decent work conditions and assisted market access but also government policies and guidance with regard to environment protection and capacity building, as it happened in the case of Thai rice farmers.

Recognising and addressing the complexity of fair trade standards, the administrative logistics required, the understanding of accountability and procurement and the respective labour arrangements are necessary steps to enable small producers performance to accrue benefits from the fair trade model. The returns to society can be that regional heterogeneity within countries finds through the fair trade model for food products a sustainable development pathway for agriculture and community development. And this means more than economic growth and WHS standards for rural and remote areas of peripheral countries in the globe. Here follow policy recommendation topics for government's intervention to increase the chances of agriculture development related to this alternative trade model:

– Facilitate access and storage of crop seeds to small farmers.
– Act lowering input costs for small producers.
– Balance asymmetrical power relations between small-scale farmers, medium-size farmers and large-scale agri-food enterprises.
– Agreements and surveillance to support the small farmers associations and cooperatives.
– Facilitate access to credit so small farmers can undertake long-term investments.
– Facilitate the access to processing plants.
– Provide training so small farmers develop capacity building skills.
– Provide transportation facilities to remote small farm lands.
– Technical assistance to avoid socioenvironmental disruptions.

Each of the listed topics can turn into a bottleneck for small farmers if not well managed and also for communities where they live. However, partnerships for sustainable development can turn small producers' steps into international trade feasible with social responsibility guidelines. This is the lesson Thai fair trade certified rice farmers demonstrated, producing great outcomes with support of government policies. Let alone all the work that need to be undertaken by the small farmers, but within the insightful social entrepreneurship initiative of fair trade, a stable income and decent work conditions are secured, a great achievement for small-scale producers' livelihoods considering that Fairtrade standards attendance constitutes a prerequisite to join the fair trade system and to maintain the fair trade certification label.

Although fair trade high standards face criticism, it is not a sound solution to lessen the robust fair trade working conditions requirements. On another hand, it is not fair that small producers are held solely accountable for internalising the expenses to dignify their work and add value to their production when they are submitted to meso structures out of their control. Difficulties of producers that are related to macroeconomic structures should be addressed through partnerships for sustainable development as per the above recommendations, with emphasis on their capacity building. This important aspect was already remarked by the Food and Agriculture Organization of the United Nations (FAO), when it stated that "provision of education in rural areas is essential if smallholders are to participate in markets, as small farmers cannot trade in sophisticated chains if they are neither literate nor numerate" (FAO 2012, p. 30). Training is usually provided by some fair trade organisations, but this also relies on meso structures from the localities where the products are grown. Once more, the value of partnerships for sustainable development can have a synergic effect.

9.7 Impact Sustainability: Final Remarks

Fair trade was presented as an example of innovation for attaining sustainable development goals that deserved to be revisited and have its opportunities related to economic growth and income generation in rural areas of developing economies brought into evidence under an evolutionary economic analysis. As from the review of empirical contributions, the success story relied on multistakeholder partnerships, and their outcomes cannot be disregarded as an opportunity to address those left behind within the economic dynamics. The innovative fair trade model has been insightful in demonstrating the range and the reach of ethical consumerism in our contemporary society, and its attempts to address inequalities evidenced by globalisation.

Notwithstanding the benefits of fair trade, the initiative can be improved to increase the outcomes. If all the criticisms are to be definitive about the fair trade movement, if fair trade is just about a niche market strategy and not a greater movement for social change, if the original fair trade model is turning into a free trade and if fair trade initiative is giving too much room for corporatisation, those constitute a broad exploratory area for empirical research and development. If all those critics make sense, fair trade organisations have the chance to review its pathways and reinvent themselves to keep up with the social license conquered, in a continuous process of creative destruction. Or the movement will tend to be replaced for a model that readdresses its pitfalls.

The Thai rice case demonstrated the grounds to recognise the potential of fair trade to create more jobs and to broaden decent work conditions for small producers from developing countries involved in rice certification schemes, as well as the associated environmental benefits, although it can be argued that these achievements are not universal when another fair trade food product as quinoa showed diverse effects. The argument of inequalities growth associated with the quinoa case works as a word of caution given the understandable motivation to bring high-quality products into mainstream markets. With this in mind, there is need for more systematic studies with different methodological approaches to assure that research findings are available as a resource and tool for further sustainable development actions and to assist in achieving the aims for social change brought by the fair trade movement. In this sense, strengthening the link between actions and research is strongly recommended.

A valuable remark is that as an alternative trade model, fair trade sheds light into the viability of an agriculture development model that does not have to surrender solely to the large agribusiness model to achieve economic growth. There is room for sustainable development actions through inclusive models for small farmers with the potential to increase job offerings and steady income sources within rural and remote areas, as the fair trade initiative has proven. This is of great relevance for governments' challenge of addressing the drain of population from rural areas into urban settings seeking desperately for income source. This is important for intra-estate and local wealth generation. The level of fairness benefits to be obtained can be supported through government policies to enable agriculture not only for large agribusiness corporations but also for small farmers. Considering that trade has a relevant role for wealth and income generation, trade agreements either on regional or multilateral levels must be prioritised so the sustainability goals are not bypassed. In conclusion, on a decision maker's level, the achievements or pitfalls of the fair trade movement may be managed to contribute further to social entrepreneurship, assuring the SDG 8 with spillover effects for attaining other sustainable development goals such as SDG 1, SDG 2, SDG 5, SDG 10, SDG 12 and SDG 17.

Acknowledgements I acknowledge the Australian Government for the research training program scholarship provided and my supervisors at the University of Tasmania, Dr. Fred Gale and Dr. Hannah Murphy-Gregory.

References

Barratt Brown M (2007) 'Fair trade' with Africa. Rev Afr Polit Econ 34(112):267–277
Becchetti L, Conzo P, Gianfreda G (2012) Market access, organic farming and productivity: the effects of fair trade affiliation on Thai farmer producer groups. Aust J Agric Resour Econ 56(1):117–140
Carimentrand A, Ballet J (2010) When fair trade increases unfairness: the case of quinoa from Bolivia. Cahier FREE 5
Carlisle L (2016) The terrace keepers. Stanf Soc Innov Rev 14(4):13–14
Choat S (2010) Marx through post-structuralism: Lyotard, Derrida, Foucault, Deleuze. Continuum, London
Cimoli M, Katz J (2003) Structural reforms, technological gaps and economic development: a Latin American perspective. Ind Corp Chang 12(2):387–411
Dopfer K (2011) Mesoeconomics: bridging micro and macro in a Schumpeterian key. In: Sectors matter!. Springer, Berlin/Heidelberg, pp 103–114
Englund O, Berndes G (2015) How do sustainability standards consider biodiversity? Wiley Interdiscip Rev Energy Environ 4(1):26–50
Fairtrade (2019) Fairtrade standard for small-scale producer organisations. Retrieved from https://www.fairtrade.net/standards
FAO, WFP, IFAD (2012) The state of food insecurity in the world in 2012. Economic growth is necessary but not sufficient to accelerate reduction of hunger and malnutrition. FAO, Rome
FLO (2003–2004) Fairtrade Labelling Organization Annual Report 2003–2004
FLO (2016–2017) Fairtrade Labelling Organization Annual Report 2016–2017
Foster J, Metcalfe JS (2012) Economic emergence: an evolutionary economic perspective. J Econ Behav Organ 82(2–3):420–432
Garrett RD, Carlson KM, Rueda X, Noojipady P (2016) Assessing the potential additionality of certification by the round table on responsible soybeans and the roundtable on sustainable palm oil. Environ Res Lett 11(4):045003
Green Net. Fairtrade: a world of difference. Producer group profile – Green Net. Retrieved from www.traiscraftschools.co.uk

Lagarde C (2017) The agenda: Christine Lagarde on globalisation. The World Economic Forum, The Economist

Malerba F, Nelson R, Orsenigo L, Winter S (1999) History-friendly'models of industry evolution: the computer industry. Ind Corp Chang 8(1):3–40

Murray DL, Raynolds LT (2007) Globalisation and its antinomies: negotiating a fair trade movement. In: Fair trade. Routledge, London, pp 19–30

Ribeiro-Duthie AC, Gale F, Murphy-Gregory H (2018) The alignment of fair trade with the sustainable development goals: a review of case studies and reports. FTIS 2018, Portsmouth

Rice RA (2001) Noble goals and challenging terrain: organic and fair trade coffee movements in the global marketplace. J. Agric. Environ. Ethics 14(1):39–66

Riisgaard L (2015) Fairtrade certification, conventions and labor. In: Handbook of research on fair trade. Edward Elgar Publishing, Northampton, pp 120–138

Schumpeter JA (1939) Business cycles: a theoretical, historical and statistical analysis of the capitalist process. McGraw Hill, New York

Schumpeter JA (1942) Capitalism, socialism and democracy. Harper & Brothers, New York

Schumpeter JA (1952/1997) History of economic analysis. Routledge, London

Sharma RK, Abidi N, Khan KM (2018) Comparison of conventional and fair trade systems on dimensions of sustainability: a study of basmati rice procurement in India. Int. J. Innov. Sustain. Dev. 12(4):446–468

Shiesari C, Gruninger B (2014) Assessing the benefits of Fairtrade orange juice for Brazilian small farmers. BSD Consulting Report, April, 92 pp.

Smith AM (2014) Cross-border innovation in south-north fair trade supply chains: the opportunities and problems of integrating fair trade governance into northern public procurement. In: Vazquez-Brust DA, Sarkis J, Cordeiro JJ (eds) Collaboration for sustainability and innovation, vol 3. Springer, Dordrecht

Taotawin N (2011) The transition from conventional to organic rice production in Northeastern Thailand: prospect and challenges. Adv Glob Change Res 45:411–435

Thavat M (2011) The tyranny of taste: the case of organic rice in Cambodia. Asia Pac Viewp 52(3):285–298

Udomkit N, Winnett A (2002) Fair trade in organic rice: a case study from Thailand. Small Enterp Dev 13(3):45–53

WFTO – World Fair Trade Organization (2016) Working together for a fairer world. Annual report 2016

Witt U (1992) Evolutionary concepts in economics. East Econ J 18(4):405–419

Chapter 10
SDG 9 Industry, Innovation, and Infrastructure

Community Capacity-Building for Sustainable Resource Governance in the Small-Scale Mining Industry

Isabel B. Franco, Franz Gonzalez Arduz, and Jairo Andres Buitrago

Abstract There is a general concern in the scholarly literature pertaining to sustainable resource governance and the implications for industries such as mining. Yet there are many unanswered questions in this area, and the research reported in this chapter increases our understanding in this regard. Our investigation has shown there are essential capacity-building areas to trace a roadmap for sustainable resource governance in small-scale mining (herein SSM). Improving identified areas seems from this research to be the most effective way to enhance the ability of small-scale miners and stakeholders to cope with pressing sustainability issues over time. This chapter presents a capacity-building roadmap for SSM and pays attention to the nature and importance of these areas to achieve the Sustainable Development Goal 9 (herein SDG), Industry, Innovation, and Infrastructure. What makes this research essential is its focus on the extent to which these areas are valuable for stakeholders involved in the SSM industry. It does this through a case study in Bolivia. This research is based on a qualitative strategy for data collection and case study methodology.

I. B. Franco (✉)
Institute for the Advanced Study of Sustainability, United Nations University Shibuya-ku, Tokyo, Japan

Australian Institute for Business and Economics, The University of Queensland, Brisbane, Australia
e-mail: connect@drisabelfranco.com

F. G. Arduz
Universidad Autonoma Tomas Frias, School of Economics, Finance and Administrative Sciences, Potosí, Bolivia

J. A. Buitrago
Universidad Nacional de Colombia, School of Engineering, Bogotá, DC, Colombia

© Springer Nature Singapore Pte Ltd. 2020
I. B. Franco et al. (eds.), *Actioning the Global Goals for Local Impact*, Science for Sustainable Societies, https://doi.org/10.1007/978-981-32-9927-6_10

Keywords SDG 9 · Industry, Innovation, and Infrastructure · Small-scale mining industry · Governance · Sustainability · Bolivia

10.1 Introduction

Global governance arrangements in the resource sector have imposed new responsibilities on the small-scale mining (herein SSM) industry, yet they do not hold the capacities to actively participate in such arrangements and achieve SDG 9, Industry, Innovation, and Infrastructure. International mandates more often encourage stakeholders to follow sustainable practices in the regions within they operate (ICMM 2005; ISO 2010; PDAC n.d.; RJC 2011). Compliance with both international and national regulatory mandates to protect the environment and foster local development through tax revenues and royalties are some of the emerging accountabilities imposed on the SSM industry. However, small-scale miners wonder if these governance arrangements are valuable for them and local communities or if they should comply with a more context-based governance framework that targets priority areas for them (CANAMIL 2015). The absence of a sustainable SSM governance framework increases existing SSM challenges threatening productivity and therefore livelihood options for locals. A situation that escalates discontent and tensions among stakeholders (miners, civil society, governments, and large-scale companies) (Franco 2014; Guzmán and Manuel 2015; Makki 2015). Based on a governance analysis of international precepts, domestic mining policies, perceptions of small-scale miners, and selected stakeholders, the research presented here proposes a capacity-building roadmap for sustainable SSM governance in Bolivia. This roadmap can be applied to other developing resource economies in Latin America and elsewhere.

Mining represents the largest source of income for small-scale miners, particularly in developing resource economies. Research indicates that ungoverned SSM escalates socio-environmental issues such as loss of livelihood options, unsustainable use of mercury, and loss of royalty revenues, among others (Cardenas 2011; Franco 2014). So far, stakeholders in isolation or through collaborative processes have implemented solutions to address these issues, namely, formal agreements between small-scale miners and large-scale mining companies, financial support, access to credit, capacity-building, and technical assistance, among others. These initiatives have somehow contribute to build small-scale miners' capacity, forge sustainable livelihoods, and increase taxes, royalties, and income generation (Molina-Escobar and Restrepo-Baena 2010; Ponce 2010; Veiga et al. 2001; Hilson 2006; Hilson and Banchirigah 2009; Franco and Ali 2016). However, these initiatives remain insufficient, particularly in the Bolivian context where issues around SSM are greater and small-scale miners' voices have not been heard.

The capacity-building roadmap here presented builds upon the international experience and a governance analysis of the Bolivian context. Policies, stakeholders (government, small-scale miners, private sector and civil society organizations), and existing collaborative governance arrangements have been examined for the purpose of this research. However, small-scale miners wonder if these governance arrangements are valuable for them.

10.2 Literature Review

10.2.1 A Review of Resource Governance

Miners have proactively fostered the creation collaborative coalitions for effective sustainable SSM governance. This emerged in response to global governance restructuring which has pushed miners to meet unrealistic accountability mechanisms, a situation that has elicited tensions within the SSM industry in developing countries such as Bolivia. Global concerns regarding the achievement of the SDG 9 and global agendas that guide mining performance are increasingly encouraging miners to become accountable for local operations. Lack of capacity at the local level to meet international demands has placed miners in a disadvantaged position to meet global requirements and account for the adverse impacts of industrial activities. This has led miners to rethink existing governance arrangements at the local level and collaborate with other stakeholders to be able to meet international and national mandates through relevant capacity-building initiatives.

The notion of governance as a multi-stakeholder collaboration scenario is frequently based on the assumption that stakeholders participate on an equal basis in decision-making processes and collaborate in the achievement of common goals (Clarkson 1995; Gibson 2000; Tracey et al. 2005). However, this unrealistic multi-stakeholder collaboration approach can be challenged due to two reasons. On the one hand, disadvantaged stakeholders such as miners lack the capacity to negotiate and benefit from collaborative governance processes. On the other hand, "because the relationships amongst actors are very often driven by factors such as unequal power relations, lack of clarity in their roles and responsibilities, and tensions that limit the possibilities of effective collaboration" (Franco 2014: 42). In Bolivian resource regions, these multi-stakeholder relationships are becoming more complicated as miners are very often tasked with unrealistic top-down agendas. This situation is becoming unsustainable for the SSM industry as well as for surrounding communities whose only livelihood option is mining.

The SSM industry in Bolivia is often challenged by the neoliberal rhetoric that posits large corporations as central actors of the global and national economies. An argument that puts more pressure on SSM as the latter has limited capacity to compete with mining giants. The neoliberal argument has largely been explored in schools of thought like economics and business ethics. In the economic realm,

Friedman's work (1970) has been one of the most significant. It also undermined the idea of social cohesion since they were fueled by corporate self-interest (Stilwell, 2006). Freidman's work (1970) emphasizes the role of corporate governance to increase shareholders' investment returns but neglects the responsibility to develop better relationships and collaboration between corporations and external stakeholders (Franco 2014). Friedman's contributions therefore undermined the idea of social cohesion since they were fueled by corporate self-interest (Stilwell 2006). The reaction against this neoliberal economic agenda led to local-level transformations that encouraged private and public corporations to develop better relationships with their stakeholders and look after their well-being (Beck 2007). However, with the arrival of Japanese, Chinese, Canadian, and other international corporations in Bolivia in the past years, Friedman's neoliberal argument is becoming the norm.

Along with the proliferation of neoliberal policies driven in part by globalization, the need to make mining actors more responsive to their external stakeholders and establish collaborative relationships with them increased. In developing resource economies, these responsibilities have increased not only for large multinationals but also for small-scale miners who very often belong to disadvantaged communities.

These governance shifts allowed miners, and the communities they belong to, to become more active rather than remaining passive actors and foster closer relationships between the community and other stakeholders like governments, NGOs, and the private sector. NGOs were involved in governance shifts which not only increased NGOs' responsibilities at the global and the local level in terms of public service provision for community sustainability but also in terms of assistance to governments to govern natural resources and to demand corporate accountability (Bell and Hindmoor 2009; Edwards et al. 1999).

Influenced by NGOs, small-scale miners in Bolivia have embraced emerging governance shifts through increased accountability. They have proactively initiated collaboration processes with external stakeholders to propose a sustainable SSM framework relevant for all. However, in informal conversations with miners, they perceive they still remain in a critical position. They are not only accountable for aligning with national and international governance precepts but are also pressured to account for the well-being of their families who in most cases belong to disadvantaged communities. Miners instinctively follow global governance mandates on social accountability. However, there is a limited understanding of the potential of SSM and the opportunities this sector could bring for all stakeholders involved.

In the literature, scholars in the business ethics domain embraced recent governance changes and challenged neoclassical theories that neglected the role of external stakeholders in governance scenarios (Clarkson 1995; Freeman 1984; Gibson 2000; Tracey et al. 2005). One of the major representatives of this governance approach was Freeman (1984) and his work on stakeholder theory. This theory made room for new approaches like stakeholder collaboration and corporate social responsibility. Freeman's work posits the notion that corporations have social obligations and therefore need to engage with external stakeholders. In the mining sector, this idea has been recently corroborated by scholars who agree that companies have obligations to their external stakeholders who are also entitled to benefit from the surpluses coming from corporate profit (Clarkson 1995; Gibson 2000; Tracey et al. 2005). In part, this arises from the fact that companies are extracting resources that can be conceived of as part of the local community. "Minerals, for example, are envisaged as public goods owned by the state on behalf of the country's citizenry, and so the companies have a moral obligation to enhance the welfare of the citizens whose resources they are extracting" (Franco 2014). In the Bolivian case, for example, natural resources are a public resource, and so, communities need to be compensated for resource extraction.

Recent shifts in governance processes in developing resource regions have led to active engagement among stakeholders (Mate 2001; Porter et al. 2013; Hamann et al. 2005). This approach is based on the assumption that stakeholders (including corporations, governments, and civil society) equally participate in decision-making processes and collaborate in the implementation of actions for the benefit of all (Clarkson 1995; Gibson 2000; Tracey et al. 2005). While this latter understanding is appealing from a theoretical standpoint, there are major challenges in practice, particularly in developing context like Bolivia. Scholars from schools of thought like corporate social responsibility, development, and environmental management argue that in reality, corporations engage with their stakeholders to pursue their own interests rather than in an effort to achieve sustainable initiatives for communities (Hilson 2006; Jenkins and Yakovleva 2006; Dartey-Baah et al. 2011). However, this is not the case for small-scale miners coming from unprivileged backgrounds. They do believe that collaboration is posited as a win-win relationship based on permanent consensus. Field observations also showed that collaboration is possible if stakeholders communicate their aspirations effectively. It also showed that tension in the Bolivian case occurs due to lack of clarity about priority areas for SSM. Hence, this article traces a roadmap to identify capacity-building priority areas relevant for miners and stakeholders involved. The propose roadmap will hopefully overcome existing barriers in a multi-stakeholder scenario for sustainable SSM governance and contribute to the achievement of SDG 9.

10.3 Methodology

10.3.1 Methods and Techniques

The research presented in this chapter is based on a qualitative methodological approach for data collection and analysis. Data obtained from primary and secondary sources were analyzed through a qualitative research strategy; nevertheless secondary quantitative techniques have been used to assess and validate available data where possible. This research dealt with diverse and multiple sets of data requiring the application of the case study method. The case studies allow for detailed and comprehensive information to be collected about a more focused issue, in this case

SSM. The collected data came from literature and policy reviews, small-scale miners' perceptions, and field observations.

Qualitative information was obtained through focus group interviews with ten (10) regional mining chambers' representatives and selected stakeholder respondents. Following respondents' requests, identities have been protected, and personal details will remain anonymous. Three focus groups were undertaken with miners and selected stakeholders at CANAMIL's (National Mining Chamber) office. Focus groups involved participants from Potosí, La Paz, Oruro, Santa Cruz, Chuquisaca, and Tarija mining regions. Small-scale miners who actively participated in group interviews were invited by CANAMIL. A governance policy analysis was also applied to facilitate the examination of some of the data collected, such as regulatory frameworks, royalty reports, corporate financial reports, and other quantitative secondary data meaningful to achieve the objective of this research. These data were collected from well-known and reliable sources.

10.3.2 Case Selection

Bolivia was selected as a case study due to the existing governance shifts in SSM. Bolivia is a developing resource country located on the Andes mountain range in South America. The Bolivian mining industry consists of three subsectors: state, private, and cooperative. State mining is represented by COMIBOL mining company, producer of tin, copper, zinc, and lead. The private mining sector integrates medium private companies, domestic and foreign investment companies, and producers of zinc, lead, silver, gold, copper, and antimony. SSM is part of the private mining industry and has a diversified mining production, as it produces all minerals and metals exported by Bolivia. SSM is mainly undertaken at COMIBOL's surrounding mining areas.

Private mining produces 70% of the GVA followed by 21% of cooperative mining and 9% of state mining. Official data of SSM production and exports are not available; however it is estimated that it represents 31.5% of the private mining production (CANAMIL 2015). There is not a clear methodology for SSM production calculations, particularly of cooperatives operating nearby COMIBOL. However, miners state that levels of production and exports have increased since 2006 (CANAMIL 2015).

According to Law 535 of Mining and Metallurgy, SSM miners are considered private agents clustered in the National Chamber of Mining (CANAMIL), an organization with autonomous management, equity capital, resources, and assets generated by members' contributions and governed under the National Constitution (Congreso de la Republica de Bolivia 2014a, b). Founded in 1940, CANAMIL produces a wide variety of minerals and metals such as tin, silver, zinc, lead, antimony, wolfram, copper, gold, and other rare-earth metals. CANAMIL's production represents 25% of the GVA. The organization currently comprises approximately 3747 miners (see Table 10.1).

Table 10.1 Regional mining chambers and number of small-scale miners (Autoridad Juridica y Administrativa Minera (AJAM) (2015)

Description	Number of small-scale miners
Potosi Regional Chamber	729
South Chichas Regional Chamber	561
Uyuni Regional Chamber	129
North of Potosi Regional Chamber	118
La Paz Regional Chamber	945
Oruro Regional Chamber	471
Cochabamba Regional Chamber	181
Chuquisaca Regional Chamber	159
Santa Cruz Regional Chamber	386
Tarija Regional Chamber	31
Beni Regional Chamber	35
Regional Chamber Pando	2
Total small-scale miners	3757

Table 10.2 Mining exports. Based on data from Ministerio de Minas y Metalurgia de Bolivia

Mineral/metal	MTF (multilateral trading facility)
Zinc	80,000
Tin	80
Gold	3
Silver	90
Antimony	450
Lead	22,000
Wolfram	86
Copper	2216
Iron	1,760,000
Manganese	718,660
Ulexite	146, 953, 923
Boric acid	14,985,462
Salt	2,261,845
Arsenic trioxide	120,000
Baritone	30,476,000
Amethyst	182,424
Ametrine	1751
Tantalite	46,558
Others	9,000,000

The number of miners, both individuals and cooperatives, is expected to increase up to 4000 miners over the next 2 years. SSM currently employs approximately 35,000 people and contributes to $US 20 million to the national royalty system (Ministry of Mining and Metallurgy of Bolivia 2015). SSM exports reached 650 million in 2017. Miners produce an average volume of 300 gross tons of ore per month (Ministerio de Minas) (Table 10.2).

Focus groups with small-scale miners showed that a capacity-building roadmap for sustainable SSM governance has the potential to increase SSM production by 15%, increase exports and revenues, and contribute to the achievement of SDG 9. Evidence also shows that miners value a bottom-up approach to capacity-building rather than a top-down agenda imposed by governments or international organizations. Such a bottom-up approach will be presented and discussed in forthcoming sections.

10.4 Discussion and Results

10.4.1 SSM Governance in Bolivia

10.4.1.1 Governance: Policy Analysis

The SSM sector in Bolivia is supported on a strong governance framework. However, it does not necessarily mean the existing governance environment is equally strong. Governance comprises policies and multi-stakeholder collaborative processes (Rakodi and Lloyd-Jones 2002; Davies 2005; Minnery 2007). These two pillars should be strong enough to secure sustainable SSM governance. Findings show that despite the proliferation of policies and regulations in place, multi-stakeholder collaboration is still fragile and needs to be enhanced in priority areas valuable for locals. Table 10.3 shows the existing policy framework that regulates the existing SSM governance:

Table 10.3 Policy framework for SSM governance (the Authors 2017)

Policy	Description
National Constitution of Bolivia Articles 369, 370, 371, and 372 (Congreso Nacional de la Republica de Bolivia 2009)	The constitution regulates mining operations carried out by state-owned and private companies and SSM cooperatives
Law no. 1777, 1997 Mining Code (Congreso Nacional de la Republica de Bolivia 1991a)	Former mining code
Law 535 of Mining and Metallurgy (Congreso Nacional de la Republica de Bolivia 1991a)	Existing mining code Replaces Law No. 1777
Decree 29272, 2007 (Congreso Nacional de la Republica de Bolivia 2014a)	Approves the existing development plan and encourages better governance through four actions: a) Modernization of existing legal mining framework b) Strengthening the role of the government as a key actor for mining development c) Development and diversification of Bolivia's mining potential d) Strengthening of SSM

The policy framework shown in Table 10.3 is supported by various institutional stakeholders. Policy analysis and field observations showed there is an active involvement of key organizations such as the Geological Mining Service (SERGEOMIN), the Mining Corporation of Bolivia (COMIBOL), the National Service for Minerals Commercialization's Control and Registration (SENARECOM), universities, and NGOs. Although miners have initiated collaborative processes with these institutions, they find it difficult to convey their governance aspirations in a multi-stakeholder scenario. Despite a strong policy framework, SSM governance is threatened mainly by lack of miners' capacity to communicate SSM priority areas for capacity-building and incorporate them into the national governance agenda. Unsustainable SSM practices, rudimentary infrastructure and transport, lack of energy sources and water, and precarious administrative structures are areas in which locals need to build their capacity. Miners agree these challenges are likely to be resolved through capacity-building that target priority areas for sustainable SSM governance (CANAMIL 2015).

10.4.1.2 Governance: Stakeholder Collaboration

At the beginning of the 1990s, the government closed the National Mining Bank (BANIM). BANIM was accountable for allocating loans to SMM. SSM loans were paid to BANIM through mineral production and were subsequently exported and discounted from SSM debt. This banking system became unsustainable as miners lacked financial planning education to be accountable for debts and revenue management. BANIM was bankrupted and SSM became a high-risk debtor. In response, the government issued Decree 22862 in 1991 to proceed with BANIM's closure (Congreso Nacional de la Republica de Bolivia 1991b). Subsequently, the government proposed the creation of the mining development fund for SSM. The fund was aimed to provide finance and technical assistance. However, the fund was hesitant to engage with SSM, and no concrete projects were developed as debt collection issues remained critical. At that time, SSM was still considered a high-risk debtor. In addition to poor SSM reputation, SSM marketing and financing systems were transferred to the private sector. This situation also jeopardized international cooperation's funding for SSM.

There was no way out for miners, and international and domestic precepts kept making them accountable for the sustainability of SSM operations. In response, miners tapped on stakeholder collaboration for capacity-building in priority areas to secure their only livelihood option. However, there was uncertainty about priority areas for SSM governance and to what extent they were valuable for miners. Various focus groups were carried out to better understand miners' priorities and assist them in fostering sustainable solutions to become active participants in Bolivia's resource governance (CANAMIL 2015).

In focus groups with leaders in the SSM industry, it was agreed that multi-stakeholder collaboration for SSM sustainability was essential to achieve SDG 9. Miners agreed that institutions such as SERGEOMIN, COMIBOL, and SENARECOM, higher education institutions, and NGOs have helped the SSM industry overcome financial and technical challenges and therefore secure their only livelihood option. Interestingly, miners perceived that collaboration between SSM and NGOs is beneficial for accessing financial, legal, and management capacity-building. CANALMIN, a non-for-profit organization, for instance, signed a cooperation agreement with the SSM sector in the mining areas of Potosi, La Paz, Oruro, Llallagua, Atocha, and Uyuni. Likewise, the SSM sector has initiated ongoing collaboration processes with the Ministry of Mining and Metallurgy, the Ministry of Development Planning, state and local governments, and indigenous communities. Slowly, the SSM industry has overcome past barriers and initiated a transformation toward sustainable SSM governance. A better understanding of priority capacity-building areas for SSM governance will also contribute to the achievement of SDG 9.

10.4.2 *Roadmap for Sustainable SSM Governance in Bolivia*

An analysis of SSM leaders' perceptions yielded interesting findings in relation to a capacity-building roadmap for sustainable SSM in the Bolivian context. While the idea of a roadmap for sustainable SSM is tempting from a theoretical standpoint, it is only possible if stakeholders involved work in collaboration with miners to target priority areas for SSM sustainability. This research identified nine priority areas to trace the path for sustainable SSM governance in Bolivia: SSM formalization, risk prevention, sustainable development, investment and technology transfer, economy diversification, geology and minerals processing, multi-stakeholder collaboration, marketing, and capacity-building. The roadmap here proposed seeks to create the foundation for sustainable SSM governance in Bolivia. Increased emphasis by stakeholders on these priority areas has the potential to foster SSM growth and overall sustainability in the Bolivian context: "If mining were not sustainable, it would have already finished. This is not possible because industrial development depends largely on minerals and metals and their transformation into goods and finished products" (Focus Group with Miners).

10.4.2.1 SSM Formalization

Miners suggest that SSM formalization should be undertaken through concession agreements regulated under the Law 535 of Mining and Metallurgy (Congreso Nacional de la Republica de Bolivia 2014a, b). In this regard, miners also agree that:

this new law establishes the regulatory framework for the formalization of mining... however, the government does not have the institutional capacity to formalize SSM or the funding to support such process. Yet, a few failed attempts towards formalization have been proposed. These have not materialized as the government does not have enough resources to support this process and miners were not convinced this was something necessary. (Focus Groups with Miners)

10.4.2.2 Risk Prevention

Participants agree that risk prevention is a priority area for sustainable SSM governance. Hence, it is miners' responsibilities to adopt the procedures and protocols for risk prevention in the use of equipment and machinery, supplies, and materials. Risk prevention activities should be regulated under the Law 602 on risk management issued in 2014 (Congreso Nacional de la Republica de Bolivia 2014a). In this respect, miners added that:

The most critical issue is that the life of workers is not valued. They are treated as a disposable resource. This is something that comes from the time of the colony. Recent trends show that having a healthy and safe worker is more productive. Just a few multinational companies apply this principle partially. The ideal scenario would be to train workers on industrial security providing adequate resources and technology. The SSM sector does not benefit from this (Focus Group with Miners).

10.4.2.3 Promoting Sustainable Development

Data analysis also showed that miners are somehow familiar with concepts such as sustainability and social responsibility. However, this does not mean they have wide knowledge about the operationalization of these concepts: "...the government should participate in all SSM activities…to make sure all initiatives are sustainable over time...the law should also be considered as an essential element of sustainable development." (Focus Group with Miners)

10.4.2.4 Investment and Technology Transfer

Miners agree that establishment of alliances with investment firms to access funding and knowledge transfer opportunities can foster SSM productivity. SSM leaders expect that both private and public actors collaborate to set the institutional conditions to access credit and funding opportunities: "Mining activities cannot be carried out without investing in technology (both, machinery and equipment) and financial resources, particularly in the production stage. Introducing the use of technology will improve overall productivity" (Focus Group with Miners).

10.4.2.5 Economic Diversification

Findings show that SSM has the potential to foster regional sustainability if it is considered a complimentary economic activity. Evidence shows there is a need to foster alternative livelihood options along with SSM to achieve overall sustainability. More importantly, increasing the participation of indigenous communities to discuss SSM governance priorities is a key factor for SSM sustainability: "Revenues derived from mining, either SSM or large mining, are important to help miners plan for reinvestment on mining and alternative activities such as transport, agriculture, particularly in the context of mining regions" (Focus Group with Miners).

10.4.2.6 Geology and Mineral Processing

Miners perceive that capacity-building in geology and minerals processing will assist them in becoming more effective to unleash the geological potential in Bolivia. Similarly, they agree that stakeholders should contribute to build miners' capacity on "the design of production plans that integrate calculations on mineral production, prices in domestic and international markets; taxes and royalty payments; costs of supplies, materials, tools, machinery and equipment"(Focus Group with Miners).

10.4.2.7 Strengthening Multi-stakeholder Collaboration

Multi-stakeholder collaboration is essential to guarantee sustainable SSM governance. Collaboration should involve universities, research institutions, indigenous communities, as well as SSM leaders. "Such a collaborative approach should follow the principles of inclusiveness and participation in alignment with the Constitution" (Focus Group with Miners).

10.4.2.8 Marketing

Miners agree that one of the factors that hinders the success of SSM has to do with poor commercialization and marketing channels. Designing better ways to commercialize SSM production in both local and international markets will increase production as well as the reputation of the SSM industry. "SSM produces small amounts of mineral material, therefore it is important to conduct marketing research to identify competitive prices and emerging markets at the national level" (Focus Group with Miners).

10.4.2.9 Capacity-Building

Capacity-building and technical assistance can also help the SSM industry in fostering sustainability. In focus groups with miners, capacity-building became a recurrent topic of conversation. Miners think they do not hold the capacities to face the challenges impose by local and global demands which have somehow threatened their ability to produce on sustainable basis. There is a common agreement among participants that miners in isolation do not hold the financial resources to cover training or technical assistance costs. Capacity-building in finance, use of technology, marketing, entrepreneurship, and sustainability are also some of the priority capacity-building areas for SSM: "One of the major weakness of SMM is that it is based on empirical knowledge and practical experience….requiring miners to build their capacities on drilling, mineral processing and use of technology" (Focus Group with Miners) (Table 10.4).

Table 10.4 Roadmap for sustainable SSM governance proposed by small-scale miners in Bolivia

Priority area	Actions
SSM formalization	Environmental and social licensing approvals Sustainability planning, addressing priority areas for SSM such as geology, production, and investment
Risk prevention	Management systems for risk prevention in SSM Increasing awareness for risk prevention management in SSM Identifying a model for SSM production for mineral traceability Investment projects that include key areas for risk management such as industrial safety, environmental and social sustainability, and quality production Developing agreements in risk prevention and in partnership with foreign and national investors
Promoting sustainable development	Inclusion of good safety practices and effective community relations with indigenous communities Enhance human capital and employment generation for miners and surrounding communities Foster entrepreneurship for SSM production Create accountability mechanisms to comply with national guidelines, institutional commitments, and international standards
Investment and technology transfer	Guarantee SSM formalization Technical assistance provided by external stakeholders and investors in mining and technology transfer Capacity-building in minerals processing and use of new technologies for SSM development
Economic diversification	Fostering connections between mining and agriculture supply chains under the mandate of the Ministry of Rural Development and Land Establishing small and medium enterprises for local procurement Creation of community councils to monitor the relationships between SSM and native indigenous communities

(continued)

Table 10.4 (continued)

Priority area	Actions
Geology and mineral processing	Design SSM plans that set clear exploration, mining, mineral processing, and marketing goals Design a prospecting and exploration plan that involves most miners in SSM Capacity-building on SSM supply chains. This will contribute to SSM growth and diversification Designing a new administrative and accounting model for SSM Capacity-building to increase SSM accountability
Strengthening stakeholder collaboration	Establishing agreements and collaboration between the Ministry of Mining and Metallurgy, SERGEOMIN, and SENARECOM Establishing positive relations between SSM and indigenous communities Increase SSM participation in social consultation processes Developing collaborative initiatives with public and private stakeholders in the mining supply chain
Marketing	Developing marketing plans to enhance the commercialization of lead, zinc, and silver Including marketing as a constituent component of SSM Capacity-building on local and international market prices and exports
Capacity-building	Capacity-building should also be targeted to assist miners and surrounding communities to find alternative livelihood options different from mining Capacity-building on sustainable production across the mining supply chain Capacity-building on entrepreneurship for economic diversification Capacity-building on finance and accounting Capacity-building on industrial safety Capacity-building on use of new technologies

10.5 Impact Sustainability: Final Remarks

Proposing a capacity-building roadmap for sustainable SSM governance valuable for miners and local communities in the Bolivian context has the potential to contribute to the achievement of SDG 9. It will also increase SSM accountability by maximizing royalty and taxes revenues as well as foster sustainable and innovative mining practices. Primarily supported by the Law No 535 and the National Development Plan, the implementation of the roadmap here proposed requires the support of various stakeholders, namely, civil society organizations, miners, government, higher education institutions, and multinational corporations. Multi-stakeholder collaboration can make a strong contribution by providing technical assistance, financial support, and capacity-building. Likewise, it is essential to define stakeholders' roles and responsibilities for the effective implementation of the roadmap. This will lead to a collaborative governance environment in which SSM can play an active role for sustainable SSM governance. Some of the roles and responsibilities for the implementation of the capacity-building roadmap are as follows.

In collaboration with the government and international cooperation agencies, this study recommends that CANAMIL monitors the implementation of the roadmap. Regional chambers and civil society organizations should also be consulted on the implementation of the nine priority areas for capacity-building, namely, SSM formalization, risk prevention, sustainable development, investment and technology transfer, economic diversification, geology and minerals processing, stakeholder collaboration, marketing, and capacity-building.

Local communities and civil society organizations operating in Bolivian mining regions should also become active participants in proposing SSM capacity-building areas relevant for community sustainability. Although civil society participation has not been widely discussed in this chapter, SSM has a direct impact on local communities. In most cases miners come from unprivileged communities; hence they are directly accountable for overall community sustainability. Interestingly, findings show that miners are aware of local development expectations and have proposed alternative livelihood options to diminish community dependency on mining. However, research findings show that economy diversification is the result of active community engagement and government leadership.

Governments should assume a more active role in SSM capacity-building and overall sustainability. Local authorities in developing resource regions are usually criticized for their weak institutional capacity. This research has found that the Bolivian government has wide expertise on issuing legislation for SSM governance and overall sustainability. However, the implementation of such legislation to foster multi-stakeholder collaboration in this area fails. This inadequate involvement in collaborative governance arrangements is hampering the opportunities to advance toward SSM sustainability. However, the implementation of more participatory governance mechanisms led by the government can have a major positive effect on capacity-building for SSM and overall sustainability. The adoption of more participatory governance approaches in which the government is the key player can also lead to improvements in government accountability and community participation toward the achievement of SDG 9.

Regarding the roles and responsibilities of SSM in the implementation of the roadmap for SSM capacity-building, this investigation has found that miners have undertaken several actions to move SSM forward and improve sustainability practices, yet there is still room for improvement. Strengthening collaboration with the government and civil society can assist SSM in overcoming this challenge. Research also shows that SSM continues to being trapped in conventional governance mechanisms hindering SSM sustainability. Context-based governance arrangements such as the roadmap for capacity-building proposed in this chapter can lead to more sustainable SSM and capacity-building initiatives valuable for miners and local communities. More influential stakeholders should also facilitate the execution of the roadmap here proposed, particularly in regard to SSM capacity-building and multi-stakeholder collaboration for its effective implementation.

This study highly recommends SSM capacity-building in nine priority areas. Capacity-building should also be accompanied by ongoing research undertaken in collaboration with universities or think tanks. This will assist stakeholders in identifying early barriers and hindering factors in the future implementation of the roadmap. In this context, the role of higher education and training institutions is essential to equip miners with the capacity to ensure overall sustainability across the nine priority areas. This transit toward sustainable SSM governance can increase overall productivity, boost the Bolivian economy, and contribute to the achievement of SDG 9.

References

Autoridad Juridica y Administrativa Minera (AJAM) (2015) Estadisticas Camaras Regionales Mineras. AJAM, La Paz
Beck U (2007) Modernizacion Reflexiva. Retrieved from http://www.criterios.es/pdf/archplus-beckmoder.pdf
Bell S, Hindmoor A (2009) Rethinking governance: the centrality of the state in modern society. Cambridge University Press, Sydney
CANAMIL (2015) Focus group discussions with stakeholders between September – November 2015. CANAMIL, La Paz
Cardenas M (2011) Población Guajira, Pobreza, Desarrollo Humano y Oportunidades Humanas para los Niños en La Guajira. Master Dissertation, Universidad Nacional de Colombia. Retrieved from http://www.bdigital.unal.edu.co/3573/1/Tesis__Mauricio__Cardenas.pdf
Clarkson MB (1995) A stakeholder framework for analyzing and evaluating corporate social performance. Acad Manag Rev 20(1):92–117
Congreso Nacional de La República de Bolivia (1991a) Ley 1777 Código de Minería. Congreso Nacional de La República de Bolivia, La Paz
Congreso Nacional de La República de Bolivia (1991b) Decreto Supremo No 22862. Congreso Nacional de La República de Bolivia, La Paz
Congreso Nacional de La República de Bolivia (2009) Constitución Politica de Colombia. Congreso Nacional de La República de Bolivia, La Paz
Congreso Nacional de La República de Bolivia (2014a) Ley 535 de Minería y Metalurgia de 2014. Congreso Nacional de La República de Bolivia, La Paz
Congreso Nacional de La República de Bolivia (2014b) Ley 602 de Gestion de Riesgos. Congreso Nacional de La República de Bolivia, La Paz
Dartey-Baah K, Amponsah-Tawiah K, Sekyere-Abankwa V (2011) Leadership and organisational culture: relevance in public sector organisations in Ghana. Business Manag Rev 1(4):59–65
Davies JS (2005) Local governance and the dialectics of hierarchy, market and network. Policy Stud 26(3–4):311–335
Edwards M, Hulme D, Wallace T (1999) NGOs in a global future: marrying local delivery to worldwide leverage. Public Adm Dev 19:117–136
Franco IB (2014) Building sustainable communities: enhancing human capital in resource regions. PhD dissertation. The University of Queensland, Brisbane, Australia

Franco I, Ali S (2016) Decentralization, corporate community development and resource governance: a comparative analysis of two mining regions in Colombia. Extractives Industries and Society. In press

Freeman RE (1984) Strategic management: a stakeholder approach. Cambridge University Press, New York

Friedman M (1970) The social responsibility of business is to increase its profits. Corporate ethics and corporate governance. In: Zimmerli WC, Holzinger M, Richter K (eds) . Springer, Berlin/Heidelberg, pp 173–178

Gibson K (2000) The moral basis of stakeholder theory. J Bus Ethics 26(3):245–257

Guzmán C, Manuel G (2015) Transforming Andean space: local experiences of mining development in Peru. PhD Thesis. Sustainable Minerals Institute, The University of Queensland

Hamann R, Kapelus P, Sonnenberg D (2005) Local governance as a complex system: lessons from mining in South Africa, Mali and Zambia. J Corp Citizensh 18:61–73

Hilson G (2006) Championing the rhetoric? "Corporate social responsibility" in Ghana's mining. Greener Manag Int 53:43–56

Hilson G, Banchirigah SM (2009) Are alternative livelihood projects alleviating poverty in mining communities? Experiences from Ghana. J Dev Stud 45(2):172–196

ICMM (International Council on Mining and Metals) (2005) Community development toolkit. Retrieved from, https://www.icmm.com/news-and-events/news/articles/icmm-presents-updated-community-development-toolkit

ISO (Organizacion Internacional de Normalizacion) (2010) ISO 26000 Guia de Responsabilidad Social, vol 26000. ISO, Suiza, p 106

Jenkins H, Yakovleva N (2006) Corporate social responsibility in the mining industry: exploring trends in social and environmental disclosure. J Clean Prod 14(3–4):271–284

Makki M (2015) Coal seam gas development and community conflict: a comparative study of community responses to coal seam gas development in Chinchilla and Tara, Queensland. PhD Thesis. School of Communication and Arts, The University of Queensland

Mate K (2001) Capacity-building and policy for sustainable development networking. Miner Energy Raw Mater Rep 16:3–25

Minnery J (2007) Stars and their supporting cast: state, market and community as actors in urban governance. Urban Policy Res 25(3):325–345

Molina-Escobar JM, Restrepo-Baena OJ (2010) Colombian mining sustainability. Dyna-Colombia 77(161):149–151

PDAC (Asociación de Prospección y Desarrollo de Canadá). (N.D). E3 A Framework for Responsible Operations- Principle and Guidance. Rerieved from http://www.pdac.ca/e3plus/English/pg/pdf/e3plus-pg-full.pdf

Ponce A (2010) Panorama del Sector Minero. Unidad de Planeación Minero Energética, Bogota

Porter M, Franks DM, Everingham J (2013) Cultivating collaboration: lessons from initiatives to understand and manage cumulative impacts in Australian resource regions. Resources Policy

Rakodi C, Lloyd-Jones T (2002) Urban livelihoods: a people Centred approach to reducing poverty. Earthscan Publications, London

RJC (Consejo de Joyeria Responsable) (2011) Guia de estándares. Retrieved from http://www.responsiblejewellery.com/downloads/boxed_set_2009/G002_2009_RJC_Standards_Guidance.pdf

Stilwell F (2006) Political economy: the contest of economic ideas. Oxford, New York

Tracey P, Phillips N, Haugh H (2005) Beyond Philanthropy: community Enterprise as a basis for corporate citizenship. J Bus Ethics 58:327–344

Veiga M, Scoble M, McAllister ML (2001) Mining with communities. Nat Res Forum 25(3):191–202

Chapter 11
SDG 10 Reducing Inequalities

Reducing Inequalities (SDG 10) in Australia's Superannuation System: A Multidimensional Approach to Achieving Female Financial Equality in Retirement

Caitlin Power

Abstract It is well-evidenced that women earn less than men; however, the incessant effects of the gender pay disparity egregiously continue into retirement. Women in Australia retire with approximately half the superannuation balance of men; consequently, this chapter is preoccupied with understanding the reasons for the gender disparity in retirement (superannuation) savings in Australia. Using the framework of the sustainable development goals (SDG), notably SDG 10 (reducing inequalities), this chapter critically engages with how superannuation policy can be ameliorated and reformed to facilitate the diverse career trajectories and primary care responsibilities many women face. In line with SDG sub-target 10.2, the empowerment and promotion of economic inclusion, irrespective of age, sex, race etc., this chapter stresses the integral role that financial education and literacy play in enabling women to better understand the vital function of superannuation savings in preparing for a sustainable retirement. Finally, the latter section of this chapter explores how superannuation policy can be structurally amended to consider the oftentimes broken and disparate career trajectories women face.

Keywords Reducing inequalities · SDG 10 · Superannuation policy · Retirement · Female financial equality · Defined benefit schemes · Australia

---

C. Power (✉)
Faculty of Humanities and Social Science, The University of Queensland,
St Lucia, QLD, Australia
e-mail: caitlin.power@uqconnect.edu.au

11.1 Introduction

It is undoubtable that the current structure of Australia's superannuation system is systemically biased towards women. To this end, and as a symptom of systematic policy inadeqauicies, women are retiring with 47% less superannuation than men (Trust 2015; Campo et al. 2015; Agency 2017; Hetherington and Smith 2017; Riach 2018; Wood 2017; Feng et al. 2019). At present, women in Australia retire with approximately half the superannuation (super) balance of men (Black 2015; Committee 2016; Coates 2018; Koukoulas 2018; Agency 2017). According to financial data, men aged between 60 and 64 retire with an average superannuation balance of $270,710. Conversely, women in the same age bracket retire with only $157,050 in super (see Table 11.1) (Agency 2017; Coates 2018). Further horrifically, according to the Association of Australian Superannuation Funds (2018) 50% of women approaching retirement in 2014 had a low super balance of nil to $49, 999.[1]

Cameron (2013, p. 2) makes the salient point that 'nowhere is the extent of gender inequality more starkly revealed than in the lifetime earnings and superannuation savings of men and women'. It is thus axiomatic that addressing the gender gap in superannuation balances is vital: single and elderly women in Australia are at the greatest risk of absolute poverty and housing stress in retirement (Coates 2018). The 2016 senate inquiry, entitled 'a husband is not a retirement plan', noted that 'older and single women are one of the fasted growing cohorts of people living in poverty' (Committee 2016). In a similar vein, a report into the lived experiences by women on the precarious cusp of poverty in retirement revealed the extent of structural inequalities which women in Australia are subjected to (Parkinson et al. 2013). In the report, women spoke of the stress of being unable to retire with one woman acknowledging that:

> We had retired at one stage and had no intention of returning to work…I was on carer's pension for quite a while but it just wasn't meeting the need. We just couldn't survive. (Parkinson et al. 2013, p. 11)

The report also highlighted the impoverished financial circumstances in which many elderly women in Australia live; these women used free showers at sporting clubs, and many women adjusted their meals to suit their meagre budget:

> Sometimes your food all day is a cup of tea and bread and a little butter and sometimes fruit but you can't buy meat and all the things you need. (Parkinson et al. 2013, p. 10)

It is indisputable that wealth inequality in retirement harms women. In this regard, the reasons behind female financial inequality in retirement are multifaceted, however, interrupted career trajectories, the gender wage gap, and taking primary responsibility for unpaid care work are critical explanatory factors that facilitate the gender disparity in retirement and superannuation savings. COTA Australia saliently emphasises that 'the wheels of retirement income security are set in motion many

---
[1] For men in the same age bracket, 33% had a low balance of nil to $49,999; see Funds (2018, p. 9).

Table 11.1 Average superannuation balances by age (2013–2014)

Age group	Women's average superannuation	Men's average superannuation	Difference	Gender superannuation gap
20–24	$3941	$6265	$2324	37.1%
25–29	$14,812	$18,072	$3360	18.0%
30–34	$25,549	$36,373	$10,825	29.8%
35–39	$34,812	$55,279	$20,467	37%
40–44	$53,536	$83,565	$30,029	35.9%
45–49	$67,805	$119,500	$51,695	43.3%
50–54	$84,228	$146,608	$62,380	42.5%
55–59	$115,046	$227,765	$112,719	49.5%
60–64	$138,154	$292,510	$154,356	52.8%
65–69	$117,113	$194,633	$77,489	39.8%
70–74	$101,960	$146,165	$44,205	30.2%
75–79	$25,692	$114,937	$89,245	77.6%
75–79	$17,468	$30,026	$12,558	41.8%
80–84	$4281	$26,226	$21,845	83.7%
Total	$54,916	$98,535	$43,619	44.3%

This table has been sourced from Agency (2017, p. 5)

decades before most people even begin to think about how well they are placed to manage financially in later life' (McGrath 2015, p. 5). This means that many elderly women in Australia are reliant on the age pension, which has been deemed inadequate and insufficient (Pattern 2016; Society et al. 2016). In a report into the adequacy of the age pension, one retiree highlighted that 'after paying major bills, we have $180 a fortnight to live on' (Society et al. 2016, p. 34). A similar report by the Association of Superannuation Funds of Australia found that to maintain a modest style of living in retirement, a single person household requires $524.30 per week (Australia 2018). However, this estimate figure excludes rent or mortgage payments and thus assumes that retirees own their principle place of residence outright. Moreover, this 'modest' estimate is 13% more than the current age pension which is approximately $458.15 per week (Services 2019a). It is evident that relying solely on the age pension in retirement places elderly retirees in a precarious financial situation. Consequently, the focus of this chapter lies in understanding the contributing factors behind the financial disparity in retirement between genders; critically, this chapter will pay particular attention to how sub-goals of sustainable development (SDG) 10 can facilitate a narrowing in the gender gap in super balances between men and women.

The structural biases innate within the foundations of Australia's superannuation system impede upon the achievement of SDG 10 – the reduction of inequality. Crucially, a key sub-goal of SDG 10 is to eliminate discriminatory policies in favour of programmes which enable social protection and thus the achievement of equality. Sub-targets 10.3 and 10.4 demand equal opportunity via the abolition of discriminatory laws, policies, and practices; moreover, sub-target 10.2 advocates for the promotion of empowerment and the economic inclusion of all, irrespective of age, gender or other status (Nations 2019). Recognising the appeals of SDG 10 and subtarget 10.2, 10.3 and 10.4, this chapter is principally concerned with understanding how Australia's superannuation policy can be modified to enable the economic equality of women in retirement. These considerations are particularly salient in light of the conclusions arising from the 2016 senate inquiry into financial security in retirement, which is that women are at a greater risk of experiencing poverty, housing stress and homelessness in retirement (Committee 2016, p. 13). These financial difficulties stem largely from the systematic inequalities embedded within superannuation policy in Australia. For example, super payments are intrinsically tied to paid work; consequently, women are at a fundamental disadvantage as they are more likely to assume primary responsibility for unpaid care work over the course of their careers (Bulbeck 2005; Agency 2017). As this chapter will investigate, the accumulation of larger superannuation balances demands full-time work and an interrupted career trajectory. In a closely related vein, the gender pay gap, gender wealth gap, and occupational segregation further obstruct the accumulation of adequate superannuation balances for retirement (Agency 2015, 2017; Committee 2016). These observations lead Riach (2018) to make the salient argument that 'women's superannuation balances are determined by relationships and cultural expectations, among them gender inequality in family care of disabled family members, and the division of household labour'. This quote pivotally highlights the structural deficits present within superannuation policy. As such, it is crucial to question how these structural biases within Australia's superannuation policy can be dismantled.

This chapter will explore how SDG 10 can assist with mitigating the burden of financial insecurity that women experience in retirement in three central parts. The first section of this chapter will explore Australia's multi-pillar pension framework. Crucially, this section will delve into the three pillars which support retired Australians: the pension, superannuation, and investments or voluntary savings. Importantly, this section will emphasise that the gradual shift towards a more privately funded (superannuation) model of retirement can be inequitable to women, as the accumulation of a larger superannuation balance is intrinsically linked to a full-time employment and pay. The second section explores the societal and gendered impediments that prevent women from accumulating sufficient superannuation for retirement. Significantly, this section emphasises that there is a confluence of barriers which inhibit women from accumulating the same superannuation balance as men; however, there are four overarching variables which critically affect the ability of women to accumulate super: the gender pay gap, gendered occupational segregation, the care penalty, and the gender wealth gap. The critical conclusion arising from this section is that there is a confluence of distinct factors which impinge upon female financial security in retirement, thus, tackling or alleviating gendered financial insecurity in retirement demands a multifaceted approach. Consequently, the third section of this chapter will consider how SDG 10 can be employed to dismantle the inequitable barriers which prevent women from achieving financial equality in retirement. With particular reference to sub-target 10.2, this chapter argues that there is sufficient scope to expand financial education and literacy programmes, specifically so that they are more tailored to the diverse career trajectories that women face. Under the recommendations of target 10.2, I will further argue that financial education should be embedded within a national school curriculum framework. Finally, this chapter will consider how sub-target 10.3 and 10.4 can be employed to dismantle the structural and systemic barriers embedded within superannuation policy in Australia. The key conclusion arising from this chapter is that within the framework of the sustainable development goals, there is ample scope to use the direction of SDG 10 to develop a national agenda which will ameliorate financial security for women in retirement.

11.2 Surveying the Literature: The Architecture of Australia's Retirement Policy

Australia's retirement system is embedded within a multi-pillar structure (see Table 11.2) (Agency 2017; Cerise 2009; T. Treasury 2009). Australia's three-pillared system stems from the World Bank's Pension Conceptual Framework, which recommends a flexible, multi-pillar model (T. W. Bank 2001, 2006; W. Bank 2008). This chapter will briefly outline Australia's three-pillar model for supporting and ensuring retirement, beginning with the age pension.

Table 11.2 Australia's three-pillar retirement income stream

The Age Pension	Superannuation	Voluntary savings and assets
• Social Security Payment (publicly funded) • Means tested • Provides a minum 'safety net' level of income in retirement	• Employers are required to pay a designated portion (9.5%) of an employees earnings into a fund • Employer contributions have increased from 3% in 1992 to the current level of 9.5%	• E.g. shares, managed funds, or cash with in a bank account • Investment properties • Family home

This figure has been compiled using information from Agency (2017), Coates (2018), T. Treasury (2019) and D. o. t. Treasury (2009)

11.2.1 The Age Pension

The first pillar is the public age pension, provided by the government (Agency 2017; Coates 2018). The age pension is designed as a 'safety net' income stream (Agnew et al. 2013; Coates 2018). Eligibility for the age pension is determined by comprehensive means testing, which includes assessing eligible income and assets (excluding the family home) (Agnew et al. 2013). Eligible assets that are assessed include, but are not limited to, investment properties, financial investments, superannuation investments, business assets, motor vehicles, collectable items etc. (Services 2019b). At present, the current maximum pension payment per fortnight for a single retiree (including the energy supplement) is $916.30, and for a couple, it is $1381.40 (Services 2019a).

11.2.2 The Superannuation Guarantee

The second pillar is the mandatory superannuation guarantee, or a defined contribution (DC) scheme. While Australia's history of occupational superannuation schemes can be traced back to federation, the mandatory superannuation guarantee was legislated by the federal government in 1992[2] (Brunner and Thorburn 2008). The superannuation guarantee mandates for a percentage (9.5%) of an employee's ordinary time earnings to be paid to a nominated super fund[3] (SuperFunds 2019). Funds paid into a superannuation scheme can only be accessed at retirement age (preservation age).[4] When superannuation legislation was first introduced, the mandatory level of employer contributions was 3%; however, the required minimum level has gradually increased to 9.5%. By July 2025 it is expected to increase to 12% of ordinary time earnings (Office 2019). Critically, the superannuation guarantee is not extended to workers who earn less than $450 in salary or wages, from a single employer, during a calendar month (Office 2018c). The total amount of super available at retirement depends upon a number of factors, such as contributions made during an employee's working life (concessional contributions); private (non-concessional) contributions; the fund's investment returns; and finally, fees and tax paid on contributions (MoneySmart 2019).

---

[2] *Superannuation Guarantee (Administration) Bill 1992*

[3] There are various distinct types of super funds, including industry schemes, corporate schemes and self-managed funds; see Brunner and Thorburn (2008, p. 15) for a typology.

[4] The current minimum preservation age (the age your super must be 'preserved' until) is between 55 and 60 depending on the year of birth; see Office (2015).

11.2.3 Voluntary Savings and Assets

The final pillar of Australia's retirement system is private savings and assets. Examples of items which are incorporated within this category include managed funds; owner-occupied properties; investment properties; share portfolios; precious metals; and cash in a bank account or term deposit (Agency 2017; Coates 2018; D. o. t. Treasury 2009). As emphasised previously, the three pillars of Australia's retirement system are modelled on the World Bank's recommendation of flexible multi-pillar pension system (T. W. Bank 2001, 2006, p. xxii).

Vis-à-vis gender inequities women face in retirement, in recent decades, there has been a gradual shift away from a predominantly state-funded pension model towards a more privately funded system of retirement income, which can be limiting to women. Brunner and Thorburn (2008, p. 43) note there has been a shift towards a greater emphasis on self-provision in retirement, either in full or part through the superannuation guarantee. They further note that 'people today have a clear understanding that superannuation and private savings will be needed if they are to have a comfortable retirement'. The stronger emphasis on the mandatory superannuation guarantee can be limiting to women. As the subsequent section of this chapter will explore, the current superannuation system in Australia limits women from accumulating an equivalent super balance as men, this is largely because the system is designed to reward full-time work and uninterrupted career trajectories. However, as emphasised in the introduction, women are oftentimes handicapped from accumulating the same superannuation savings as men, as they are more likely to take on unpaid care work, work reduced hours or leave the workforce (Bulbeck 2005; P. Commission 2013; Agency 2017). Accordingly, the following section will explore these elucidatory variables in greater depth.

11.3 Superannuation and the Gender Gap in Retirement Income

There are several key interrelated factors that hinder the ability of women to accumulate equal superannuation balances as men (Funds 2018; Hetherington and Smith 2017). These include the gender pay gap, interrupted career trajectories, family and care commitments, gender segregation in the labour market and societal and cultural factors (McGrath 2015; Committee 2016; Agency 2017; Feng et al. 2019). Hetherington and Smith (2017) argue that the confluence of variables and diverse circumstances mean that women's retirement income in Australia is a wicked problem (see Fig. 11.1).

It is evident that the variety of barriers to achieving equal financial security in retirement is multifaceted. Consequently, this section will focus on four overarching factors which impinge upon the ability of women to accumulate sufficient super

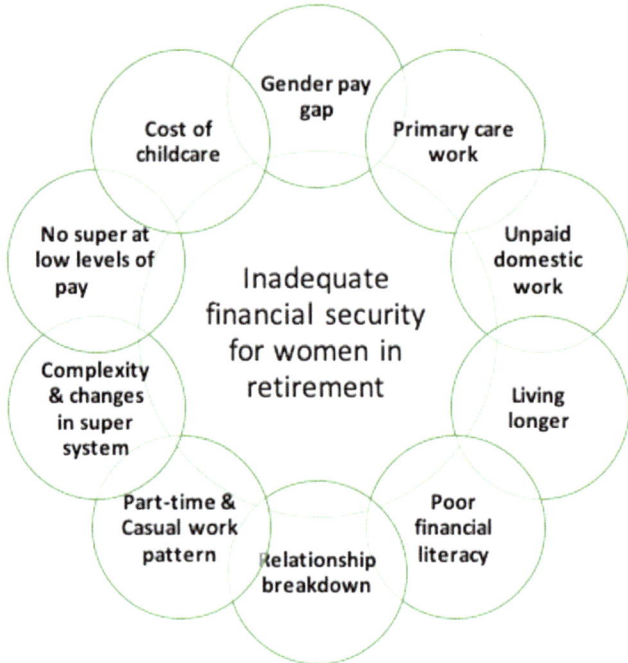

Fig. 11.1 The multifaceted sources which contribute towards female financial insecurity in retirement (This figure has been sourced from (Hetherington and Smith 2017))

balances for retirement: the gender wealth gap, the gender pay gap, gendered occupational segregation in the labour market and the care penalty.

11.3.1 Gender Wealth Gap

The gender wealth gap is the difference between men and women's accumulation of assets (Ravazzini and Chesters 2018). While the gender pay gap is oriented towards illustrating the wealth gap in terms of wages, the gender wealth gap depicts a *more* fuller picture of the economic disparity between genders. In this regard, the gender wealth gap aims to be more all-encompassing and focuses on assets, bonds, property, possessions, etc. (Austen et al. 2013). In Australia, the gender wealth gap between men and women increased from 10.4% to 22.8% between 2002 and 2010 (Committee 2016). This equates to an approximate gap of $46,900 (Austen et al. 2013).[5] A single male under 35 possesses assets worth $120,200 which is approxi-

---

[5] In this study, Austen et al. (2013, p. 20) note that the increase in the gender wealth gap was primarily driven by the higher rate of increase in the median value of primary home assets held by single male households (SMHs) compared to single female households (FMH). The key conclu-

mately $56,700 (89%) more than the average single female in the same age cohort (Austen et al. 2013; Wade 2014). According to Professor Austen, 'the data suggests we are going to see substantial gender wealth inequalities in old age' (as cited in (Wade 2014). Critically, reasons for the gender disparity in wealth accumulation are multifaceted and diverse; for example, the gender wealth gap can be attributed to women having lower average incomes, the gender pay gap, and biases within the labour market such as the glass ceiling and occupational segregation (Austen et al. 2013; Ravazzini and Chesters 2018). The fundamental point is that the explanatory factors associated with the gender wealth gap also affect female financial security in retirement. In this vein, it is necessary to consider the gender wealth gap in conjunction with mitigating variables such as the gender pay gap and gendered occupational segregation.

11.3.2 Gender Pay Gap

The gender pay gap is a critical contributor towards the gender disparity in superannuation balances. The pay gap represents the difference between men and women's average weekly full-time equivalent earnings, expressed as a percentage of men's earnings (Agency 2019b). At present, women earn approximately $239.80 per week less than men, with this figure being even greater when women's part-time and casual earnings are considered (Agency 2019a; Cerise 2009, p. 10). The Human Rights Commission (2017) makes the salient point that if women were to work full-time over a 45-year career and take no time off for parenting or unpaid care commitments, they would still earn approximately $700,000 less than men. In this regard, the gender pay gap is a punishment to women throughout their working life (Committee 2016). This is because the accumulation of super is intrinsically linked to paid work (Hetherington and Smith 2017). It is estimated that the gender pay gap in average full-time earnings for full-time employees results in a 19.3% shortfall in superannuation contributions for women compared to men (Black, 2015). However, the gender pay gap is more multidimensional than the chasm between average weekly full-time earnings. In this regard, the gender pay gap assumes different forms: for example, women are less likely to occupy positions of upper management, women are more likely to face interruptions to their careers and women are more likely to work in an occupation or industry which attracts lower wages (Wingrove and Ferrier 2016; Hetherington and Smith 2017). The gender pay gap assumes a more insidious posture when it is considered alongside gendered occupational segregation and unpaid care work. Folbre (2017) makes the critical point that when women earn less than their partners, they are more likely to take on increased responsibility for family care. As such, it is necessary to consider how gendered occupational segregation affects female earning capacity.

---

sion from their study is that the differential growth rates in the value of the primary asset type drive the gender wealth gap.

11.3.3 Gendered Occupational Segregation

Female-dominated industries and occupations attract lower wages (Committee 2016; Agency 2017; A. H. R. Commission 2017). Jobs in the field of health care, social assistance, education and training attract lower average remuneration than male-dominated industries, such as construction and mining services (Woman 2017). In a concerning vein, women constitute 60–80% of workers in lower-paid industries such as health care, social assistance, education and training (Committee 2016; A. H. R. Commission 2017). KPMG's report into pay inequality in Australia revealed that 22.4% of women work in administrative positions, compared to 7.4% of men (Wingrove and Ferrier 2016). Further to this point, there is a gender disparity between women and men in managerial roles across all sectors; 19.5% of men occupy managerial roles compared to 11% of women (Wingrove and Ferrier 2016). Gendered occupational segregation and lower pay in female-dominated industries mean that women are less likely to accumulate an equal superannuation balance as men (Committee 2016). Thus, it is clear that gendered occupational segregation contributes to, and fuels, the financial insecurity faced by women in retirement. Closely related to gendered occupational segregation is the 'care penalty' faced by women: male-dominated workplaces attract a smaller proportion of part-time employees, suggesting that women continue to sustain the responsibility for unpaid care work (Folbre 2017). It is noted that expectations surrounding high hours of work can deter women from entering traditionally male-dominated industries (Folbre 2017). In sum, occupational segregation can have a damaging impact on a female's ability to accumulate sufficient superannuation savings for retirement (A. H. R. Commission 2017).

11.3.4 The Care Penalty

There is an increased likelihood that women will take on a greater share of unpaid care work throughout her career (Unions 2016; Agency 2017). Feng et al. (2019) note that in comparison to OECD countries, Australian women work increasingly in part-time employment; 52.8% (76.3%) of women with dependent children under age 5 (ages 6–14) were employed in 2006/2007, compared to 94% (92%) of men. This evidence supports the conclusion that mothers largely remain the primary child caregiver. Critically, unpaid care work translates into lower superannuation savings. Research into superannuation contributions emphasise the deleterious effects of *missing* contributions in the early years of retirement savings (Warner 2014). In a study by the Melbourne Institute, research demonstrates that when the decreased earning capacity of unpaid care work is balanced against superannuation contributions, it translates into a $126 million gap in earnings during the period that a mother takes maternity leave (Baker 2011b).[6] Notably, as superannuation contributions are innately linked to paid work, while a mother is taking parental leave, oftentimes she

---

[6] Baker (2011b, p. 12) research uses sample of 80, 725 women who stopped working to have a child and returned to work within 12 months.

is making no or reduced superannuation contributions during this period. Moreover, when this is considered in conjunction with the *wage penalty* or the decreased earning capacity faced by women who take parental leave, it amplifies the absent superannuation contributions made during a period in which a woman takes unpaid care leave. Saint-Martin and Venn (2010) highlight that when women take leave to care for a newborn child, they are subject to a *wage penalty*, which in a deleterious manner translates into a reduced earning capacity. They point out that 'long periods of part-time work could damage individuals' career prospects and increase their (women's) risk of poverty in retirement' (Saint-Martin and Venn 2010). In a similar vein, Baker (2011a) highlights that following maternity leave, women often return to work in a reduced capacity, with more than eight in ten (82%) of women returning to work on a part-time basis after the birth of a child, however, this results in a wage penalty: 'women returning to work within 12 months of taking leave suffered a wage penalty during the first year back at work of almost 7%, this increases to 12% in the following year'. Critically, decreased superannuation contributions during maternity or unpaid care leave have a powerful augmented effect over later superannuation balances, as the power of compounding returns magnifies into larger retirement pots over time (Feng et al. 2019). Finally, societal stigma towards working mothers and stay at home fathers and the assumption that the decision to work in a more reduced or flexible capacity is considered an individual choice further contribute towards the insidious barriers that prevent women from achieving financial security in retirement. As the Human Rights Commission report into gender and workplace stigmatisation emphatically encapsulates: '[a] women's decisions to take time out of paid work, to trade salary for flexibility or to work in a low paid job are often viewed as a matter of individual choice and responsibility. Yet, these choices are very often constrained by a range of external factors such as inflexible workplace structures, family dynamics, cultural pressures and gendered stereotypes…the sum of these factors could leave her in a financially improvised retirement' (A. H. R. Commission 2017, pp. 9–10). This quote saliently emphasises the central crux of female financial insecurity in retirement that financial vulnerability is the result of multifaceted variables, and in this regard, it can be described as a wicked problem (Hetherington and Smith 2017). This conclusion leads into the subsequent section of this chapter, which addresses SDG 10 within the context of the gender gap in super. The following section will use SDG 10 – reducing inequalities – to highlight that mitigating or alleviating female financial insecurity in retirement requires a multilevel and multifaceted approach.

11.4 Utilising the Methodological Understanding of SDG 10: Reducing Inequalities and the Gender Gap in Super

It is undeniable that the superannuation system is not working for women (Trust 2015). Considering this conclusion, it is both pertinent and necessary to interrogate the role of SDG 10 in alleviating the burden of female financial insecurity in retirement. Sustainable Development Goal (SDG) 10 is concerned with the reduction in inequality within and amongst countries. It is a critical preoccupation of SDG 10 to eliminate discriminatory policies in favour of programmes that enable social equality. This is in line with sub-targets 10.3 and 10.4 which respectively call for equal opportunity via the elimination of discriminatory laws, policies and practices; moreover, it is also in line with sub-target 10.2 which advocates for the promotion and empowerment of the economic inclusion of all, irrespective of age, gender or other status (Nations 2019). As emphasised in the previous section of this chapter, the current structure of superannuation policy is prejudicial to women and the achievement of financial security in retirement. The accumulation of larger superannuation demands a full-time work pattern and continual employment. However, while achieving financial equality in retirement is the desirable ultimata, because of the complexity and wicked nature of the issue, the realisation of equality in superannuation balances needs a multifaceted approach (Hetherington and Smith 2017; Jericho 2018; Riach 2018). As a consequence, this section will consider how sub-targets 10.2, 10.3 and 10.4 can be utilised to provide a platform for the reduction of inequality in superannuation savings between men and women. The first section will promulgate the argument that within the boundaries of sub-target 10.2, there is a greater need to empower women and young adults through targeted financial literacy and education programmes. Finally, the latter section will consider how sub-targets 10.3 and 10.4 can be utilised to remove discriminatory policies which hinder women from achieving equal superannuation.

11.4.1 SDG 10.2: Empowering Women and Young Adults Through Financial Education and Literacy

The functioning of Australia's three-pillar retirement system requires individuals to be well informed and possess a sufficient level of financial literacy (Agnew et al. 2013). In a study of Australian financial literacy levels, Agnew et al. (2013, p. 8) found that women scored lower than men in a financial literacy test. They concluded that financial illiteracy is more prevalent amongst 'younger individuals, women, those with less education and those who are not employed'. Their findings are linked to the argument that financial literacy and education play a crucial role in ensuring adequate preparation for retirement. On a larger scale and similar vein, the OECD, in conjunction with the International Network on Financial Education, estimated that young adults are amongst the lowest levels of financial literacy (OECD 2012). For example, in Denmark, 73% of young adults have little or no knowledge of interest rates (OECD 2012). Similarly, Ali et al. (2015) found that women have lower levels of knowledge about the superannuation system. In a questionnaire designed to test respondents' knowledge in regard to super, 20% of women were able to answer seven or more questions correctly. It is evident that there is a decreased level of financial comprehension amongst women and young adults, and in this regard, it is evident that there needs to be a greater emphasis on providing tailored financial literacy.

The Victorian Women's Trust (2015) emphasised that amongst young Australian women, there is a lack of confidence and consequently knowledge about how to effectively plan for retirement. In their recommendations to the Senate Standing Committee on Economics, the trust called for more targeted communication for women to encourage them to engage with super. Critically, they emphasised that there is a strong impetus for super funds to customise their communication to reflect the lived experience of Australian women, stressing that information should be tailored to address barriers such as career breaks and unpaid care responsibilities. To this end, information should be easy to understand and devoid of financial jargon. Ali et al. (2015, p. 101) concur with this suggestion, arguing that there is a need for more tailored information which reflects the life experiences and knowledge levels of different demographics. Similarly, Women in Super noted that financial planning training across Australia does not include gender awareness nor address the topic of how gender can affect retirement planning (Wood and Buckley 2015). They argued for greater gender-specific education which addresses insurance needs, asset allocation and the benefits of saving. It is clear there is a greater need to provide targeted information to women and young adults vis-à-vis superannuation and how to understand and engage with their super. Consequently, in 2014, The Australian Taxation Office (ATO) launched a campaign called 'five-step super check', which encouraged women to undertake five simple steps to ensure that they increase their super savings. The campaign used social media, proactive media, and paid advertising: it particularly targeted women aged 25–49 (Committee 2016). However, targeted educational programmes need to be embedded within a broader financial education and literacy framework. In this regard, several suggestions to the Senate Standing Committee on Economics stressed that there needs to be a stronger emphasis on financial literacy programmes in Australian schools (Committee 2016). To assist with financial education in schools, the OECD (2012) proposes a number of recommendations. For example, the financial literacy programme should be embedded within a broader 'coordinated national strategy', and there should be a 'learning framework' which articulates key goals, learning outcomes, content and pedagogical approaches. Finally, the OECD suggests that financial literacy should form a 'core part of the school curriculum'. In this regard, instead of teaching financial literacy as a stand-alone subject, it is recommended that financial concepts are integrated into core subjects such as mathematics, economics, social science or citizenship (OECD 2012). The payoff in investing in developing financial literacy frameworks and programmes is 'substantial' (Behrman et al. 2012). In a US-based study, Behrman et al. (2012, p. 303) show that financial literacy enhances the likelihood that individuals will make contributions to their retirement savings. Similarly, in an Australian context, Professor Carsten Murawski stressed that 'early intervention education is one of a number of avenues to systematically change peoples' behaviour'. It is evident that within the scope sub-target 10.2, which calls for the empowerment and promotion of the economic inclusion of all, there is sufficient space to pay a greater heed towards developing a sustainable and comprehensive financial education programme, particularly, a programme which focuses specifically on empowering young adults and women in key areas of retirement planning.

However, it is important to recognise that financial literacy programmes are sufficient to a point (Riach 2018). In order to improve equality between women and men in retirement, it is also necessary to address the systemic barriers innate within the current superannuation system. In this regard, it is beneficial and necessary to consider how sub-target 10.3 and 10.4 can be employed to reduce the economic inequalities in retirement between men and women.

11.4.2   SDG 10.3 and 10.4: Removing Systemic Barriers to Meliorate Superannuation Policy for Women

This section will focus on the systemic barriers of superannuation policy in Australia; crucially, these systemic obstructions place women in a precarious disadvantage of being unable to accumulate adequate super savings for retirement. As investigated in the first section of this chapter, this largely stems from the preference superannuation policy given to full-time and continuous work. However, as women are more likely to take parental leave, unpaid care leave or work part-time, this inflicts a handicap upon women in accumulating the same superannuation savings as men. As such, this section will consider how the superannuation system can be reformed to account for the different work and career trajectories that women face. The mechanisms that this chapter will address include increasing the low income super tax offset (LISTO) contributions to $1000; paying superannuation during paid parental leave abolishing the $450 minimum threshold; and finally, establishing an independent review panel. Crucially, these mechanisms align with sub-targets 10.3 and 10.4 of SDG 10, which, respectively, call for the adoption of policies that achieve greater equality and secure equality of opportunity (Nations 2019).

11.4.2.1   Taxation of Concessional Contributions: Increasing LISTO Co-contributions

Concessional contributions[7] are taxed at a flat rate of 15%; however, this is a punitive tax for low-income earners (Funds 2018). For example, an individual earning up to the tax-free threshold ($18,200) per annum has an income tax rate of 0%. Similarly, an individual earning $37,000 per annum has an income tax rate of 9.65%. It is evident that the flat rate tax of 15% on concessional contributions penalises low-income earners, as they lose 15% of their contribution; thus, it can be

---

[7] Concessional contributions are contributions made into a super fund before tax. This includes employer contributions (compulsory contributions), additional employer concessional contributions, salary sacrifice payments and contributions that are allowed as an income tax deduction. Importantly, once the concessional contributions are in your super fund, they are taxed at a 15% rate. Concessional contributions are distinguishable from non-concessional contributions; see Office (2018a, b).

argued that low-income earners are not encouraged or incentivised to make additional contributions to their super. Industry Super Australia stresses that the regressive nature of tax on concessional contributions provides an inadequate incentive for low-income earners to make additional contributions to their super (Campo et al. 2015). Industry Super Australia stresses that the regressive nature of tax on concessional contributions provides an inadequate incentive for low-income earners to make additional contributions to their super (Campo et al. 2015). The Association of Superannuation Funds (2018, p. 3) suggests that consideration should be given to the idea of a 'top-up' payment or super co-contribution to incentivise low-income earners to engage with their super. More substantially, it has been recommended that the low income super tax offset (LISTO)[8] contribution be increased to $1000 (Scheerlinck and Webb 2017). It is estimated that increasing the LISTO contribution to $1000 would produce a 14.7% increase in superannuation balances for low-income earners, many of whom are female. Women in Super estimate that a woman aged 25 with a starting salary of $25,000 p.a. and a projected super balance of $205,210 would be able to increase this projected balance through the proposed $1000 government contribution by 14.7%. They estimate that through this scheme, the projected retirement balance would be $235, 347 (Wood 2017).

11.4.2.2 Superannuation Payments During Parental Leave and Removal of the $450 Threshold

Paying superannuation payments while on Commonwealth Paid Parental Leave or employee Paid Parental Leave would assist in narrowing the gender gap in superannuation balances (Campo et al. 2015; Wood and Buckley 2015; Committee 2016; Dyrenfurth 2018). While all employees are generally entitled to 18 weeks' government-assisted paid parental leave, these payments do not attract superannuation. Conversely, sick leave, annual leave, and long service leave all attract super contributions. It is estimated that paying super during government paid parental leave would increase retirement savings by 1.7% (Campo et al. 2015). In conjunction to this, removing the $450 threshold exemption would boost the superannuation of women who work casually or part-time across different employers and jobs (Committee 2016). Women in Super noted that there are many females working part-time or casually across two or three different jobs, none of which allows them to reach the $450 monthly threshold. Across the two or three distinct jobs, they may cumulatively earn more than $450 a month, yet receive no super contributions from any employer (Committee 2016). Campo et al. (2015) estimate that 250,000 individuals are excluded from superannuation contributions due to not satisfying the $450 threshold.

---

[8] The low income super tax offset contribution (LISTO) provides low-income earners with an adjusted taxable income of up to $37,000 with a contribution equal to 15% of their total concessional (pre-tax) super contributions. It is currently capped at $500; see Office (2017).

11.4.2.3 Independent Review Panel

Finally, COTA Australia stresses that to understand the complexities that underpin retirement income in Australia, independent and holistic commission should be established (McGrath 2015). They argue that a retirement income review should be composed of an independent chair and expert members: the committee should be tasked with reviewing all aspects of superannuation and retirement policy. The review committee should further be tasked with evaluating and optimising retirement income policy in Australia (McGrath 2015).

In sum, this chapter has considered how SDG 10 can be utilised to give space and mobility to the dismantling of systemic barriers that exist within superannuation policy in Australia. The central conclusion arising from this section is that there is scope to employ sub-targets 10.2, 10.3 and 10.4 of SDG 10 to achieve sufficient financial equality in retirement for women in Australia. Critically, this conclusion is intrinsically connected to the central argument of this chapter, that is, within the framework of the sustainable development goals, there is ample scope to use the direction of SDG 10 to develop a national agenda which will ameliorate financial security for women in retirement.

11.5 Impact Sustainability: Final Remarks

11.5.1 Cultivating a Sustainable Impact: Improving Gender Equality in Retirement Through SDG 10

This chapter was primarily concerned with understanding female financial security in retirement – particularly with regard to the gender gap in superannuation balances. The first section of this chapter outlined Australia's retirement and pension system. The critical conclusion arising from this section was that the shift towards a greater emphasis on self-provision in retirement is limiting to women. This is because the current structure of superannuation policy in Australia privileges a full-time and uninterrupted career trajectory; however, women are often handicapped from achieving an equal superannuation balance as men as they are more likely to take on unpaid care work, work in a reduced capacity (fewer hours), or leave the workforce all together. Critically, this conclusion leads to the second section of this chapter which offered a more in-depth explanation as to why women are disadvantaged in achieving sufficient superannuation balances for retirement. Pivotally, this section emphasised that the confluence of associated variables means that the gap in women's retirement income can be considered a wicked problem. This section of the chapter investigated four overarching factors that inhibit the ability of women to accumulate equal super balances: the gender wealth gap, the gender pay gap, gendered occupational segregation and the care penalty. The conclusion arising from this section was that female financial insecurity is multifaceted and thus demands a multilevel approach. This conclusion broached into the final section of this chapter, which used the SDG framework to explore how superannuation policy in Australia can be ameliorated in order to mitigate the gap in retirement savings. With particular reference to SDG 10 – reduced inequalities – the third section emphasised that tackling female financial security in retirement demanded a multi-pronged approach. To this end, the final section utilised sub-target 10.2, 10.3 and 10.4 to explore mechanisms by which the gender gap in superannuation savings can be narrowed. Utilising SDG 10.2 this section found that there is a greater need and impetus to empower women and young adults through targeted financial literacy and education programmes, particularly, with information that will take into account the unique and diverse life circumstances that women face. Finally, the latter section of the chapter drew on SDG 10.3 and 10.4 to argue that discriminatory policies embedded within the architecture of the superannuation system need to be addressed and dismantled. Policy mechanisms that require amendment include increasing the low income super tax offset (LISTO) contributions to $1000; paying superannuation during paid parental leave; abolishing the $450 minimum threshold; and finally, establishing an independent review panel. The central conclusion arising from this chapter is that within the framework of the SDGs, there is both ample and sufficient scope to use the direction of SDG 10 to develop a national agenda on improving female financial security in retirement.

References

Agency WGE (2015) Submission – senate inquiry into economic security for women in retirement. Retrieved from https://www.wgea.gov.au/sites/default/files/WGEA-Submission-Economic_Security_for_Women_in_Retirement.pdf

Agency WGE (2017) Women's economic security in retirement: insight paper. Retrieved from https://www.wgea.gov.au/sites/default/files/insights-paper-womens-economic-security-in-retirement.pdf

Agency WGE (2019a) Australia's gender pay gap statistics, 22 February 2019. Retrieved from https://www.wgea.gov.au/data/fact-sheets/australias-gender-pay-gap-statistics

Agency WGE (2019b) What is the gender pay gap? Retrieved from https://www.wgea.gov.au/addressing-pay-equity/what-gender-pay-gap

Agnew J, Bateman H, Thorp S (2013) Financial literacy and retirement planning in Australia. Retrieved from Sydney: https://financialcaptability.gov.au/files/report_financial-literacy-and-retirement-planning-in-australia_2013.pdf

Ali P, Anderson M, Clarke M, Ramsay I, Shekhar C (2015) No thought for tomorrow: young Australian adults' knowledge, behaviour and attitudes about superannuation. Law Financ Mark Rev 9(2):90–105. https://doi.org/10.1080/17521440.2015.1052667

Austen S, Ong R, Bawa S, Jefferson T (2013) Trends in the gender wealth gap among single households in Australia 2002–2010. Retrieved from https://bcec.edu.au/assets/Trends-in-the-Gender-Wealth-Gap-Among-Single-Households-in-Australia-2002-2010.pdf

Australia TAoSFo (2018) ASFA retirement standard budgets review. Retrieved from https://www.superannuation.asn.au/ArticleDocuments/269/2018-ASFA-Retirement-Standard-Budgets-Review.pdf.aspx?Embed=Y

Baker D (2011a) Maternity leave and reduced future earning capacity. Retrieved from https://aifs.gov.au/publications/family-matters/issue-89/maternity-leave-and-reduced-future-earning-capacity

Baker D (2011b) The wage-penalty effect: the hidden cost of maternity leave. Retrieved from Melbourne: https://melbourneinstitute.unimelb.edu.au/assets/documents/hilda-biography/other-oublicstions/2011/Baker_The_Wage_Penalty_Effect.pdf

Bank TW (2001) Social protection sector strategy: from safety net to springboard. Retrieved from Washington D.C.: http://documents.worldbank.org/curated/en/299921468765558913/pdf/multi-page.pdf

Bank TW (2006) Pension reform and the development of pension systems: an evaluation of World Bank assistance. Retrieved from Washington D.C.: http://inweb90.worldbank.org/oed/oedoclib.nsf/DocUNIDViewForJavaSearch/43B436DFBB2723D085257108005F6309/$file/pensions_evaluation.pdf

Bank W (2008) The World Bank pension conceptual framework. Retrieved from Washington, D.C.: http://siteresources.worldbank.org/INTPENSIONS/Resources/393544311121194657824/PRPNoteConcept_Sept2008.pdf

Behrman JR, Mitchell OS, Soo CK, Bravo D (2012) The effects of financial education and financial literacy. Am Econ Rev 102(3):300–304. https://doi.org/10.1257/aer.102.3.300

Black M (2015) Pay gap leads to 19.3% annual super shortfall for full-time women [Press release]. Retrieved from https://www.wgea.gov.au/media-releases/pay-gap-leads-193-annual-super-shortfall-full-time-women

Brunner GG, Thorburn C (2008) The market for retirement products in Australia. Retrieved from https://openknowledge.worldbank.org/bitstream/handle/10986/6934/WPS4749.pdf?sequence=1&isAllowed=y

Bulbeck C (2005) Gender policies: hers to his. In: Walter PS a J (ed) Ideas and influence: social science and public policy in Australia. UNSW Press, Sydney, pp 141–158

Cameron P (2013) What's choice got to do with it? Women's lifetime financial disadvantage and the superannuation gender pay gap. Retrieved from http://www.tai.org.au/sites/default/files/PB_55_Whats_choice_got_to_do_with_it.pdf

Campo R, Goodwin A, Engert L (2015) Inquiry into economic security for women in retirement. Retrieved from http://www.industrysuperaustralia.com/assets/Submission/Inquiry-into-economic-security-of-women-in-retirement-ISA-Submission.pdf

Cerise S (2009) Accumulating poverty? Women's experiences of inequality over the lifecycle. Retrieved from Sydney, NSW

Coates B (2018) What's the best way to close the gender gap in retirement incomes? Retrieved from https://grattan.edu.au/wp-content/uploads/2018/02/ESA-Gender-Economics-Symposium-Brendan-Coates-8-February-2018-FOR-GRATTAN-WEB.pdf

Commission P (2013) An ageing Australia: preparing for the future. Retrieved from Canberra: https://www.pc.gov.au/research/completed/ageing-australia/aging-australia.pdf

Commission AHR (2017) Gender segregation in the workplace and its impact on women's economic equality. Retrieved from Sydney, NSW: https://www.humanrights.gov.au/sites/default/files/AHRC_Submission_Inquiry_Gender_Segregation_Workplace2017.pdf

Committee ER (2016) 'A husband is not a retirement plan' achieving economic security for women in retirement. Retrieved from Canberra: https://www.aph.gov.au/Parliamentary_Business/Committees/Senate/Economics/Economic_security_for_women_in_retirement/Report

Dyrenfurth N (2018) Super ideas: securing Australia's retirement income system. Retrieved from https://static1.squarespace.com/static/587e1296579fb39e3199b6e9/t/5a825449652deaaa6dd9a0a7/1518490792145/Super+Ideas%3A+Securing+Australia%E2%80%99s+Retirement+Income+System

Feng J, Gerrans P, Mouland C, Whiteside N, Strydom M (2019) Why women have lower retirement savings: the Australian case. Fem Econ 25(1):145–173. https://doi.org/10.1080/13545701.2018.1533250

Folbre N (2017) The care penalty and gender inequality. In: Averett SL, Argys LM, Hoffmann SD (eds) The Oxford handbook of women and gender inequality. Oxford University Press, Oxford

Funds TAoS (2018) Women's economic security in retirement

Hetherington D, Smith W (2017) Not so super, for women: superannuation and Women's retirement outcomes. Retrieved from https://percapita.org.au/our_work/not-so-super-for-women/

Jericho G (2018) The gender gap in retirement savings isn't just about super. The Guardian. Retrieved from https://www.theguardian.com/business/grogonomics/2018/deb/13/the-gender-gap-in-retirement-savings-isnt-just-about-super

Koukoulas S (2018) Defining the concept of economic security for all women: policy recommendations to boost women's economic security. Retrieved from https://www.security4women.org.au/wp-content/uploads/20180625-eS4W_White-Paper_Defining-the-Concept-of-Economic-Security-for-Women.pdf

McGrath S (2015) Submission to the senate standing Committee on economics: inquiry into economic security for women in retirement

MoneySmart (2019) Getting your super. Retrieved from https://www.moneysmart.gov.au/superannuation-and-retirement/how-super-works/getting-your-super

Nations U (2019) Sustainable development goal 10. Retrieved from https://sustainabledevelopment.un.org/sdg10

OECD (2012) Financial education in schools. OECD

Office AT (2015) Preservation of super, 1st December 2015. Retrieved from https://www.ato.gov.au/super/self-managed-super-funds/paying-benefits/preservation-of-super/

Office AT (2017) Low income super tax offset contribution, 28th November 2017. Retrieved from https://www.ato.gov.au/super/self-managed-super-funds/super-changes-for-self-managed-super-funds/low-income-super-tax-offset-contribution/

Office AT (2018a) Concessional contributions. Retrieved from https://www.ato.gov.au/Individuals/Super/In-detail/Growing-your-super/Super-contributions%2D%2D-too-much-can-mean-extra-tax/?page=2

Office AT (2018b) Non-concessional contributions. Retrieved from https://www.ato.gov.au/Individuals/Super/In-detail/Growing-your-super/Super-contributions%2D%2D-too-much-can-mean-extra-tax/?page=3

Office AT (2018c) Working out if you have to pay super. Retrieved from https://www.ato.gov.au/business/super-for-employers/working-out-if-you-have-to-pay-super/

Office AT (2019) Super guarantee percentage, 4th March 2019. Retrieved from https://www.ato.gov.au/Rates/Key-superannuation-rates-and-thresholds/?page=24

Parkinson D, Weiss C, Cara C, Duncan A, Judd K (2013) Living longer on less: women speak on superannuation and retirement. Retrieved from https://www.whealth.com.au/documents/work/living-longer-on-less.pdf

Patten S (2016) What life is really like on the age pension. The Sydney Morning Herald. Retrieved from https://www.smh.com.au/money/what-life-is-really-like-on-the-age-pension-20160426-gofbs8.html

Ravazzini L, Chesters J (2018) Inequality and wealth: comparing the gender wealth gap in Switzerland and Australia. Fem Econ 24(4):83–107. https://doi.org/10.1080/13545701.2018.1458202

Riach K (2018) Snakes and ladders: why women's superannuation is complex. ABC News. Retrieved from https://www.abc.net.au/news/2018-10-02/superannuation-women-work-mothers-parenting-finances/10328884

Saint-Martin A, Venn D (2010) Does part-time work pay? Retrieved from http://oecdobserver.org/news/archivestory.php/aid/3311/Does_part-time_work_pay_.html

Scheerlinck E, Webb R (2017) Mind the gap! Resolving the undercoverage of Australia's world class superannuation system. Retrieved from http://www.aist.asn.au/media/1100159/20171215_submission_pre-budget_v.1.0_final.pdf

Services DoH (2019a) Age pension – payment rates. Retrieved from https://www.humanservices.gov.au/individuals/services/centrelink/age-pension/eligibility/payment-rates

Services DoH (2019b) Assets. Retrieved from https://www.humanservices.gov.au/individuals/enablers/assets/30621

Society B, Hub TLI, Percapita (2016) The Adequacy of the Age Pension in Australia: An assessment of pensioner living standards. Retrieved from http://percapita.org.au/wp-content/uploads/2016/09/Pension-Adequacy_Final.pdf

SuperFunds I (2019) Super rules: superannuation obligations for employers. Retrieved from https://www.industrysuper.com/for-employers/super-rules/

Treasury Dot (2009) Australia's future tax system – the retirement income system: report on strategic issues. Retrieved from Barton, ACT: http://taxreview.treasury.gov.au/content/downloads/retirement_income_report_strategic_issues/retirement_income_report_20090515.pdf

Treasury T (2019) Australia's retirement income system. Retrieved from https://treasury.gov.au/programs-and-initiatives-superannuation-charter-of-superannuation-adequacy/report/part-1/

Trust VWs (2015) Economic security for women in retirement: submission 33. Retrieved from https://www.aph.gov.au/Parliamentary_Business/Committees/Senate/Economics/Economic_security_for_women_in_retirement/Report

Unions ACoT (2016) The gender pay gap – over the life cycle. Retrieved from https://www.actu.org.au/media/886499/the-gender-pay-gap-over-the-life-cycle-h2.pdf

Wade M (2014) Rich man, poor woman: the gender wealth gap widens. The Sydney Morning Herald. Retrieved from https://www.smh.com.au/national/rich-man-poor-woman-the-gender-wealth-gap-widens-20141107-11igay.html

Warner R (2014) Retirement savings gap as at 30 June 2014. Retrieved from https://ricewarner.com/wp-content/uploads/2015/12/Retirement-Savings-Gap-as-at-30-June-2014.pdf

Wingrove G, Ferrier S (2016) She's price(d)less: the economics of the gender pay gap. Retrieved from https://home.kpmg/content/dam/kpmg/au/pdf/2016/gender-pay/gap-economics/full-report.pdf

Woman NFfA (2017) Gender segregation in the workplace and its impact on Women's equality. Retrieved from http://www.5050foundation.edu.au/assets/reports/documents/2017-Gender-segregation-in-the-workplace-and-its-impact-on-womens-economic-equality.pdf

Wood C (2017) 2018–19 Pre-Budget Submission. Retrieved from https://clarety-wis.s3.amazonaws.com/userimages/WIS_Submissions/Women_in_Super_2018-19_Pre-Budget_Submission.pdf

Wood C, Buckley S (2015) Submission to the senate inquiry into the economic security of women in retirement. Retrieved from https://www.aph.gov.au/Parliamentary_Business/Commitees/Senate/Economics/Economic_security_for_women_in_retirement/Submissions

Chapter 12
SDG 11 Sustainable Cities and Communities

SDG 11 and the New Urban Agenda: Global Sustainability Frameworks for Local Action

Hitesh Vaidya and Tathagata Chatterji

Abstract Recent global policy discourses orchestrated under the aegis of the United Nations, such as the Agenda for Sustainable Development (2030) and the New Urban Agenda of UN Habitat, stress upon the need for concerted focus at the city and the community scale – not only to achieve long-term developmental objectives but also to make direct tangible benefits to the quality of lives of the people. The world at large is gradually taking an urban turn, as more and more people are moving to the cities. Cities account for 55% of the population, produce 85% of the global GDP but also 75% of the greenhouse gas emissions. The issues of global sustainability cannot be addressed, without strongly addressing sustainability at the urban scale. This chapter focuses on SDG 11 as the analytical framework to explore how the transformative force of urbanization represents opportunity and challenge to meet several other sustainability challenges, such as SDG 1 (poverty reduction), SDG 4 (education), SDG 5 (gender equality), SDG 6 (clean water and sanitation), SDG 7 (affordable and clean energy), SDG 8 (economic growth) and SDG 13 (climate action). The research highlights research and action points for urban governance systems to mainstream sustainability concerns through their local planning and development mechanism.

Keywords SDG 11 · New Urban Agenda · Urban local bodies · India

---

H. Vaidya (✉)
India Country Manager, UN Habitat, New Delhi, India
e-mail: hitesh.vaidya@un.org

T. Chatterji
Xavier School of Human Settlements, Xavier University Bhubaneswar,
Kakudia, Odisha, India
e-mail: tathagata@xub.edu.in

© Springer Nature Singapore Pte Ltd. 2020
I. B. Franco et al. (eds.), *Actioning the Global Goals for Local Impact*, Science for Sustainable Societies, https://doi.org/10.1007/978-981-32-9927-6_12

12.1 Introduction

While launching the Sustainable Development Goals (SDG), Ban Ki-Moon, the former Secretary-General of the United Nations, noted that 'Cities are where the battle for sustainable development will be won or lost' (Fabre 2017, p. 4). Urbanization has become a defining phenomenon of the twenty-first century, as we are increasingly living in an urban world. In 1950, the world was predominantly rural, as global urbanization level was 30%. By 2050, the scenario is projected to reverse, as urbanization level is expected to reach 70% (UN 2018).

Cities are the hubs of innovation, employment and wealth generation. Urban areas already account for 55% of the global population and produce 85% of the global GDP. But the way the processes of urbanization are unfolding is also deeply problematic. Urban wastes are polluting our air, water and soil resources, and cities account for 75% of the greenhouse gas emissions. Therefore, the issues of global sustainability cannot be addressed, without strongly addressing the question of urban sustainability.

Recognizing the importance of the cities in contemporary world, the UN General Assembly in 2015 decided to adopt 'sustainable cities and communities' as a distinct goal (SDG 11) under Agenda for Sustainable Development (2030). The overarching aim of SDG 11 is to make cities and human settlements inclusive, safe, resilient and sustainable. Recent global policy discourses orchestrated under the aegis of the United Nations, such as COP 24, Paris Agreement (UN Framework Convention on Climate Change) and the New Urban Agenda of UN Habitat stress upon the need for concerted focus at the city and the community scale – not only to achieve long-term developmental objectives but also to make direct tangible benefits to the quality of lives of the people.

This chapter focuses on SDG 11 as the analytical framework to explore how the transformative force of urbanization represents opportunity and challenge to meet several other sustainability challenges. The research highlights research and action points for urban governance systems to mainstream sustainability concerns through their local planning and development mechanism. The Sustainable Development Goals have adopted a comprehensive systems approach. The targets and indicators of the 17 SDGs are tied in such a manner that pursuance of one goal often leads to cascading benefits. City as a spatial platform offers opportunities to address sustainability concerns in a range of sectoral infrastructure domains, such as transportation, energy, water, education and healthcare through regulatory, fiscal, planning and managerial instruments. Consequently, SDG 11 has robust linkages with several other SDGs. For example, LED street lighting, which reduce urban municipal expenditure, also contribute towards energy efficiency (SDG 7) and promote sustainable production and consumption (SDG 12).

The global goals set under SDG 11 vitally depend on their integration with the local developmental agenda for effective implementation and to make tangible differences in the everyday lives of people. SDG is closely aligned with the New Urban Agenda (NUA), which was adopted at the Habitat III summit in Quito, Ecuador, in

October 2016. The NUA is the guiding document for the UN system's urban engagements over the next 20 years. The signatory countries are required to frame respective National Urban Policies as a framework to drive urban sustainability targets. Thus, the National Urban Policy can serve as the qualitative toolbox to guide and monitor accomplishment of the Sustainable Development Agenda 2030. It provides a template to national government agencies to benchmark progress at different cities and prioritize its funding mechanisms accordingly. Furthermore, it is also a roadmap for urban local bodies (ULBs) and allied agencies of urban governance, to incorporate the SDG 11 targets through their local planning and governance frameworks.

The contemporary global urbanization is being driven by newly industrializing countries of Asia and Africa. The developing countries often lack adequate state capacities to comprehensively address urban issues, and civic infrastructure gaps are being perpetuated. Supply of new infrastructure is being quickly outpaced by population pressure caused by natural growth as well as migration. Between 2000 and 2014, the proportion of urban population living in slums declined by 20% (from 28.4 to 22.8%) across the globe, whereas the rate of new home construction lagged far behind the rate of urban population growth, and the number of people living in slums actually increased from 807 to 883 million over this period (UN Habitat 2016).

Here, in this paper we focus on India to understand how SDG 11 is being implemented at the national and ULB level. India had been one of the fastest-growing major economies of the world over the past couple of decades. With 370 million people living in the cities, it is also the world's second largest urban system. According to World Urbanization Prospects (2018) published by the UN, India's urban population is projected to further increase by 145 million between 2018 and 2030. By 2050, the urban population would increase by another 416 million – which would 50 million more than the combined population of the United States and Canada (UN 2018). The Government of India has aligned much of its developmental support mechanism with the SDG and also formulated a framework for integration of the same through local governance entities.

The rest of the chapter is organized as follows. The second section discusses the concept of SDG as a systems approach by focusing between SDG 11 and other SDGs. The third section analyses how the universal goals of SDG 11 are being operationalized through national and local governance agencies. The fourth section concludes the chapter by pointing out areas for further research.

12.2 Literature Review: A Systems Approach Towards Urban Management

SDG 11 is a spatially organized developmental framework, which seeks to address several interrelated issues by focusing on urban sustainability. Like other SDG goals and targets, it is based on the systems approach with an emphasis on cross-linkages with other developmental priorities and aims to achieve the desired level of outcome by 2030, through a set of attainable targets and indicators designed to guide policy actions at local levels.

The concept of 'sustainable development' rests on achieving balance between economic, social and environmental objectives. Rather than maximizing gain through one specific system, trade-offs between multiple systems and objectives are encouraged. As Babier and Burges (2017) put it, the theory of sustainable development is fundamentally different from *capital approach* which guides much of the developmental policies in practice. Urban developmental policies, in particular, are frequently designed to improve functional efficiency of civic infrastructure and attract new investments. Such policies may at times meet the economic objectives of reducing poverty and generate jobs but also end up widening social polarization and increasing pollution. On the other hand, a sustainable economic development approach would not only strive for efficiency, equity and poverty reduction but at the same time take ecological footprints and social impacts into consideration. Such comprehensive processes inter alia would involve trade-offs.

However, it has to be kept in mind that sustainability as a conceptual model faces practical limitations in terms of applicability in policy design. The point of trade-off between conflicting goals and priorities is negotiated at various levels of governance hierarchy and power equations between the stakeholders involved. Therefore, priorities assigned to economic environmental and equity concerns vary according to local context. It is the place-specific political-economic scenario which dictates the urban form through land use regulations and the planning norms. It is the land supply mechanism, which shapes urban growth pattern, whether a city would expand outwards through monetization of peri-urban agricultural land to accommodate new economic opportunities; or the building bye-laws could be tweaked to accommodate high-rise, high density to encourage transit-oriented development and compact urban form. It is a difficult choice to make at a city level, as urban local governments are often ill equipped to make informed decisions due to technical and financial capacity constraints. Moreover, social (i.e. livelihood securities, polarization, gentrification, gender) and environmental (i.e. loss of green cover, filling up of wetlands and water bodies, deforestation, air pollution, waste dumping) impacts of development are seldom considered in depth. Needless to say, the problems are most acute in rapidly urbanizing cities of the Global South.

The adaption of the Sustainable Development Goals including the stand-alone urban goal of *making cities safe, inclusive, resilient and sustainable* (SDG11) firmly places urbanization at the forefront of international development policy. This recognition goes beyond viewing urbanization simply as a demographic phenomenon but rather as a transformative process capable of galvanizing momentum for many aspects of global development. However, these goals are interconnected – often the key to success on one will involve tackling issues more commonly associated with another.

Thus Goal 11 is most relevant and seeks to *make cities and human settlements inclusive, safe, resilient and sustainable* through eliminating slumlike conditions, providing accessible and affordable transport systems, reducing urban sprawl, increasing participation in urban governance, enhancing cultural and heritage preservation, addressing urban resilience and climate change challenges, better management of urban environments (pollution and waste management), providing access to safe and secure public spaces for all and improving urban management through better urban policies and regulations. Specific targets from other goals also fall within its purview, such as equal access to resources by all people (Goal 1), halving the number of road accident fatalities by 2030 (Goal 3), ensuring access to safe water and sanitation for all (Goal 6), reducing waste generation (Goal 12) and strengthening resilience and adaptive capacity to climate hazards (Goal 13).

In this scenario, SDG 11 can be seen as a comprehensive template to guide urban policy makers. The targets and indicators (as indicated in Table 12.1) are useful in measuring welfare gains and support evidence-based policy design. It also fits in to systems approach towards sustainable development and has substantial implications for several other SDG aims. For example, issues related to affordable housing and inclusive planning under SDG 11 are closely linked to the objectives SDG 1 on eradicating poverty (Fig. 12.1).

Table 12.1 SDG 11 targets and indicators

Target	Proposed indicators
12.1 By 2030, ensure access for all to adequate, safe and affordable housing and basic services and upgrade slums	12.1.1 Proportion of urban population living in slums, informal settlements or inadequate housing
12.2 By 2030, provide access to safe, affordable, accessible and sustainable transport systems for all, improving road safety, notably by expanding public transport, with special attention to the needs of those in vulnerable situations, women, children, persons with disabilities and older persons	12.2.1 Proportion of population that has convenient access to public transport, by sex, age and persons with disabilities
12.3 By 2030, enhance inclusive and sustainable urbanization and capacity for participatory, integrated and sustainable human settlement planning and management in all countries	12.3.1 Ratio of land consumption rate to population growth rate
	12.3.2 Proportion of cities with a direct participation structure of civil society in urban planning and management that operate regularly and democratically
12.4 Strengthen efforts to protect and safeguard the world's cultural and natural heritage	12.4.1 Total expenditure (public and private) per capita spent on the preservation, protection and conservation of all cultural and natural heritage, by type of heritage (cultural, natural, mixed and World Heritage Centre designation), level of government (national, regional and local/municipal), type of expenditure (operating expenditure/investment) and type of private funding (donations in kind, private non-profit sector and sponsorship)

(continued)

Table 12.1 (continued)

Target	Proposed indicators
12.5 By 2030, significantly reduce the number of deaths and the number of people affected and substantially decrease the direct economic losses relative to global gross domestic product caused by disasters, including water-related disasters, with a focus on protecting the poor and people in vulnerable situations	12.5.1 Number of deaths, missing persons and persons directly affected by disaster per 100,000 people
	12.5.2 Direct economic loss in relation to global GDP, damage to critical infrastructure and number of disruption of basic services, attributed to disasters.
12.6 By 2030, reduce the adverse per capita environmental impact of cities, including by paying special attention to air quality and municipal and other waste management	12.6.1 Proportion of urban solid waste regularly collected and with adequate final discharge out of total urban solid waste generated, by cities
	12.6.2 Annual mean levels of fine particulate matter (e.g. PM2.5 and PM10) in cities (population weighted)
12.7 By 2030, provide universal access to safe, inclusive and accessible, green and public spaces, in particular for women and children, older persons and persons with disabilities	12.7.1 Average share of the built-up area of cities that is open space for public use for all, by sex, age and persons with disabilities
	12.7.2 Proportion of persons victim of physical or sexual harassment, by sex, age, disability status and place of occurrence, in the previous 12 months
12.a Support positive economic, social and environmental links between urban, peri-urban and rural areas by strengthening national and regional development planning	12.a.1 Proportion of population living in cities that implement urban and regional development plans integrating population projections and resource needs, by size of city
12.b By 2020, substantially increase the number of cities and human settlements adopting and implementing integrated policies and plans towards inclusion, resource efficiency, mitigation and adaptation to climate change, resilience to disasters and develop and implement, in line with the Sendai Framework for Disaster Risk Reduction 2015–2030, holistic disaster risk management at all levels	12.b.1 Number of countries that adopt and implement national disaster risk reduction strategies in line with the Sendai Framework for Disaster Risk Reduction 2015–2030.
	12.b.2 Proportion of local governments that adopt and implement local disaster risk reduction strategies in line with national disaster risk reduction strategies
12.c Support least developed countries, including through financial and technical assistance, in building sustainable and resilient buildings utilizing local materials	12.c.1 Proportion of financial support to the least developed countries that is allocated to the construction and retrofitting of sustainable, resilient and resource-efficient buildings utilizing local materials

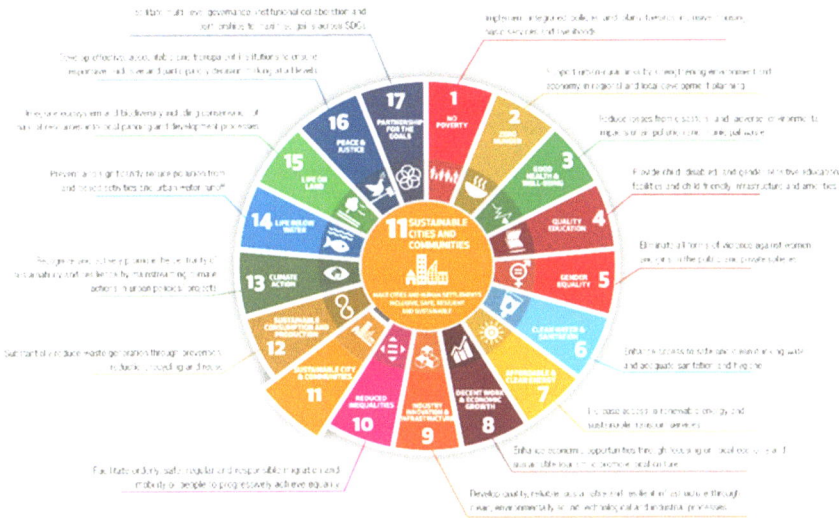

Fig. 12.1 Interconnections between SDG 11 and other SDGs. (Source: UN Habitat)

12.3 Discussion and Results: Localizing Global Development Goals

While SDG 11 sets broad objectives, it is important to contextualize and mainstream the universal goals into local development process, to make tangible differences in the lives of people – and here urban local bodies have crucial role to play. To make timebound progress to achieve sustainability goals, the targets and indicators of SDG 11 need to be aligned and benchmarked with infrastructure delivery process through the local development plans and budgetary priorities. However, incorporation of global goals into local planning and policy discourse faces considerable governance and economic challenge. This section looks at how these issues are being negotiated through the example of India.

India had been one of the fastest-growing major economies of the world over the past couple of decades. India is also the second largest urban system in the world, with an urban population of 377 million at an urbanization level of 31.14%, and has three megacities with ten million plus population (Census 2011). Although in percentage terms India's urbanization rate is less compared to big Asian countries like China or Indonesia, it has started to accelerate. For the first time in history, net population growth rate in urban areas exceeded rural population growth rate – during the 2001–2011 census decade. It is projected that the urban population would reach 600 million (40%) by 2031 and 850 million (50%) by 2051 (ibid).

Table 12.2 Socio-economic indicators of Indian cities (in percentage terms)

Indicators	2001	2011
Literacy rate	73.08	84.1
Urban sex ratio	841	929
Infant mortality rate	42	29
Population below poverty line	25.5	13.7
Household with safe drinking water	90	91.4
Household with electricity	91.6	97.9
Household with septic tank and flush	70.7	81.6

Source: Handbook of urban statistics (2019)

Table 12.3 Indicators selected for SDG India index

SDG global targets		Indicators selected for SDG India index	National target value for 2030
12.1 By 2030, ensure access for all to adequate, safe and affordable housing	1	Houses completed under Pradhan Mantri Awas Yojana for houses as a percentage of net demand assessment for houses	100
	2	Percentage of urban households living in slums	0
12.6 By 2030, reduce the adverse per capita environmental impact of cities, including by paying special attention to air quality and municipal and other waste management	3	Percentage of municipal wards with 100% door to door waste collection	100
	4	Percentage of waste processed	100

Source: SDG India Baseline Index (2018)

Urban areas face major challenges regarding provision of affordable housing and basic infrastructure services. During census 2001–2011, although the population of slum household percentage reduced 18.3 to 17.4, in absolute terms there was net increase of 13,920,191 (Census 2011). Similarly, as Table 12.2 shows, socio-economic and access to basic amenities had improved, but the shortfalls still remain.

In recent years, India's national government has rolled out several city-centric centrally funded missions which are in sync with the SDG 11 objectives regarding sustainable urban development, such as Jawaharlal National Urban Renewal Mission (JNNURM), Atal Mission for Rejuvenation and Urban Transformation (AMRUT), Smart Cities Mission, Pradhan Mantri Awas Yojana (PMAY – Housing for All) and Swachh Bharat Mission (Clean India Mission).

Two of the SDG 11 targets in particular, focusing on affordable housing (PMAY), sanitation and access to clean water supply, are specifically covered by national missions (Swachh Bharat Mission), as indicated in Table 12.3.

However, it is important to note here that under the multilevel governance structure of the Indian constitution, responsibilities for urban development mainly lie with the state governments and urban local bodies. Roles and responsibilities of the national government are limited – setting directions and providing funds. Direct implementation responsibilities lie with the institutions and agencies associated with urban local governance, including the elected municipal governments, city development authorities and parastatal agencies under the state governments. Elected municipal governments enjoy constitutional status with defined roles and responsibilities vis-à-vis the state government, following the 74th Amendment Act of 1992. Under the constitutional devolution, the municipal governments are mandated to perform 18 functions, which contribute towards urban sustainability.

Recently the Ministry of Housing and Urban Affairs (MoHUA) has developed a guidance for the ULB to leverage SDG 11 as a lens to address several other interlinked SDG goals, as outlined in Table 12.4. The ULBs, which are in the process of preparing the statutory master plan or urban mission-linked city development plans, have the opportunity to synchronize their plans with SDGs.

Table 12.4 Framework for sustainable urban development at ULB

What ULB can do?	Action points for urban local bodies to achieve sustainability goals
SDG 1: No poverty – *end poverty in all its forms everywhere*	
	Facilitate universal access to housing and homeless shelters, water and sanitation, social protections, etc.
	Building of institutions like self-help groups
	Resilience to extreme climate events and other environmental shocks
	Strategies targeted towards women's employment, access to finance and ownership and control of resources
SDG 2: Zero hunger – *end hunger, achieve food, security and improved nutrition and promote sustainable agriculture*	
	Enrol – families under Public Distribution Scheme (PDS), elderly people under old age pension scheme
	Ensure – growth monitoring of children under 6 years, coverage of pregnant and lactating women, adolescent girls under ICDS supplementary nutrition programme, quality midday meals
SDG 3: Good health and well-being – *ensure healthy lives and promote well-being for all at all ages*	
	Maintain and Monitor – quality of healthcare services, overall cleanliness to combat malaria, water-borne diseases
	Ensure – improving air quality and municipal waste management, effective functioning of health sanitation and nutrition committees and Rogi Kalyan Samiti, linkage to referral centre and 24/7 availability of emergency services delivery infrastructure
SDG 4: Quality education – *ensure inclusive and equitable education and promote lifelong learning opportunities for all*	
	Facilitate – safe and reliable transport for children, access to entitlements like scholarships, uniforms, text books, midday meals, etc., building of girls' toilets to ensure retention of girls evening and mobile schools

(continued)

Table 12.4 (continued)

What ULB can do?	Action points for urban local bodies to achieve sustainability goals
SDG 5: Gender equality – *achieve gender equality and empower all women and girls*	
Promote – universal birth registration, awareness against gender discrimination and gender-based violence, gender-responsive planning and gender-sensitive budgeting, women's participation and leadership in ULBs	
Facilitate – formation of ward, zonal and ULB cooperative samitis, comprising grassroots level functionaries and women's representatives from the community, gender status studies and dissemination of findings in ULB	
SDG 6: Clean water and sanitation – *ensure availability and sustainable management of water and sanitation for all*	
Ensure – identification of households without access to toilets and piped water, facilities for solid and liquid waste management, water-use efficiency by rationalizing water use	
Promote – hygiene education, introduction of efficient water-use technologies, ground water recharge and permeable surfaces, awareness of the additional vulnerability of women with disabilities and religious minorities	
SDG 7: Affordable and clean energy – *ensure access to affordable, reliable, sustainable and modern energy for all*	
Ensure – universal access to affordable, reliable and modern energy services	
Facilitate solar and other renewal energy, energy audit for all electrical installations under ULB and utility undertakings, promote use of LED lighting, promote solar and other renewable energy in urban infrastructure	
SDG 8: Decent work and economic growth – *promote sustained, inclusive and sustainable economic growth, full and productive employment and decent work for all*	
Map – micro, small and medium enterprises within the ULBs, informal enterprises within the urban areas	
SDG 9: Industry, innovation and infrastructure – *build resilient infrastructure, promote inclusive and sustainable industrialization and foster innovation*	
Map – existing and potential industry clusters in the area and hazardous industries and promote appropriate risk mitigation strategies	
Identify and track – start-ups that could lead innovation in the area	
SDG 10: Reduce inequality – *reduce inequality within and amongst countries*	
Identify – the vulnerable groups of people, spaces and incidence of discrimination against women, SCs, STs, minorities and persons with disabilities	
Promote – enhanced representation and voice for the most marginalized groups in decision-making, universal access to affordable housing, social services and public utilities	
SDG 11: Sustainable cities and communities – *make cities and human settlements inclusive, safe, resilient and sustainable*	
Map – access of affordable housing and basic infrastructure for all, particularly slum dwellers and access of safe, affordable, accessible and sustainable transport systems for all	
Identify and track – potential areas for upgrading, redevelopment and greenfield development, environmentally sensitive development, link between infrastructure with urban land use and vulnerability assessment	
SDG 12: Sustainable consumption and production – *ensure sustainable consumption and production patterns*	
Ensure – identification of major solid waste and industrial waste generators, reduction, reuse and recycle of wastes, green building construction	
Promote – circular economy, 'use, treat and reuse' approach, eco-labelling for goods and services, sustainable and responsible tourism	

(continued)

Table 12.4 (continued)

What ULB can do?	Action points for urban local bodies to achieve sustainability goals
SDG 13: Climate action – *take urgent action to combat climate change and its impact*	
	Integrate – sectoral climate change considerations into local planning
	Facilitate – conducting of vulnerability assessments, make environment/climate change impact assessment for all major projects
SDG 14: Life below water – *conserve and sustainably use the oceans, seas and marine resources for sustainable development*	
	Ensure – revitalization of waterfronts and make these assets attractive and open to public, prevention and significant reduction of marine pollution of all kinds, in particular from land-based activities, including industrial waste and wastewater
SDG 15: Life on land – *restore and promote sustainable use of terrestrial ecosystems, sustainably manage forest, combat desertification, halt and reserve land degradation and halt biodiversity loss*	
	Ensure – integrated planning and curb urban footprint into ecological systems, balance rural-urban linkages with minimum impacts to the terrestrial ecosystems, promote integrated planning and curb urban sprawl, focus on mixed use and compact cities
	Facilitate – urgent and significant action to reduce the degradation of natural habitats caused by poor planned urbanizations
SDG 16: Peace, justice and strong institutions – *peace, justice and strong institution: promote peaceful and inclusive societies for sustainable development, provide access to justice for all and build effective, accountable and inclusive institutions at all levels*	
	Map – accountability and transparency in urban governance
	Facilitate- awareness on citizens' rights, capacity building of municipal staff and elected representatives, informed, active and meaningful participation of all social groups in decision-making
SDG 17: Partnership for the goals – *strengthen the means of implementation and revitalize the global partnership for sustainable development*	
	Ensure – knowledge sharing and build capacities of government officials in urban local bodies, develop context-specific knowledge products adapted to the absorption capacity of target audience
	Facilitate – promoting information exchange and experience sharing (policy reforms, new technologies, performance monitoring and innovative service delivery options) for sustainable urban management

Source: Prepared by Authors – based on publicly available data sourced from the Ministry of Housing and Urban Affairs and UN Resident Commissioner (2018)

12.4 Impact Sustainability: Final Remarks

12.4.1 Pointers for Further Research and Action

The inclusion of SDG 11 in the Sustainable Development Agenda 2030 recognizes the importance of the cities in the global platform. The quest for urban sustainability is not a stand-alone goal but rather an opportunity to address several other objectives, such as climate change, poverty eradication, access to safe water supply, energy demand management, social inclusion and spatial justice.

SDG 11 is also closely tied with the New Urban Agenda, which was adopted in 2016. The NUA requires the signatory countries to frame respective National Urban Policies – as a framework to drive urban sustainability targets. The National Urban Policy is expected to guide and monitor urban development priorities to accomplish Sustainable Development Agenda 2030.

Success of SDG 11 will depend on the extent to which they are contextualized to local situations and mainstreamed within local urban planning frameworks and budgetary constraints. The interlinked nature of SDG 11 would require an integrated approach towards urban development, overcoming sectoral boundaries. Government functionaries need to be equipped to build synergies within institutional boundaries and work on new partnerships while strengthening existing institutions.

Prime responsibility regarding SDG 11 implementation lies with the institutions of urban local governance – which includes the elected ULB, city development authorities and various parastatal bodies providing water supply or running transportation systems. In India, the lever of urban development lies with the state governments and most of the agencies responsible for delivery of urban infrastructure function directly under the state governments and often serve multiple administrative domains. In most cases the elected ULB are weak, despite the constitutional mandate. Therefore, the crucial onus lies at the level of state bureaucracy. In recent years, the national government started to pay greater attention towards the cities and channelize funds through mission-based programmes. While the projects carried out by the mission-based programmes do fulfil several SDG priorities, they are rolled out by specific project management units and special purpose vehicles.

Achieving inclusive, safe, resilient and sustainable cities for all, as outlined in Sustainable Development Goal 11 and the New Urban Agenda, requires officials to recognize the interdependence of global goals and local actions and to follow an integrated vision for development and urban resilience building. Such an approach demands a paradigm shift in the way governance structures are managed not only between national and local governments but also between local governments and communities. To this end, decentralization and fiscal federalism that brings local governments and users in proximity to urban services and development must be enabled. The trust between communities and local leadership must be strengthened through transparency and accountability seated in advocacy, data and evidence to monitor the progress towards a common vision for the city. Local leadership must be encouraged to innovate and leapfrog from business-as-usual tools and technologies, even if it means taking calculated risks. Metropolitan and urban local bodies in India, thus empowered, can take on a strong leadership role in the achievement of the Sustainable Development Goals.

City administrators and political leaders, mayors, municipal councillors and higher bureaucracy require to be sensitized about the conceptual and operational framework of the SDGs, interlinkages between the SDG 11 and the New Urban Agenda and other SDGs. There is a strong need for creating awareness and building the required capacities within the administration at the middle and upper management level. Integrated planning involving such diverse set of stakeholders at various governance hierarchies would require a strong institution with skilled technical expertise in planning (economic, social, physical), management (project, finance, operational, organizational) and governance (stakeholder engagement, inter-agency coordination). As the SDG 11 has multiple sets of targets and indicators, database needs to be developed to monitor progress and benchmark services being delivered to facilitate decision-making based on ground-level inputs. Urban observatories may be developed for dissemination of data and information sharing.

References

Babier EB, Burges JC (2017) The sustainable development goals and the systems approach to sustainability, Discussion Paper. Economics 11(2017.28):1–23

Census (2011) Population Census 2001–11, Registrar General and Census Commissioner. Government of India, New Delhi

Fabre EA (2017) Local implementation of the SDGs and the New Urban Agenda – towards a Swedish National Urban Policy. Global Utmaning, Stockholm

Ministry of Housing and Urban Affairs (2019) Handbook of Urban Statistics. Government of India, New Delhi

Ministry of Housing and Urban Affairs and UN Resident Commissioner (2018) Sustainable Development Goals and Urban Local Bodies. Office of the UN Resident Commissioner, New Delhi

Niti Ayog (2018) SDG India Base Line Index. Government of India, New Delhi

UN Habitat (2016) Sustainable development goal – 11: a guide to assist national and local governments to monitor and report SDG goal 11+ indicators. UN Habitat, Nairobi

United Nations (2018) World urbanization prospects, United Nations Department of Economic and Social Affairs, viewed 31 October 2018. https://population.un.org/wup/

Chapter 13
SDG 12 Responsible Consumption and Production

Sustainable Community Development Through Entrepreneurship: Corporate-Based Versus Wellbeing-Centred Approaches to Responsible Production

Isabel B. Franco and Lance Newey

Abstract This chapter aims to build new theory about the links between sustainable community development, entrepreneurship, community wellbeing and interlinkages between the sustainable development goal 12 (hereafter SDG 12), sustainable consumption and production, and the sustainable development goal 3 (hereafter SDG 3) – good health and wellbeing. New theory is needed because multidimensional wellbeing has not been used as an outcome variable with which to assess relative merits or understand the intricacies of how development approaches achieve synergies or fragmentation between the varying components of wellbeing. The research presented here is based on a case study qualitative methodology strategy. Evidence shows that resource-rich regions of Latin America are sites for sustainable community development and responsible production by international companies. Through a comparative case study of two resource towns in Colombia – Antioquia and Risaralda – we find contrasting approaches with different outcomes. A top-down corporate-based approach to sustainable community development occurred in Antioquia but bound the future of the community to resource extraction (mining) with limited attention to other aspects of community wellbeing. This reduced the overall resilience and wellbeing opportunities for the community. In Risaralda, by contrast, a

---

I. B. Franco (✉)
Institute for the Advanced Study of Sustainability, United Nations University Shibuya-ku, Tokyo, Japan

Australian Institute for Business and Economics, The University of Queensland, Brisbane, Australia
e-mail: connect@drisabelfranco.com

L. Newey
The University of Queensland, Business School, Brisbane, Australia
e-mail: l.newey@business.uq.edu.au

© Springer Nature Singapore Pte Ltd. 2020
I. B. Franco et al. (eds.), *Actioning the Global Goals for Local Impact*, Science for Sustainable Societies, https://doi.org/10.1007/978-981-32-9927-6_13

more responsible, wellbeing-conscious approach was adopted based on local entrepreneurship. Entrepreneurship here was not only focused on economic development and the future self-sufficiency of the community apart from mining but was also conscious of producing responsibly and building a greater range of wellbeing components other than just economic. We use these results to articulate a wellbeing-centric approach to development called Entrepreneurship for Community Wellbeing.

Keywords SDG 12 responsible consumption and production · SDG 3 good health and wellbeing · Entrepreneurship · Wellbeing · Sustainability · Corporate social responsibility

13.1 Introduction

This chapter asks the question: how can corporate responsible producers help underdeveloped yet resource-rich towns undertake development in a way that leads to community wellbeing? Based on its rich, multidisciplinary development, in this chapter we define community wellbeing as the capacity of a community to flourish, including being able to fulfil important goals and persist in the face of obstacles (cf. Gough et al. 2007). In other words, wellbeing is a resilient capacity to achieve sustainable development aspirations (Burroughs and Rindfleisch 2002) and SDG 3 (UN 2018). Wellbeing is multidimensional and for our purposes includes eight components: economic, social, cultural, environmental, psychological, spiritual, material and physical (Keyes 1998; OECD 2013; Peterman et al. 2002).[1] Although the term "community" has various meanings, in this paper we focus exclusively on communities as bounded geographic spaces and where there is a desire to achieve collective goals. Community wellbeing refers to the wellbeing of a distinct geographic space of people whose fortunes are bound together (Peredo and Chrisman 2006). We also focus specifically on noncore peripheral communities/towns outside the main metropolitan core.

Our research question then poses how communities can improve their wellbeing across the eight components: economic, social, cultural, environmental, psychological, spiritual, material and physical. Such a view recognises that these components are interlinked and either benefit or damage occurs depending on whether actors, particularly corporations, are mindful of these interactions. Although a number of scholars note how "harmonic" (Misoczky 2011) and local entrepreneurship-based (McMullen 2011; McWade 2012; Peredo and Chrisman 2006) approaches to development aim to create a broader range of outcomes for communities, we lack a concept which unites these various outcomes into a parsimonious and composite measure (Dana et al. 2014; Gray et al. 2014; Huggins and Thompson 2014; Korsgaard and Anderson 2011). We advance "wellbeing" as a concept which fills this gap and can offer coherence to understanding how different community development strategies achieve multiple outcomes.

---

[1] In this chapter, when we use the term "multidimensional wellbeing", we are referring to all eight components listed here.

In order to advance the merits of a wellbeing-based approach, we compare and contrast two leading methods of sustainable community development in terms of their outcomes for community wellbeing. On the one hand, resource-rich towns can place their hopes in the hands of multinationals who foster a local economy through centring it on a key industry such as mining, agriculture or tourism. On the other hand, towns may instead base their development less on any one industry and more through the variety offered by local entrepreneurship.

Our research objective is to build new theory about the links between sustainable community development, entrepreneurship and community wellbeing. New theory is needed because multidimensional wellbeing has not been used as an outcome variable with which to assess relative merits or understand the intricacies of how development approaches achieve synergies or fragmentation between the varying components of wellbeing.

Two case studies in the resource-rich but underdeveloped Colombian mining regions of Antioquia and Risaralda offer research sites for a comparison between economic-focused industry-centred approaches and more wellbeing-centric approaches to sustainable community development. Both are noncore regions, defined as "outside the principal metropolitan areas" (Lagendijk and Lorentzen 2007: 459). We find that the more wellbeing-centric approach adopted in Risaralda led to better overall community wellbeing outcomes across the eight components compared with Antioquia. Moreover, the latter's entrepreneurship and development were centred largely on the fortunes of the mining industry thus making its resilience susceptible to the vagaries of the mining cycle. By contrast, in Risaralda, entrepreneurship and sustainable community development were more focused on local empowerment and leadership to chart a direction of development where mining was only a part and not the core.

Our main contribution is to advance empirical evidence for an eight-component model of wellbeing as an approach to sustainable community development through entrepreneurship and towards the achievement of SDG 12 (UN 2018). Consistent with research showing a diminishing returns effect between economic wealth and life satisfaction (Easterlin 1974; Gasper 2005; Patrizii et al. 2017), the eight-component model focuses on a wider range of elements that regions and towns must pay attention to for the life fulfilment of their populations and responsible production of companies operating in the region. Achieving outcomes on these eight wellbeing components requires a mix of entrepreneurship types within a community including commercial, social and institutional (Baumgartner et al. 2013). This broad wellbeing perspective contributes to that emerging literature focused on more "harmonic" (Misoczky 2011) and local entrepreneurship-based (McMullen 2011; McWade 2012; Peredo and Chrisman 2006) approaches to development that aim to create a broader range of value types in towns (Dana et al. 2014; Gray et al. 2014; Huggins and Thompson 2014; Korsgaard and Anderson 2011).

Wellbeing thus serves as an overarching goal around which communities can develop entrepreneurship strategies. This shifts the focus of noncore regional strategies beyond just economic growth and product/service innovation to consider how these strategies affect other components of wellbeing. This also questions the assumption that any growth is good, responsible growth. We also know that noncore regions need tailor-made entrepreneurship policies on account of their peculiar conditions including high factor costs and remoteness (Anderson 2000). Our research indicates that noncore regions also benefit from entrepreneurship policies which support an overarching set of wellbeing goals. Wellbeing offers an end game for other contextual features advocated in the literature to support entrepreneurship in small towns including SME innovation and R&D, firm clusters, knowledge providers and innovation support institutions (Toedtling and Kaufmann 2001: 1215).

Attending to wellbeing shifts the focus of communities away from just economic development to also build their long-term resilience through other components of wellbeing such as psychological, spiritual, cultural and social. This wider set of wellbeing components helps buffer noncore communities from the ups and downs of economic cycles. Despite its increasing academic interest and movements towards measuring the wellbeing of nations, the concept of wellbeing is yet to make its mark on understanding sustainable development through entrepreneurship. The concept is mentioned by development scholars, but existing wellbeing literature is inadequately consulted, leaving the term to mis-specified use. Yet, when adequately conceptualised, a multidimensional concept such as wellbeing can help to guide how communities use entrepreneurship to create more balanced communities and avoid negative externalities that can come from a failure to understand and consider all aspects of wellbeing. We thus wish to explore how wellbeing may be used to guide sustainable development through entrepreneurship.

In the next section, we review the concept of wellbeing and the two development strategies of interest: industry-centred and local entrepreneurship. Our discussion highlights how wellbeing offers a different evaluative framework for assessing the merits of different sustainable community development approaches. We then outline our methodology, including a justification for a comparative theory-building case study approach. Empirical evidence and analysis follow, illustrating how a wellbeing-centric approach led to better community outcomes compared with more industry-based approaches. The following section then abstracts from the case analysis to distil key theoretical variables demarcating industry-based and wellbeing-centric approaches including analysis of their respective benefits and limitations. Theoretical and practical implications of an entrepreneurship for community wellbeing approach are presented before concluding the paper.

13.2 Literature Review

13.2.1 Multidimensional Wellbeing

The concept of wellbeing has gained parlance in numerous disciplines such as psychology, sociology, economics, anthropology and political science in order to capture the multiple components that make up quality of life (Gasper 2004; Gough et al. 2007; McGregor 2007; Seligman 2011; Stiglitz et al. 2010). In policy circles

Table 13.1 Components of wellbeing

Component	Example definition	Example references
Economic	The stock and flow of economic resources that an individual or household receives over time	OECD (2013)
Social	How people come together as one community Includes political participation, equity of opportunity, freedom, agency, law and order	Keyes (1998)
Cultural	The extent to which people can actively engage with their culture, customs, traditions and values inherent to their specific cultural heritage	Torjman (2004)
Environmental	Refers to the health of the natural environment and whether it provides the conditions to support life goals in a location	Callicott (1996)
Psychological	Is about lives going well. It is the combination of feeling good and functioning effectively	Ryff (2014)
Spiritual	Personal search for meaning and purpose in life, connection with a transcendent dimension of existence and the experiences and feelings associated with that search	Peterman et al. (2002); Pargament and Sweeney (2011)
Physical	Quality of physical health	WHOQOL Group (1998)
Material	Whether an individual has the material assets that allow them to do and be what they seek in life	Perry (2009)

also, wellbeing is an increasing focus of national governments as a reaction against purely economic indicators (Bleys 2012). The Report of the Commission on the Measurement of Economic Performance and Social Progress initiated by the French President Nicolas Sarkozy proclaimed that "the time is ripe for our measurement system to shift in emphasis from measuring economic production to measuring people's wellbeing" (Stiglitz et al. 2010: 10).

Based on existing literature, we focus on an eight-dimensional framework of wellbeing: economic, physical, material, social, cultural, psychological, spiritual and environmental. Table 13.1 (see Appendix) offers definitions and example references for each component. These components are what are known in the literature and which have become apparent as important to wellbeing through the adverse consequences that result in communities when they are overlooked.

This multidisciplinary approach recognises that wellbeing emerges at the intersection of the interrelationship between the various components. In some ways, this interrelationship is characterised by counterbalance. If let go too far, economic wellbeing depletes natural resources and environmental wellbeing. The pull towards one whole integrated community of social wellbeing needs to also recognise the plurality of cultural wellbeing. Diseases of affluence (McKeown 1988; Novotny 2005) which accrue in wealthy economics and harm physical health require sound psychological wellbeing to rein in hedonistic tendencies. Finally, excessive materialism requires the counterbalance of inner contentment and spiritual nourishing, which in turn creates a more sustainable relationship with the natural environment.

We see here the seedbed of ideas for achieving quality of life in towns and how ignorance of the principle of counterbalance can lead to considerable wellbeing costs. There is growing recognition that economic factors alone do not contribute to wellbeing (Diener et al. 2010). While research shows a link between income and positive life evaluations, these results may be moderated by values (Burroughs and Rindfleisch 2002; Ng and Diener 2014). This means that it cannot be assumed that economic wellbeing equates to sustainable community development, at least not without taking into account the wider value structure of the community. This is often significant when Western conglomerates enter indigenous communities and values collide. Clear examples here are the indigenous communities of Latin America, which have implemented policies of buen vivir (Gudynas 2011; Monni and Pallottino 2015; Villalba 2013), which mirror closely our eight-component model of wellbeing. Community values captured in frameworks like buen vivir highlight the importance of notions of balance, harmony and multidimensionality to small indigenous towns.

Critique of Top-Down Industry-Centred Approaches

Our conceptualisation of wellbeing has important implications for understanding how communities can develop. It also provides a wider framework for evaluating whether the espoused sustainable community development strategies of multinationals will sustainably benefit communities or not. The integration of emerging regions into a global economy can involve certain "models of development" being imposed from outside, which can conflict with local priorities, values and interests upsetting social, psychological, cultural and spiritual aspects of wellbeing (Pike et al. 2007). In turn this has led to the escalation of extreme poverty as well as imbalanced community wellbeing and discontent (Cook 2006; Haan and Maxwell 1998; Harrison 2006; Kabeer 2000; Shankar and Shah 2003).

Poor and local communities often feel that they are not adequately consulted and/or compensated for their loss of livelihood options (Dana et al. 2009; Surborg 2012). Large companies can be prone to management by extraction, exclusion and expulsion (Banerjee 2011). A sense of community wellbeing at the local level can be undermined by underestimating the significance of "place" and embedded socio-cultural identity that resist commercial homogenisation (Dana et al. 2014). How the local strategies of multinationals are treated by local governments is also important as institutional strategies can help or hinder the development of entrepreneurship, thus affecting the sustainability and evenness of development of towns (Williams and Vorley 2015).

More specifically, large multinationals often base community development around an industry, such as mining, tourism or agriculture, depending on the resource supplies of a region. We refer to this as an industry-based approach to community development to connote an approach based around a key industry which forms a nucleus for economic development. This industry-based approach can determine path dependencies of the development of local communities leading them on either a self-sufficient track or one that is dependent and wellbeing imbalanced (Williams and Vorley 2014). The vagaries of industrial cycles can leave these communities vulnerable. Moreover, such development efforts can be lopsided

(Huggins and Thompson 2014) and exacerbate regional unevenness (Pike et al. 2007). Local economic gains can be offset if local entrepreneurs are not adequately educated and trained to consider unintended consequences such as environmental degradation, excessive class inequalities and perceptions of Western cultural imperialism causing affront to indigenous culture and values (Banerjee 2011; Dana, et al. 2014; Misoczky 2011).

Entrepreneurship and Community Wellbeing
Traditionally, entrepreneurship refers to the formation, evaluation and exploitation of opportunities for commercial gain (Shane and Venkataraman 2000) neglecting sustainable paths for responsible production. The entrepreneur spots opportunities to create financial value through innovation or doing the same things better. The intent is for growth, traditionally narrowly defined as consisting of economic indicators. However, such opportunity recognition and value creation motivations and skills of entrepreneurs have now been extended to also include social and environmental value creation – or what is commonly known as social entrepreneurship (Bacq and Janssen 2011; Korsgaard and Anderson 2011; Nicholls 2010; Ratten and Welpe 2011).

Entrepreneurship for community wellbeing is a further extension of this logic that goes beyond just "social" to include a more multidimensional and holistic goal for entrepreneurship. Entrepreneurship for community wellbeing is an approach where communities (1) set multidimensional wellbeing as their goal and (2) develop different types of entrepreneurship (commercial, social, institutional) in a way that seeks to achieve the community wellbeing goals. As a bottom-up approach to entrepreneurship, entrepreneurship for community wellbeing is more likely to create value for communities than existing top-down approaches. This emerging approach advocates for the transformation of vulnerable communities to higher states of organisation and the realisation of their full potential (cf. Easterly 2006).

Indeed, the improved wellbeing of entrepreneurs themselves and their families can have flow-on effects to the broader community, as occurred in Cahn's (2008) case study of fine mat weaving in a Samoan micro-enterprise. The latter enterprise blended its economic activity with the cultural way of life, cultural norms, and offered psychological benefits through status. In contrast, a village-based coconut oil production enterprise suffered from tensions between its economic structure and the cultural commitments of its members. The fine mat weaving enterprise enhanced not only economic but also social and cultural assets, which seemed to improve sustainability. Top-down approaches may miss these critical sociocultural-economic dynamics, thus fracturing the enterprise and wider society (Peredo and McLean 2013). Entrepreneurship for community wellbeing is also based on a new set of values focused on the community (Peredo and Anderson 2006), including the most marginalised and exploited. This approach is also intended to imprint communities and other stakeholders involved with strong values of egalitarianism and sensitivity for the suffering of others (Maalouf 2014).

Empirical research in Latin America shows that communities' lack of education and opportunities to work for either a company or an industry relevant to the local economy further compromises community wellbeing and escalates poverty, loss of livelihood options and conflict (Molina-Escobar and Restrepo-Baena 2010). This has been caused, amongst others, for irresponsible production patterns of multinational companies. However, accessing education to find employment opportunities in the mining industry is usually perceived by communities as a top-down approach that often does not meet their development aspirations. This top-down approach to wellbeing may in fact prevent stakeholders in the resource sector from impacting communities in any positively responsible way (Mate 2001). Hamdouch et al. (2016) discuss this issue. They examine how towns and communities make choices according to their development aspirations. Empirical research also shows that there is an untapped potential in resource regions for fostering entrepreneurship (Franco 2014) for improved community wellbeing.

Building communities' capacity for entrepreneurship for improved wellbeing will make a strong contribution to social sustainability in resource-rich towns. We argue that local communities which gain resilience through a history of entrepreneurship can successfully cope with the global challenges posited by the mining industry and improve community wellbeing at the local level (Gray et al. 2014; Williams and Vorley 2014). This in turn provides them with a greater ability to be receptive to education for entrepreneurship as an agent of change and development.

In this context, entrepreneurship for community wellbeing is an untapped agenda for sustainable development. While a number of researchers have undertaken studies exemplifying the multiple types of value created by entrepreneurship (Gray et al. 2014; Korsgaard and Anderson 2011; Williams and Vorley 2014), we feel that wellbeing can be a useful concept for the continued growth of this stream of research. From an entrepreneurship perspective, one of the main advantages is that a wellbeing approach requires a portfolio approach to community governance of entrepreneurship. Such a portfolio approach encourages the need for commercial entrepreneurship, which creates primarily economic value but also social value. However, the social value created is only that which is economically rational and maximises profit. To counterbalance this, social entrepreneurship, including community-based enterprises (Peredo and Chrisman 2006; Peredo and McLean 2006), can address a wider range of wellbeing opportunities by relaxing the assumption of profit maximisation (Zahra et al. 2014). Here, increased value created on non-economic components of wellbeing is deemed equally, if not more, valuable than financial wealth maximisation. Actors thus define and perceive entrepreneurial opportunities differently rather than just through the lens of standard economic evaluations (Peredo and McLean 2013). This portfolio approach requires community-level governance if the right types and spread of entrepreneurship are to occur to maximise wellbeing (Lichtenstein and Lyons 2001).

Bringing our arguments together, we contend that resource-rich towns have important choices to make concerning the paths of their development. The attraction of employment and riches from mining companies can lead to imbalanced community development from a wellbeing perspective. The concept of wellbeing provides a framework to guide sustainable community development through entrepreneurship. Moreover, community self-sufficiency beyond mining calls for entrepreneurship skill-building allowing locals to determine their destiny. Such entrepreneurship though needs to be combined with wellbeing education and governance if community resilience and balance are to be achieved. This "entrepreneurship for community wellbeing" approach combines and goes beyond both traditional pure-profit-maximising entrepreneurship and social entrepreneurship (Zahra et al. 2014). It includes them by fostering both in a community but also goes beyond them to consider the interaction between and goal pursuit of multiple components of wellbeing, not just economic or social. By taking a portfolio approach to entrepreneurship at the community level, it forces deeper consideration of how different entrepreneurial initiatives not only contribute to wellbeing but how they interact with each other in positive or negative ways, thus pushing for that analysis which maximises overall community wellbeing (Lichtenstein and Lyons 2010).

13.3 Methodology

As wellbeing is a new way to evaluate different approaches to community development, it is important to more closely examine how different approaches to community development – industry-centred versus entrepreneurship – achieve or not achieve community wellbeing. Case studies would permit deep interrogation of locals to canvass their subjective perceptions of how these different strategies impact their wellbeing across eight components (Eisenhardt 1989; Yin 2003). We are interested in hearing the perspectives and evaluations of various stakeholders in order to form a composite picture of the links between sustainable community development strategies and community wellbeing. Case studies offer this multi-stakeholder access and investigation (Eisenhardt 1989). Moreover, case studies will permit a longitudinal perspective to take into account long-term wellbeing.

We undertook a qualitative comparative case study of two mining towns in Colombia. The latter is an appropriate sample because of the granularity of analysis possible at the local level in the mining areas of the country that have gone through resource extraction and emerging entrepreneurship. Colombia has a fraught history of internal political conflict with weakened governance structures that have escalated local struggles like civil conflict and poverty. Nevertheless, the country has experienced escalating mining growth over the last three decades. Colombia is the main producer of coal in Latin America and the twelfth largest globally, the third major producer of nickel after Cuba and the Dominican Republic.

13.3.1 Case Studies

Our two comparative case studies occurred in the regions of Antioquia and Risaralda. We chose these case study regions because they appeared to follow different paths to sustainable community development. Antioquia pursued an industry-centric development approach based around mining, while Risaralda was more conscious of diversity and encouraged entrepreneurship, business diversity and community resilience while also pursuing mining. The two cases thus adopted a different attitude and development strategy towards mining. Different outcomes also followed with Antioquia experiencing higher levels of community dissatisfaction with their approach. These differences were made for interesting sites for research in order to better understand the links between entrepreneurship, sustainable development and wellbeing.

13.3.2 Data Collection

Using case studies as the basic methodological approach, the analysis used a literature review, document analysis (policy analysis), stakeholder analysis, face to face semi-structured interviews, group interviews, observations and field notes. Field interviewing was conducted in Risaralda, Antioquia and the capital city of Bogota. Interviews and focus group interviews with key stakeholders at the national, state and local levels were applied to map, explore and evaluate the roles of these stakeholders in existing approaches to community wellbeing.

All interviews were fully transcribed and resulted in an interview sample of 48 representatives from the private sector, government and civil society, drawn from both case study areas and Bogota city where state organisations are located. The participants were chosen as representatives of their organisation or agency. The highest proportion of participants comes from civil society and community organisations and NGOs. Twelve community members and ten representatives from civil society organisations were interviewed. This number is followed by representatives from government organisations. Top level government officers were interviewed representing twelve participants. Nine company executives and private sector senior representatives were also interviewed.

In addition to group and individual interviewing, observations and field notes also constitute key methodological techniques for data collection procedures. Observations helped support some of the researchers' perceptions during interviews. Similarly, field notes were recorded, particularly when spontaneous events that related to the argument that drive this paper happened in the Risaralda and Antioquia case study areas.

13.3.3 Data Analysis

Our qualitative case study research sought to inductively build new theory concerning the links between community development, entrepreneurship and wellbeing. Theory building is an appropriate objective as even though each of the three main constructs (sustainable community development, entrepreneurship and wellbeing) has an extensive literature behind it, there is little known about how they intersect when consciously brought together. Case studies are needed to tease out these interactive links.

As such, our data analysis followed procedures for inductive theory building from cases as prescribed by Braun and Clarke's (2006) thematic analysis, Eisenhardt and Graebner (2007) and Yin's (2003) chain of evidence method to analyse the case data (also Eisenhardt 1989). The chain of evidence method calls for clear links in moving from data to theory, recalling that our overall research question is: how can underdeveloped yet resource-rich towns undertake development in a way that leads to community wellbeing? Theory building is a process of data reduction based on cause-effect evidence. Our process was to first group data into themes based on (1) features of the community development approach adopted and (2) subjective reactions of different stakeholders. From these two sets of themes, we then induced theoretical inferences of cause-effect between features and subjective reactions. Tables 13.2 and 13.3 record the data from this inductive theory-building process.

13.4 Discussion and Results

In this section, we summarise findings from our data analysis as indicated in the first two columns of Tables 13.2 and 13.3. In the next section of the paper, we further develop the theoretical inferences that emerge from this analysis and displayed in the third column of Tables 13.2 and 13.3 and Figs. 13.1 and 13.2.

13.4.1 Antioquia

The State of Antioquia is located on the North Pacific Coast of Colombia. The region is going through a mining boom that has increased poverty, loss of livelihood options and civil war. This situation compromises the possibilities to promote community wellbeing at the local level:

> Internal conflict dynamics have been exacerbated by illegal groups including guerrillas, paramilitary and bacrim (criminal bands). These illegitimate actors own small-scale mines to launder money or support their illegal businesses. In addition, they have found the current mining boom to be the best opportunity to extort money from companies operating in the region. Community livelihoods have been heavily impacted by the indirect benefits of mineral extraction to these groups. (Buitrago Franco 2014)

Antioquia has as an active mining industry, as it holds the largest reserves of gold, silver, coal, platinum and construction materials in Colombia. The industry also has a long history in the region. Mining projects currently operated by domestic and multinational companies have significant implications for local communities (Camara de Comercio de Medellin para Antioquia 2010, p. 14; Sistema de Informacion Minero Energetico Colombiano 2010, pp. 16–17). Antioquia hosts a diverse population inhabiting urban and rural areas. Data from the actual census register (DANE 2010) shows that Antioquia had an urban population of 4,340,744 inhabitants and 1,260,763 people inhabiting rural areas in 2005. Out of the total population, 51.7% were female and 48.3% male. Antioquia also hosted a total of 1,458,193 households in the same year.

Table 13.2 Community development approach – Antioquia

Mining company community development approach (features)	Stakeholder attitudes (reactions)	Data analysis (inferences)
"This is a mining town. Mining is the only livelihood for 80% of community members. We have experiencing some difficulties in engaging with informal miners as they operate in our land title. However, we have been trying to develop better relationships by providing them with contracts to operate on legal basis" (Mining Company Corporate Representative)	"I think in the coming years this town will disappear. Companies promised us development and better opportunities for small miners. Where are the multinationals that promised us employment? There are a couple of companies that do provide us with jobs but what about the other companies? For example, there are no employment opportunities for small miners. People live here out of 'rebusque' … If it is 5 or 6 pm and my children are starving and I have not made any income during the day. What do you think I should do? I go and steal something to feed my children, right? That is 'rebusque'" (Community Member).	Mining viewed as a positive source of economic wellbeing by both the mining companies and this community stakeholder. However, only some seem to be reached in terms of benefits. Unfulfilled economic wellbeing a trigger for crime
"The idea is that locals put their skills at the company's disposal so that later on, they can get gainful employment opportunities … At the end of the day we want communities to operate the mine" (corporate representative)	"Some mining towns in Antioquia have a strong dependence on mining. This situation has diminished the possibilities of sustaining communities in the long-term" (Government representative)	Mining companies very mining-centric in terms of community development. Awareness though from the government representative that this limits long-term community viability
"We do not implement HCB initiatives different from mining because we do not have direct relationship with other sectors … we think that tourism or agriculture are sectors in which we do not fit. For this reason we focus on education for mining" (Senior Corporate representative)	"Our organisation was created from the need to build different livelihood options. Women in this town did not have any value … We were labelled either as witches and/or prostitutes. We have been provided with some training by government organisations … A company donated us three machines … that was long time ago. There are small mines we have some contracts with and we also get some contribution from the government but nothing else apart from that" (Women Community Member focus group)	Mining-centrism of the mining company triggers entrepreneurship for alternative livelihood options Women neglected in community development and employment

(continued)

Table 13.2 (continued)

Mining company community development approach (features)	Stakeholder attitudes (reactions)	Data analysis (inferences)
"Nevertheless, the company's human capacity building (HCB) approach is a top-down mining-focused approach to development. The company has implemented HCB initiatives to up-skill local miners and potentially engage them in large-scale mining projects. According to the company, it has also contributed to physical infrastructure for the development of entrepreneurship HCB initiatives" (Corporate representative)	"I am the only miner in my family. I have three children and want them to study. I am aware of the risks of working at the mine and I did not want them to be part of the industry. This is a mining town… a person who does not go to school becomes a miner. Most of us do not know anything else than mining, hence, we have to work here and sustain our family members with the little income we get" (Community Member focus group)	Tension between the aspirations of mining companies and the wishes of parents for their children's futures
"We have tried to approach NGOs and CBOs to work on human capacity building for community entrepreneurship but we have not got any positive answer so far" (Senior Corporate representative)	"I know a miner called … that earned USD$ 700,000 after finding a gold deposit. That happened a few years ago. At the present time he has a handicapped daughter and lives in very poor conditions… when miners get money they spend it in a very short time. This explains why there is much prostitution and alcoholism in the town" (Miner and Community Member)	Economic wellbeing can be short-lived unless supported by attention to education in other aspects of wellbeing
"Small businesses around mining such as restaurants, laundries and transport agencies have been created… (The) purpose (of these initiatives) is to help community members get organised and stop working on informal basis" (Senior State Government representative)	"Why do you think there is violence in the world? What do you do when you have a family to support and you do not have a job? … you do whatever to get some income to feed your children, right? Well, there are many people experiencing this situation in this region, even though this is a mining region … The situation is very complex in these towns. At least 3 to 4 people are killed every day … This has got worse during the last 3 months" (NGO director)	The need for business diversity, formal business options, creation of an economy of diverse businesses

(continued)

Table 13.2 (continued)

Mining company community development approach (features)	Stakeholder attitudes (reactions)	Data analysis (inferences)
"The way we work is the following: We meet with 16 majors and design initiatives intended to respond to the region's development plan. In doing so, we work closely with governments, companies and mining communities to agree on the development approach that the community needs. We try to develop initiatives relevant for these communities … However, all cases are different. There are times when those agreements are not very productive; however, when they take place, these initiatives have a positive impact on communities" (Government Tertiary Institution Senior Representative)	"Other industries have been displaced as companies and local traders import goods and services, instead of investing and supporting local production in mining towns. Sectors like agriculture and manufacturing are threatened as mining companies pay higher wages to local farmers who have chosen mining over their traditional livelihoods. In addition, community members who have participated in educational programmes have not been able to find gainful employment opportunities. Most of the 600 children who finish high school per year become miners and bartenders. These results lead to questions about whether existing HCB approaches can create sustainable livelihoods in Antioquia" (Academic)	Attempts at coordinated community development based on regional plans
		The need for ecosystem development – an ecosystem of diverse businesses, education and training and government planning
		Attraction to mining employment creates industry shortfalls elsewhere

Industry-Based Development in Antioquia

The current approach to development in some locations of Antioquia led by stakeholders in the region (companies, governments and civil society actors) is intended to attract and retain human capital for the mining industry. A mining company representative stated: "This is a mining town. Mining is the only livelihood for 80% of community members" (Table 13.2, Row 1). Minimal attention has been paid by stakeholders to wider aspects of community wellbeing and long-term resilience. For instance, a government representative remarked that "Some mining towns in Antioquia have a strong dependence on mining. This situation has diminished the possibilities of sustaining communities in the long-term" (Table 13.2, Row 2). In this context locals are becoming insufficiently resilient to withstand mining-induced changes and have not been able to realise their full potential. "This is a mining town… a person who does not go to school becomes a miner. Most of us do not know anything else than mining, hence, we have to work here and sustain our family members with the little income we get" (Table 13.2, Row 4), says a community member. A senior Colombian government representative stated:

> Global corporations invest on projects that are convenient for their business such as infrastructure and facilities for mine's workers rather than on more sustainable legacies for communities adjacent to their operations. This corporate approach is preventing local communities from achieving development aspirations which questions the notion of sustainable and responsible mining in the region.

13 SDG 12 Responsible Consumption and Production

Table 13.3 Community development approach – Risaralda

Community development approach	Stakeholder attitudes	Data analysis
"We do not want mining-dependent communities. Instead, we need to develop sustainable supply chains and entrepreneurship amongst community members. People usually think competitiveness is an overnight process. This is a false expectation. It is a long-term commitment that will allow us to forge more sustainable communities" (senior government representative)	"We are a group of twelve community members. We got organised to create employment opportunities in the region. We have delivered jewellery design courses for the community so they will be able to work with us in the future. By the end of the year we might be able to employ more people. Our purpose is to benefit all community members who work with us" (community leader)	Economic diversity through community entrepreneurship Community leadership as overseer of community entrepreneurship
"Mining is an activity that can cause adverse effects. However, as you see we have a strong commitment to the region … regulatory agencies have visited us and get surprised about our commitment to community members … our overall goal is to provide locals with economic sustainable benefits" (corporate representative)	"We want mining companies to operate socially responsible. We want most of their employees to be locals. Employees should also be equipped with suitable conditions to perform their jobs with high safety standards" (senior local government representative)	Social and environmental wellbeing a key community value Company focus on minimising negative externalities
	"We have outstanding relations with the company. They are supporting our projects. It does not mean we cannot perceive their environmental impacts. We are aware of that. However, our relationships have been outstanding so far" (community members focus group)	
"Locals have a strong sense of community organisation. They own agriculture-based community associations that stimulate the local economy and foster employment. For example, they own associations for blackberry and coffee production and commercialisation" (corporate representative)	"I have been working in mining since I was 7 years old. The company has provided us with some resources to participate in a dress-making course. I asked the company for HCB initiatives in which we women could get some knowledge to sustain ourselves in the long-term. The company, in partnership with a government VET institution, implemented this initiative. However, we need more of these initiatives, as this one in Risaralda"	Focus on community self-sustainability through entrepreneurship High community motivation making it happen

(continued)

Table 13.3 (continued)

Community development approach	Stakeholder attitudes	Data analysis
"We do not want mining and exploration companies operating in Risaralda to extract our resources and leave the town without any legacies for the communities. We do not ask them for money – apart from the royalties and taxes they are obliged to pay. Instead, we want them to build community capacity so that we can export our local goods internationally. This does not cost much to the company but benefits substantially the community" (senior state government representative)	"Community members have been given education in human relations, food security, family guidance and conflict resolution" (community representatives focus group)	Community leadership focused on capacity building and local business development
		Broader wellbeing-focused educational programmes to support social and psychological wellbeing
"Informal miners are resistant to change in their practices and culture. However, we have been able to set clear agreements that benefit all parties" (senior corporate representative)	"Mining companies have a very good relationship with communities, particularly with artisanal miners. Some of them keep developing their livelihoods in land owned by the multinational"	Community development approach that respects local cultural wellbeing
	"Corporations are open to discuss local issues with the whole community. This has resulted in positive consultation processes led by the company" (community leader)	
"Corporations are open to discuss local issues with the whole community. This has resulted in positive consultation processes led by the company" (community leader)	"We have outstanding relations with the company. They are supporting our projects. It does not mean we cannot perceive their environmental impacts. We are aware of that. However, our relationships have been outstanding so far" (community members, focus group)	Informal forums for the purpose of staying in touch with the psychological wellbeing of locals

The research identified three factors that are the main determinants of the inappropriateness of community wellbeing approaches in Antioquia: overinvestment in irrelevant initiatives, limited understanding of entrepreneurship for community wellbeing as an approach to development and approaches that do not reach the broader community but only the mining industry's current and potential employees. These aspects have led to the escalation of social problems such as discontent and resentment amongst locals. More importantly, it has also had adverse implications for local economic development.

There is overinvestment in poorly focused initiatives in the Antioquia case region. Communities have been equipped with training and education in areas they find do not help them realise their entrepreneurship potential or access gainful employment

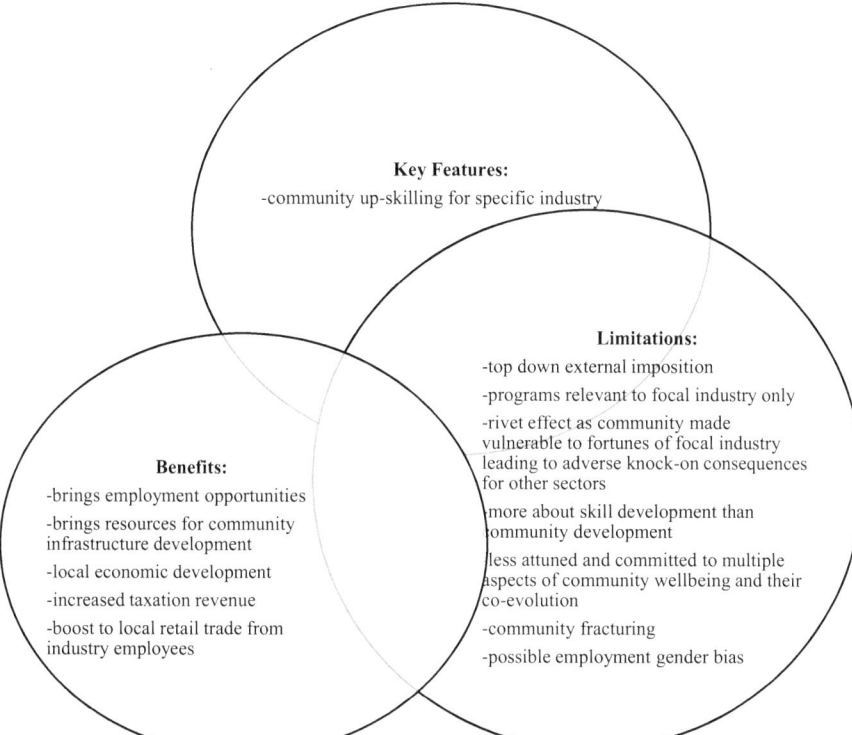

Fig. 13.1 Industry-based community development

opportunities in sectors other than mining. "What is worse is that these actions are delivered over and over, becoming a waste of resources" (Civil Society Representative, Interview, October, 2012). There is a strong perception from government representatives that there is an increasing interest coming from stakeholders, (particularly governments and companies) in up-skilling communities: "We are interested in demonstrating that we can do responsible mining … we want mining to provide us with wealth. However, it is not just about money, it is also about accessing other services such as education" (Government Representative, Interview, October, 2012).

However, the current approach is not effective for locals. The existing top-down development approach has become a reactive rather than a proactive and genuine approach. The research indicates that communities no longer need more initiatives in the form of just education but need a combination of this primary asset with other forms of human capital-like entrepreneurship, employment opportunities and income generation (Franco 2014). This will result in the realisation of communities' potential and the development of sustainable livelihood options different from mining and more accurate to Antioquia's context.

A second factor concerns the limited understanding of entrepreneurship for community wellbeing as an approach to development. Community wellbeing is about

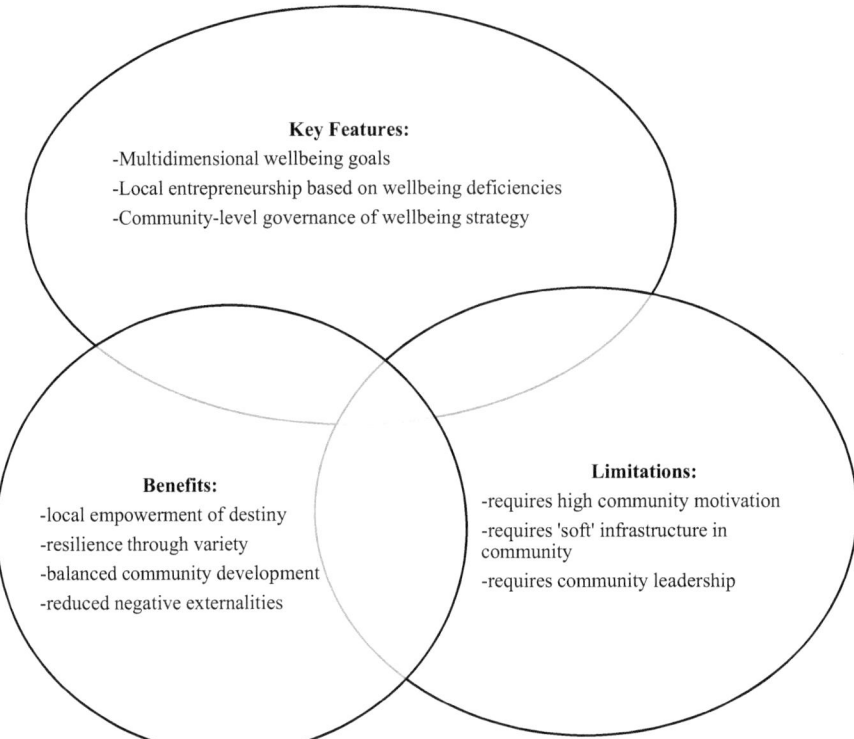

Fig. 13.2 Entrepreneurship for community wellbeing

assisting communities to meet their wellbeing needs as well as assisting them to evolve and improve their living conditions. However, the mining industry's narrow understanding of the potential of entrepreneurship for improved community wellbeing diminishes the possibilities of achieving locally relevant development: "We do not implement capacity-building initiatives different from mining because we do not have direct relationship with other sectors … we think that tourism or agriculture are sectors in which we do not fit. For this reason we focus on education for mining" (Table 13.2, Row, 3) (Senior Corporate Representative). Thus, this narrow understanding appears to be strong at the corporate level. However, in an interview with a miner and former educator, he strongly criticised the existing system and expressed his concern regarding community sustainability: "communities in the region find it very difficult to become self-sustainable. I am a miner but I do not think mining can be a driver for development" (Community leader, Interview, February, 2014). Another community member commented: "I think in the coming years this town will disappear" (Table 13.2, Row 1).

Despite the large number of programmes in place, most of them target employed miners, and only a few involve the broader community. This is a third factor that leads to the irrelevance of many approaches to community wellbeing in Antioquia. "A domestic mining company operating in the region is currently building capacity of 500 miners but these actions have been wasted or misused … the domestic company has spent much money in unnecessary mining training for us" (Local Miners, Focus Group, October, 2012). "(These) initiatives include, but are not limited to, partnerships between mining companies and tertiary institutions to develop undergraduate practicum projects and mining student tours" (Mining Engineering Students from Antioquia, Focus Group, October, 2012).

However, if these initiatives do not go beyond a mining-orientated approach that helps students become active actors in achieving their development aspirations, these initiatives will end up in dreams of development (Murray 1997). Similarly, mining companies in partnership with local governments are running initiatives to build high school students' and locals' capacity in mining practices. Although these actions have helped communities get a broader understanding of the industry, they have not met the community's expectations yet and therefore have been incapable to foster relevant community wellbeing.

Women also have been displaced and had to turn to figuring out their livelihood as mining employment seems to cater mainly for men. "Our organisation was created from the need to build different livelihood options. Women in this town did not have any value … We were labelled either as witches and/or prostitutes. We have been provided with some training by government organisations … A company donated us three machines … that was long time ago. There are small mines we have some contracts with and we also get some contribution from the government but nothing else apart from that" (Table 13.2, Row 3), says a community member.

13.4.2 Risaralda

Risaralda is a region located in the Colombian Andes mountain range. This geographical area is one of the most active producers of minerals and metals in Colombia. According to the census, Risaralda hosted a population of 859,666 people in 2005 (DANE 2005). Out of the total population, 665,104 people inhabited urban areas, whereas 194,562 were located in rural areas. 51.3% of the population was female and 48.7% male. Statistics also show that Risaralda hosted 230,532 households in urban and rural areas. With the increased mining operations, community enterprise has been more successful in gaining traction in delivering development outcomes, and communities are more receptive to private capital as an agent of development. "The company provided us with training in coffee making and coffee tasting. We have also requested support from companies and governments resulting in the implementation of jointly funded …. actions" (Community Leaders, Focus Group, October, 2012). Research findings indicate that this approach has positive results for communities (Buitrago Franco 2014). "We have been involved in several … initiatives delivered by different organisations such as Artesanias de Colombia, government-funded educational institutions and NGOs. We have experienced the benefit of these initiatives. It is reflected in the products that we design" (Community Leader, Jewellery CBO, Interview, November, 2012).

Governments, the private sector and civil society have joined in efforts to develop relevant approaches to community wellbeing that tackle key community issues. Following Mount and Mulc (2007) partnerships can foster community development. "In partnership with a government-funded educational institution we are assisting [communities] with some resources to begin their own plantations (in coffee, plantain, etc)" (Corporate Representative, Interview, October, 2012). Companies have also partnered with schools to provide children with initiatives such as reading and writing workshops (Franco 2014). Collaboration processes are characterised by active community engagement. Mining is part of Risaralda's cultural and political/economic life; however, the recent escalation of mining operations undertaken mainly by international companies has not prevented communities in Risaralda from creating alternative livelihoods and benefiting from resource development. Recent development of mining and exploration projects has positioned this activity as one of the main economic activities in the region. Operations have been undertaken by Canadian and domestic exploration and mining companies. Metals production in Risaralda represented 6.71% of Colombia's total production in 2005 (UPME 2005). Effective stakeholder collaboration processes along with a long history of entrepreneurship and economic benefits derived from mining have driven community wellbeing. Existing exploration and mining projects have helped communities to foster entrepreneurship and achieve their development aspirations:

> I have been working in mining since I was 7 years old. The company has provided us with some resources to participate in a dress-making course. I asked the company for … initiatives in which we women could get some knowledge to sustain ourselves in the long-term. The company, in partnership with a government VET institution, implemented this initiative. However, we need more of these initiatives in Risaralda. Hopefully there are more coming up. (Risaralda Community Leader, November, 2012)

This has been possible due to some governance processes and factors in place that have allowed companies to play a strong role in promoting sustainable community development leading to community wellbeing.

Entrepreneurship for Community Wellbeing in Risaralda

While many of the current approaches to community wellbeing face some challenges in Antioquia, key findings of this research showed that the approach to wellbeing in Risaralda has helped locals develop coping capacities to deal with mining-induced changes. A more coordinated and integrated approach to community wellbeing is taking centre stage in Risaralda. The existing approach is based on community needs and puts entrepreneurship at the centre of community development. The existing developmental model has also been able to meet community development aspirations and help communities evolve despite the adverse impacts derived from the escalation of mining operations. A senior Government representative stated:

> We do not want mining-dependent communities. Instead, we need to develop sustainable supply chains and entrepreneurship amongst community members. People usually think competitiveness is an overnight process. This is a false expectation. It is a long-term commitment that will allow us to forge more sustainable communities. (Table 13.3, Row 1)

Mining will eventually affect the livelihood of farmers, agricultural labourers and other rural workers whose jobs currently depend on agriculture. Farmers are choosing mining over agriculture which might displace traditional livelihood options in the long-term. However, it is mainly due to active community participation, broader understanding of community wellbeing as an approach to development and good governance that stakeholders are becoming more successful in gaining traction in delivering development outcomes. Active participation in decision-making has also helped some communities to proactively request initiatives intended to positively transform their assets and help cope with potential mining impacts. Collective decision-making and consensus are key drivers for an integrated approach (Maalouf 2014) to wellbeing. This approach has also helped many communities achieve their development goals and realise their entrepreneurial potential. It has also helped companies to produce in a more responsible manner. A community member remarked:

> Locals have a strong sense of community organisation. They own agriculture-based community associations that stimulate the local economy and foster employment. For example, they own associations for blackberry and coffee production and commercialisation. (Table 13.3, Row 3)

Based on Risaralda's community aspirations and expectations, corporations and governments have embarked on social responsibility agendas in which entrepreneurship is a key component. This has had positive impacts on locals, as these initiatives are more in tune with local circumstances. Generally, entrepreneurship is itself perceived as an approach to development by stakeholders involved. Community consultation in relation to needs and expectations is highly appreciated in the development of these initiatives, resulting in immediate benefits for locals. For example, local coffee and jewellery producers, as well as women leaders, have already experienced the benefits of these initiatives: "We have trained in jewellery design … the company has also provided us with some financial assistance to attend international fairs so that we can promote and sell our products" (Community Members, Interview, October, 2012). Evidence shows that if allocated effectively, financial aid can transform livelihoods positively (Akudugu 2011).

There is a close relationship between good governance and the effectiveness of these initiatives. Collaboration for entrepreneurship amongst stakeholders has been a driver to enhance resilient and developed community assets (Missens et al. 2007). Governance is not restricted just to the role of government but also involves other parties such as the private sector (Davies 2005) and civil society (Mayer and Knox 2010). It is part of the role of these three parties to help communities develop asset-based adaptation strategies to respond to mining impacts. Mostly, it is their responsibility to assist communities to meet their wellbeing needs and improve their living conditions. Lessons can be learnt from the Risaralda case. Approaches need to be further developed according to community expectations and needs. This bottom-up (community-focused) rather than top-down (mining-focused) approach to community wellbeing is already having significant effects in many of Risaralda's communities and equipping them with more resilient assets, helping them to realise their entrepreneurial potential and sustaining their livelihoods in the long-term. However, if existing top-down approaches keep being implemented, this will not only be detrimental for locals but also for companies operating in mining regions as it will escalate community discontent and resentment.

Driven by strong links of reciprocity and empathy, locals more often partner with other civil groups or local stakeholders to develop community associations. The importance of becoming self-sustainable reinforces their desire for a high level of economic independence and points to a need to enhance these organisations. This will not only strengthen their capacity to respond to potential mining impacts but also assist them in the generation of local employment. To date community-based organisations have contributed strongly to protecting and enhancing entrepreneurial assets in Risaralda. Community organisations provide employment opportunities to 616 households, and they expect to play a stronger role in the coming years. A representative of one of these CBOs said: "At the present time we are benefiting 11 households through employment generation … In the future it will not only be 11 but 20, 30, 50 families benefited from this organisation" (Community Leader, Jewellery CBO, Interview, November, 2012).

Three community organisations in particular have made important contributions to employment generation in the region. The Association for Coffee Production and Commercialization has benefitted 499 households with employment, while the Associations for Blackberry Production and Commercialization and Jewellery Production and Commercialization have benefitted 106 and 11 households, respectively. Community members of these organisations have participated in entrepreneurship and education initiatives led by the government. According to the Ministry of Education, 76% of former participants in these actions have been able to access gainful employment opportunities from these initiatives (Ministerio de Educacion Nacional 2014).

State actors have also taken part in fostering social sustainability by supporting these organisations. The regional government often encourages exploration and mining companies to engage with CBOs to develop entrepreneurial assets. In response, companies have committed to supporting the community endeavours; however, they also have a particular interest in linking community organisations' work with company goals. Local authorities acknowledge that the mining industry plays a key role, and actions need to be undertaken to generate mining-related employment opportunities. Nevertheless, companies willing to employ community members also need to contribute to community wellbeing: "We want mining companies to operate socially responsible. We want most of their employees to be locals. Employees should also be equipped with suitable conditions to perform their jobs with high safety standards" (Table 13.3, Row 2) (senior local government representative, November, 2012).

Fostering sustainable development (Rakodi and Lloyd-Jones 2002; Franco 2014) and wellbeing is a long-term process. Hence, if mining is to support the long-term development of work and entrepreneurship opportunities outside mining, the industry needs to come up with strategies to protect community assets and meet its expectations from the start of advanced exploration through mining development stages of the mining cycle and also after mine closure. Corporate representatives state that "during the exploration stage (they) are hiring locals to support current drilling projects ... The local community board gives (the company) potential workers' CVs and (the company) selects them through an internal process" (Corporate Representative, Interview, October, 2012).

Other corporate initiatives for employment generation are based on the implementation of bottom-up approaches for entrepreneurship. However, these initiatives are still in a very early stage, and so their impact on communities cannot yet be measured. For example, dress-making entrepreneurship programmes in which female leaders are participating are one of the existing bottom-up initiatives in place: "This training will help us get employed either by the company or educational institutions to make the company workers' or school students' uniforms" (Community Representatives, Focus Group, November, 2012).

Stakeholders in Risaralda have come to the realisation that natural resource extraction has compromised physical and environmental wellbeing and therefore communities should be compensated for by a gain in other components of wellbeing. Such compensation has mainly occurred in improvements in economic wellbeing with a focus on economic diversification. This has led to positive livelihood transformations, as locals have been able to generate income from industries other than mining.

Quinchia is a small town with a strong sense of community. Despite the long history of conflict in this town, locals have become psychologically resilient. They are supportive of one another and very often gather together to discuss issues relevant to all. These after-work informal catch-up meetings take place at the main square or in the public park. In those meetings, small groups of company staff, community members and government representatives get together to discuss topics that have an impact on the community. Observations undertaken during fieldwork show that the agreements that resulted from these informal meetings were often more effective than those formally established and contributed to improve community wellbeing in other dimensions, including psychological, material, cultural and environmental.

Investments in infrastructure have advanced material wellbeing, fostered long-term improvements in livelihoods and physically equipped educational institutions for the promotion of cultural wellbeing in the form of education: "community members have been given education in human relations, food security, family guidance and conflict resolution" (Community Representatives, Focus Group, November, 2012). Locals have a strong perception that education in social responsibility has assisted them cope with shocks like internal conflict. One of the respondents agrees that being exposed to this initiative has helped him and the community organisation he represents "prevent conflicts within (the) association" (Community Leader, Interview, November, 2012).

Abstracting from the Antioquia case, Fig. 13.1 shows the key features, benefits and limitations of a top-down industry-based approach to sustainable development. Such an approach delivers mining-related benefits in the form of skill training, employment opportunities, local boost to retail trade from miners and increased government tax revenue for more local spending. However, interviewees strongly lamented the mining industry-centred nature of this development approach and how benefits are offset by costs to other aspects of wellbeing. This approach stays away from the responsible production practices associated with SDG 12. The wider business development of the region can be neglected, long-term prosperity made vulnerable to the fortunes of mining, lack of local leadership and empowerment to be in control of their destiny and too much top-down imposition. There is also a rivet effect (Martin and Sunley 2015) as any adversity experienced by the mining sector can have knock-on adverse consequences for other dependent sectors such as retail.

Figure 13.2 then displays the key features, benefits and limitations of the entrepreneurship for community wellbeing approach. Based on the case data, three key features of this approach are salient: (1) multidimensional wellbeing as the overarching framework for sustainable community development, (2) entrepreneurship based on wellbeing deficiencies and (3) community-level governance of wellbeing strategy. The Risaralda case illustrates the greater community support for this approach and benefits in terms of resilience, community empowerment and control and development for community wellbeing rather than just mining. From a resilience perspective (Martin and Sunley 2015), an industry-based approach reduces modularity by building tight coupling of parts of a community to a mining core.

In contrast, a wellbeing approach increases modularity by increasing the diversification of business creation, thus not locking the community's eggs into one basket. Davies and Tonts (2010) argue that the more diverse a region's economy, the more socio-economically resilient it will be (also Gray et al. 2014). Importantly also, a wellbeing approach builds emotional resilience by not centring the community's psyche on economic factors alone as constituting its quality of life. This approach leads to better outcomes for community members from a perspective which recognises that wellbeing is about having, doing and being across multiple components including economic, social, cultural, environmental, psychological, spiritual, material and physical. Entrepreneurship for community wellbeing though requires high levels of community engagement to make it work. This assumes a willingness and energy in the community to implement a wellbeing development agenda in connection with the SDG 12.

13.5 Impact Sustainability: Final Remarks

Our study set about to explore how wellbeing may be used to guide sustainable development in small towns/communities. Our comparative case study showed that when communities have an overarching multidimensional vision for community wellbeing, they fare better in the longer term than those who are less mindful of the full range of components which constitute human wellbeing. Setting this overarching vision provides the framework for guiding entrepreneurship initiatives by helping to identify wellbeing priorities in the community and where entrepreneurial goods and services for wellbeing are most needed. In this way, communities develop based on the best interests of multidimensional wellbeing and learn to connect local business development to these community aims. Moreover, these communities develop capabilities in understanding how the various components of wellbeing (e.g. social, environmental, economic, spiritual) interact and thus can better synchronise different types of entrepreneurial business.

Most specifically, our research findings demonstrate that stakeholders who embark on community wellbeing development rather than industry-based capacity building are more successful in delivering sustainable outcomes in resource regions. This is similar to the findings elsewhere such as Dana et al.'s (2014) study of two French villages and the impact of tourism, Gray et al.'s (2014) case study of Women in Business Development Incorporated in Samoa and Korsgaard and Anderson's (2011) study of the sustainable living project of Friland. All cases demonstrate the need for development to adopt a multidimensional wellbeing approach.

The Risaralda case illustrates that a community wellbeing approach with a focus on entrepreneurship is helping communities become more resilient to potential mining impacts. In Risaralda, examples can be seen where mining has been combined with jewellery design and trade entrepreneurship initiatives; a livelihood option that is currently adding value to resource extraction in the region. This has been possible due to active community participation, broader understanding of entrepreneurship for community wellbeing and good governance. Conversely, the existing top-down or mining approach in some locations of Antioquia is often substituting for traditional livelihood options such as agriculture and other entrepreneurship initiatives relevant for local communities. This is not only compromising community sustainability but also jeopardising the mining industry's opportunities to make a strong contribution at the community level and produce in a responsible manner. This is mainly due to overinvestment in irrelevant initiatives, limited understanding of entrepreneurship for community wellbeing as an approach to development and approaches to community wellbeing that do not reach the broader community but only the mining industry and potential employees.

In this context, entrepreneurship for community wellbeing adds to traditional discussions heralding the benefits of entrepreneurship to developing economies (Fischer et al. 2018) and to the achievement of SDGs 3 and 12. If not undertaken with all aspects of wellbeing under consideration, then entrepreneurship can encourage business types that create negative externalities such as increased inequality, environmental degradation, indigenous colonisation, "diseases of affluence" (Ezzati et al. 2005; McKeown 1988) and the proliferation of a consumer culture. A wellbeing approach and our findings also add to discussions about regional resilience (Davies and Tonts 2010; Martin and Sunley 2015) by extending the focus beyond economic resilience to include multiple aspects of wellbeing resilience. Even if adverse economic shocks occur, the effects are not just economic but affect multiple aspects of wellbeing. A wellbeing-conscious approach can render communities less vulnerable to external shocks through having built wellbeing capital or slack that can buffer. Experience with wellbeing can aid recovery. Without this stock of wellbeing, communities are more vulnerable to more severe adverse effects across multiple wellbeing components.

Entrepreneurship is here itself important to community and regional resilience and overall sustainability. Developing a path dependence based on entrepreneurship builds innovation capacity. The latter emphasises newness, change and adaptation. Hence, a community and region improves its resilience as its history and capabilities for entrepreneurship and innovation enable positive adaptations to failed sectors. A wellbeing approach then encourages variety and a history of path creation, thereby steering away from too much path dependence (Martin and Sunley 2006).

Future research can further explore how a wellbeing approach to entrepreneurial development leads to more resilient communities and regions with less negative externalities than those regions adopting a purely economic development logic. Included is the need to develop a measure of multidimensional wellbeing across the eight components conceptualised in this paper. A score for community wellbeing on each of the eight components could be based on averaging scores of individuals. Such a test could also serve as an ideal type where a total maximum wellbeing score could serve as the barometer against which one could measure a total wellbeing score for the community. Research could also test if a portfolio approach to community entrepreneurship achieves greater multidimensional wellbeing than existing approaches which prioritise economic development. Research is also needed into the governance required to oversee the portfolio of enterprises in the community, set community wellbeing goals and develop entrepreneurship.

Social enterprises also have a role in entrepreneurship for community wellbeing (Peredo and Chrisman 2006; Peredo and McLean 2006). Social entrepreneurship can help to prioritise aspects of wellbeing in their portfolios (e.g. a focus on contributing to social wellbeing or environmental or spiritual). However, as with for-profit entrepreneurship, there are benefits to the community-level coordination of social entrepreneurship based on wellbeing deficiencies in the community. Such coordination keeps things focused on both community and wellbeing. Social entrepreneurship literature could benefit by studying how wellbeing could become an overarching goal and point of difference with traditional profit-maximisation entrepreneurship. This includes study of the interactions between various components of wellbeing and how these play out at a community level.

Corporations stand to bring useful benefits to the communities in which they operate. Importantly though, such companies would benefit from a wellbeing-based approach to development investment and responsible production. Investing in the entrepreneurship skills of locals needs to be complemented with investments in leadership and governance from a wellbeing perspective.

In this context, the implementation of an entrepreneurship for community wellbeing approach coupled with other initiatives, such as food security and farming programmes, is highly recommended due to the likelihood that mining impacts will have adverse implications for locals. This will allow activities like mining to coexist with other industries and activities relevant to the local economy and will add value to other livelihood options relevant to communities, as well as helping to ensure that improved community wellbeing remains after the mines have closed.

It is advisable to build community capacity for wellbeing-based approaches to entrepreneurship in developing resource regions of Latin America. Existing approaches to community wellbeing can cause tensions at the community level. Current processes intended to up-skill informal miners and the broader community need further development in order to help locals enhance their capacity to obtain gainful employment and therefore foster community wellbeing. However, the participation of multiple stakeholders is vital to foster entrepreneurship based on community wellbeing. This emerging view is likely to deliver relevant employment opportunities and prepare locals for adverse impacts derived from the escalation of resource extraction.

Entrepreneurship for community wellbeing is an essential asset for enhancing other human capital such as income and employment. Existing community wellbeing approaches are mainly mining oriented, particularly in the Antioquia case, which is becoming a limitation for locals whose livelihood options and development aspirations are not always directly linked to the mining industry. Thus, it is recommended that entrepreneurship for community wellbeing education be secured for locals in non-mining subjects. These subjects need to enhance locals' abilities to take up entrepreneurship or employment opportunities that are offered outside the mining sector.

The dearth of both gainful entrepreneurship options and appropriate education in this field is hindering locals in achieving their development aspirations and improving community wellbeing. This is leading to diminished opportunities for income generation and preventing them from employing other human capital assets they may have access to. Entrepreneurship for community wellbeing should be adopted by multiple stakeholders at all levels of governance and involve private organisations and civil society. This will help stakeholders to immerse themselves in the contexts of the most vulnerable communities in developing resource regions and tackle the structural disconnects (extreme poverty, inequality and conflict) that prevent communities in Colombia and elsewhere in Latin America from fostering sustainable wellbeing and following their dreams of sustainable development.

Acknowledgements One of the authors would also like to thank Les and Lee Boby and Reta and Albert Basnett for the inspiration to work with communities.

References

Akudugu MA (2011) Rural banks' financial capital and livelihoods development of women farmers in Ghana. J Enterp Commun 5(4):248–264

Anderson AR (2000) Paradox in the periphery: an entrepreneurial reconstruction? Entrep Reg Dev 12(2):91–109

AngloGold Ashanti (2008) Colombia: country report. Retrieved March 25, 2014 from http://www.anglogold.com/NR/rdonlyres/1103A1F1-D744-4220-B726-6E48242F7039/0/Colombia2008.pdf

Bacq S, Janssen F (2011) The multiple faces of social entrepreneurship: a review of definitional issues based on geographical and thematic criteria. Entrep Reg Dev 23(5–6):373–403

Banerjee SB (2011) Voices of the governed: towards a theory of the translocal. Organization 18(3):323–344

Baumgartner D, Pütz M, Seidl I (2013) What kind of entrepreneurship drives regional development in European non-core regions? A literature review on empirical entrepreneurship research. Eur Plan Stud 21(8):1095–1127. https://doi.org/10.1080/09654313.2012.722937

Bleys B (2012) Beyond GDP: classifying alternative measures for progress. Soc Indic Res 109:355–376

Braun V, Clarke V (2006) Using thematic analysis in psychology. Qual Res Psychol 3(2):77–101

Burroughs JE, Rindfleisch A (2002) Materialism and well-being: a conflicting values perspective. J Consum Res 29(3):348–370

Cahn M (2008) Indigenous entrepreneurship, culture and micro-enterprise in the Pacific Islands: case studies from Samoa. Entrep Reg Dev 20:1–18

Callicott JB (1996) Environmental wellness. Lit Med 15(1):146–160

Cámara de Comercio de Medellín para Antioquia (2010) Minería: Potencial para Iniciativas Cluster en Antioquia Documento Comunidad. Cámara de Comercio de Medellín para Antioquia, Medellín

Cook S (2006) Structural change, growth and poverty reduction in Asia: pathways to inclusive development. Pol Rev 24(1):51–80

Cummins RA, Weinberg MK (2013) Subjective wellbeing – multi-item measurement: a review. In: Glatzer W (ed) Global handbook of wellbeing and quality of life. Springer, Dordrecht

Dana LP, Anderson BR, Meis-Mason A (2009) A study of the impact of oil and gas development on the Dene first nations of the Sahtu (Great Bear Lake) region of the Canadian Northwest Territories (NWT). J Enterp Commun 3(1):94–117

Dana LP, Gurau C, Lasch F (2014) Entrepreneurship, tourism and regional development: a tale of two villages. Entrep Reg Dev 26(3–4):357–374

DANE (Departamento Administrativo Nacional de Estadísticas) (2005) Perfil Risaralda. DANE, Bogota

DANE (Departamento Administrativo Nacional de Estadísticas) (2010) Censo general 2005 – Perfil Antioquia. DANE, Bogota

Davies JS (2005) Local governance and the dialectics of hierarchy, market and network. Policy Stud 26(3–4):311–335

Davies A, Tonts M (2010) Economic diversity and regional socio-economic performance. Geogr Res 48:223–234

Diener E (2009) Assessing well-being: the collected works of Ed Diener. Social indicators research series, 39, Springer, Dordrecht

Diener E, Seligman MEP (2010) Beyond money: toward an economy of wellbeing. Psychol Sci 5(1):1–31

Diener E, Ng W, Harter J, Arora R (2010) Wealth and happiness across the world: material prosperity predicts life evaluation, whereas psychological prosperity predicts positive feeling. J Pers Soc Psychol 99(1):52–61

Doyal L, Gough I (1991) A theory of human need. Macmillan, London

Eade D (2007) Capacity building: who builds whose capacity? Dev Pract 17(4):630–639

Easterlin RA (1974) Does economic growth improve the human lot? Some empirical evidence. Nations Househ Econ Growth 89:89–125

Easterly W (2006) The white Man's burden: why the west's efforts to aid the rest have done so much ill and so little good. Oxford University Press, Oxford

Eisenhardt KM (1989) Building theories from case study research. Acad Manag Rev 14(4):532–550

Eisenhardt KM, Graebner ME (2007) Theory building from cases: opportunities and challenges. Acad Manag J 50(1):25–32

Ezzati M, Hoorn SV, Lawes CMM, Leach R, James WPT (2005) Rethinking the "diseases of affluence" paradigm: global patterns of nutritional risks in relation to economic development. PLoS Med 3:e133

Fischer BB, Queiroz S, Vonortas NS (2018) On the location of knowledge-intensive entrepreneurship in developing countries: lessons from São Paulo, Brazil Entrepreneurship & Regional Development, pp 1–27

Franco I (2014) Building sustainable communities: enhancing human capital in resource regions. PhD diss., The University of Queensland

Freire P (1970) Pedagogia do Oprimido (23 a Reimpressao ed.). Paz e Terra S/A, Sao Paulo

Gasper D (2004) Human well-being: concepts and conceptualizations. Discussion paper 2004/06. Helsinki, Finland: United Nations University, World Institute for Development Economics Research

Gasper D (2005) Subjective and objective well-being in relation to economic inputs: puzzles and responses. Rev Soc Econ 63(2):177–206

Giddens A (1999) Runaway world: how globalisation is reshaping our lives. Profile, London

Gough I, McGregor JA, Camfield L (2007) Theorising wellbeing in international development. In: Gough I, McGregor JA (eds) Wellbeing in developing countries: from theory to research. Cambridge University, Cambridge, UK

GrancolombiaGold (2012) Map of properties. Retrieved 19, July 2012, from http://www.grancolombiagold.com/Properties/Map-of-Properties/default.aspx

Gray BJ, Duncan S, Kirkwood J, Walton S (2014) Encouraging sustainable entrepreneurship in climate-threatened communities: a Samoan case study. Entrep Reg Dev 26(5–6):401–430

Gudynas E (2011) Buen vivir: today's tomorrow. Development 54(4):441–447

Haan AD, Maxwell S (1998) Editorial: poverty and social exclusion in north and south. IDS Bull 29(1):1–9

Hallerod B, Selden D (2013) The multidimensional characteristics of wellbeing: how different aspects of wellbeing interact and do not interact with each other. Soc Indic Res 113:807–825

Hamdouch A, Demaziere C, Banovac K (2016) The socioeconomic profiles of small and medium sized towns: insights from European case studies. Royal Dutch Geogr Soc 108(4):456–471

Harrison AE (2006) Globalization and poverty NBER working paper 12347. National Bureau of Economic Research, Cambridge

Huggins R, Thompson P (2014) Culture, entrepreneurship and uneven development: a spatial analysis. Entrep Reg Dev 26(9–10):726–752

Kabeer N (2000) Social exclusion, poverty and discrimination – towards an analytical framework. IDS Bull 31(4):83–97

Keyes CLM (1998) Social well-being. Soc Psychol Q 61(2):121–140

Korsgaard S, Anderson AR (2011) Enacting entrepreneurship as social value creation. Int Small Bus J 29(2):135–151

Lagendijk A, Lorentzen A (2007) Proximity, knowledge and innovation in peripheral regions: on the intersection between geographical and organizational proximity. Eur Plan Stud 15(4):457–466

Lichtenstein GA, Lyons TS (2001) The entrepreneurial development system: transforming business talent and community economies. Econ Dev Q 15(1):3–20

Lichtenstein GA, Lyons TS (2010) Investing in entrepreneurs: a strategic approach for strengthening your regional and community economy. Praeger, California

Maalouf E (2014) Emerge! The rise of functional democracy and the future of the Middle East. Select Book, New York

Martin R, Sunley P (2006) Path dependence and regional economic evolution. J Econ Geogr 6:395–437

Martin R, Sunley P (2015) On the notion of regional economic resilience: conceptualization and explanation. J Econ Geogr 15:1–42

Mate K (2001) Capacity-building and policy for sustainable development networking. Minerals Energy Raw Mater Rep 16:3–25

Mayer and Knox (2010) Small-town sustainability: prospects in the second modernity. Eur Plan Stud 18-10:1545–1565

McGregor JA (2007) Researching wellbeing: from concepts to methodology. In: Gough I, McGregor JA (eds) Wellbeing in developing countries: from theory to research. Cambridge University Press, Cambridge

McKeown T (1988) The origins of human disease. Blackwell, Oxford

McMullen JS (2011) Delineating the domain of development entrepreneurship: a market-based approach to facilitating inclusive economic growth. Entrep Theory Pract 35(1):185–193

McWade W (2012) The role for social enterprises and social investors in the development struggle. J Soc Entrep 3(1):96–112

Ministerio de Educación Nacional (2014) Articulación de la educación no formal y contínua con la educación formal y el sistema nacional de formación para el trabajo. Accessed 17 Apr 2014. http://www.mineducacion.gov.co/cvn/1665/article-107689.html

Misoczky MC (2011) World visions in dispute in contemporary Latin America: development x harmonic life. Organization 18(3):345–363

Missens R, Paul Dana L, Anderson R (2007) Aboriginal partnerships in Canada: focus on the Diavik diamond mine. J Enterp Commun 1(1):54–76

Molina-Escobar JM, Restrepo-Baena OJ (2010) Colombian mining sustainability. Dyna-Colombia 77(161):149–151

Monni S, Pallottino M (2015) A new agenda for international development cooperation: lessons learnt from the buen vivir experience. Development 58(1):49–57

Mount J, Mulc H (2007) Community economic development through partnerships: the case of the Sudbury regional business Centre. J Enterp Commun 1(4):337–351

Murray P (1997) Dreams of development: Colombia's national school of mines and its engineers, 1887–1970. The University of Alabama Press, Tuscaloosa

Ng W, Diener E (2014) What matters to the rich and the poor? Subjective well-being, financial satisfaction, and postmaterialist needs across the world. J Pers Soc Psychol 107(2):326–338

Nicholls A (2010) The legitimacy of social entrepreneurship: reflexive isomorphism in a pre–paradigmatic field. Entrep Theory Pract 34(4):611–633

Novotny TE (2005) Why we need to rethink diseases of affluence. PLoS Med 2(5):e104

OECD (2013) Economic Well-being. In: OECD framework for statistics on the distribution of household income, consumption and wealth

Pargament KI, Sweeney PJ (2011) Building spiritual fitness in the army. American Psychologist, January, pp 58–64

Patrizii V, Pettini A, Resce G (2017) The cost of well-being. Soc Indic Res 133:985–1010

Pendall R, Foster KA, Cowell M (2010) Resilience and regions: building understanding of the metaphor. Camb J Reg Econ Soc 3(1):71–84

Peredo M, Anderson R (2006) Indigenous entrepreneurship research: themes and variations. Int Res Bus Discipl 5:253–273

Peredo AM, Chrisman JJ (2006) Toward a theory of community-based enterprise. Acad Manag Rev 31(2):305–328

Peredo AM, McLean M (2006) Social entrepreneurship: a critical review of the concept. J World Bus 41(1):56–65

Peredo AM, McLean M (2013) Indigenous development and the cultural captivity of entrepreneurship. Bus Soc 52(4):592–620

Perry B (2009) Non-income measures of material wellbeing and hardship: first results from the 2008 New Zealand living standards survey with international comparisons. Ministry of Social Development, Wellington

Peterman AH, Fitchett G, Brady MJ, Hernandez L, Cella D (2002) Measuring spiritual well-being in people with cancer: the functional assessment of chronic illness therapy – spiritual well-being scale (FACIT-Sp). Ann Behav Med 24(1):49–58

Pike A, Rodriguez-Pose A, Tomaney J (2007) What kind of local and regional development and for whom? Reg Stud 41(9):1253–1269

Rakodi C, Lloyd-Jones T (2002) Urban livelihoods: a people Centred approach to reducing poverty. Earthscan Publications, London

Ratten V, Welpe IM (2011) Community-based, social and societal entrepreneurship. Entrep Reg Dev 23(5–6):283–286

Ryff CD (2014) Psychological well-being revisited: advances in the science and practice of eudaimonia. Psychother Psychosom 83:10–28

Seligman MEP (2011) Flourish: a visionary new understanding of happiness and wellbeing. Free Press, New York
Sen A (1979) Equality of what? Standford University, Oxford
Shane S, Venkataraman S (2000) The promise of entrepreneurship as a field of research. Acad Manag Rev 25(1):217–226
Shankar R, Shah A (2003) Bridging the economic divide within countries: a scorecard on the performance of regional policies in reducing regional income disparities. World Dev 31(8):1421–1441
Sharmer O, Kaufer K (2013) Leading from the emerging future, 1st edn. Berrett-Koehler Publishers, San Francisco
Sigler T (2015) Case study areas [map]. Brisbane, January, 2015
Sistema de Información Minero Energético Colombiano (2010) Distritos Mineros. Accessed 26 Mar 2014. http://190.90.10.157/Distritos%20Mineros/
Stiglitz JE, Sen A, Fitoussi J-P (2010) Mismeasuring our lives: why GDP doesn't add up. The report by the commission on the measurement of economic performance and social progress. The New Press, New York
Surborg, B. 2012 *The Production of the World City: Extractive Industries in a Global Urban Economy*. PhD diss., the University of British Columbia. Accessed from https://circle.ubc.ca/bitstream/handle/2429/40719/ubc_2012_spring_surborg_bjoern.pdf?sequence=1.
Toedtling F, Kaufmann A (2001) The role of the region for innovation activities of SMEs. Eur Urban Reg Stud 8(3):203–215
Torjman S (2004) Culture and recreation: Links to well-being. Caledon Institute of Social Policy, Ottawa
UN (2018) About the United Nations sustainable development goals. Retrieved from https://www.un.org/sustainabledevelopment/sustainable-development-goals/2018, September 5
UNDP (United Nations Development Program) (2011) Capacity development. Retrieved June 23, 2011 from, http://www.beta.undp.org/undp/en/home/ourwork/capacitybuilding/approach.html
UPME (Unidad de Planeación Minero Energética) (2005) Distritos mineros: exportaciones e infraestructura. Bogota: Accessed 26 Mar 2014. http://www.upme.gov.co/Docs/Distritos_Mineros.pdf
UPME (Unidad de Planeación Minero Energética) (2014) Indicadores de la Minería en Colombia. Accessed 14 Jan 2015. http://www.upme.gov.co/Docs/Plan_Minero/2014/Indicadores%20de%20la%20Miner%C3%ADa%20en%20Colombia.pdf
Villalba U (2013) Buen Vivir vs development: a paradigm shift in the Andes? Third World Q 34(8):1427–1442. https://doi.org/10.1080/01436597.2013.831594
WHOQOL Group (1998) Development of the world health organization WHOQOL-BREF quality of life assessment. Psychol Med 28(3):551–558
Williams N, Vorley T (2014) Economic resilience and entrepreneurship: lessons from the Sheffield city region. Entrep Reg Dev 26(3–4):257–281
Williams N, Vorley T (2015) The impact of institutional change on entrepreneurship in a crisis-hit economy: the case of Greece. Entrep Reg Dev 27(1–2):28–49
World Bank (1996) World development report. Oxford University Press, Washington, DC
Yin RK (2003) Case study research: design and methods. Sage, Thousand Oaks
Zahra SA, Newey LR, Li Y (2014) On the Frontiers: the implications of social entrepreneurship for international entrepreneurship. Entrep Theory Pract 38:137–158

Chapter 14
SDG 13 Climate Action

Climate Education: Identifying Challenges to Climate Action

Isabel B. Franco, Rosemarie Tapia, and James Tracey

Abstract There is an increasing consensus that stakeholders at education institutions, specifically educators and students, face enormous challenges in translating climate education into effective action. Therefore, a better understanding of these issues – both within and beyond education institutions – is paramount. This chapter addresses such issues within education institutions, including teachers and students misconceptions regarding climate change, behavioural issues and social considerations such as peer pressure. This investigation worked to critically analyse issues with the current climate education scheme, whilst identifying key areas in which improvements should be made to effectively promote climate action and contribute to the achievement of the sustainable development goal related to climate action (SDG 13).

Keywords Climate change · Sustainability · SDG 13 climate change · Education · Impact

---

I. B. Franco (✉)
Institute for the Advanced Study of Sustainability, United Nations University Shibuya-ku, Tokyo, Japan

Australian Institute for Business and Economics, The University of Queensland, Brisbane, Australia
e-mail: connect@drisabelfranco.com

R. Tapia
Exceptional Women in Sustainability – eWisely, Sydney, Australia

J. Tracey
Faculty of Engineering, University of New South Wales, Sydney, Australia

14.1 Introduction

Globally, awareness of climate education as a key component of sustainable development is increasing. Yet, there is still limited understanding regarding global climate precepts and challenges which impede many key actors, including educators and students from promoting climate education (Mahat et al. 2016). The transformation of the current education system to fully integrate climate education into curriculums and practice is essential. Such a change will hopefully aid future generations in realizing sustainable development, through inspiring the next generation to become active participants in forging a more sustainable future for all. In this context, climate education needs to be taken seriously to prepare students and educators to face pressing climate challenges. In doing so, the international community has embarked on a global campaign to 'embed education into plans and efforts to address sustainable development and integrat[e] sustainable development into education institutional priorities, curricular and pedagogy' (Tilbury 2015).

As a result of this investigation, it is clear that both educators and students should create incentives which foster sustainable consumption and behaviours through enabling new technologies and climate considerate extracurricular activities. Furthermore, they should also work to develop curriculums focused on scientific research and local climate needs whilst developing teaching methodologies for climate issues which provide students with relevant practical experience. Finally they should work to foster collaboration between schools and both public and private stakeholders for the development of a climate-focused curriculum and the allocation of resources to support the transition towards climate action. This chapter identifies a series of challenges faced by education institutions in translating climate education into action and provides policy recommendations for addressing these challenges.

14.2 Literature Review

14.2.1 Climate Education to Action: Hindering and Fostering Factors

Yet, how students and key actors can move forward from climate education to action remains an issue inside and outside education institutions. Teachings on 'responsible consumption' and '3R reutilization, reuse and recycle' integration in the curriculum, for example, have delivered poor results in practice (Mahat et al. 2016; Sidiropoulos 2014). Other themes such as 'disaster risk management' have mainly been included in business curriculum overlooking other fields equally relevant (Apronti et al. 2015; Brundiers 2017; Herrera 2016; Naoufal 2014; Shaw 2014). Curriculum exploring water stream-climate nexus targets mainly coastal areas, and issues around 'cleaner production' and 'greenhouse gas emissions' (Colliver 2017) are still perceived as abstract themes. Due to increasing climate challenges and scarce clarity on climate change, this chapter increases our understanding of factors needed to operationalize climate education.

Table 14.1 Perceptions on climate education/action

Type	Factor	Actor	Hindering factors	Fostering factors	Policy recommendations
Internal factors	Lifestyles	Students	Resistance to sacrifice unsustainable lifestyles and behaviours (e.g. diet, clothing)	Creating an identity around sustainable lifestyles (e.g. second-hand shopping and vegetarian food consumption)	Promoting incentives to change sustainable lifestyles and behaviours
			Lack of control to make sustainable decisions		
	Peer pressure	School peers	Pressure from fashion trends (friends and music, amongst others)	Linking climate policy to young culture and values	Use of technology for climate change, appealing to the youth (e.g. Tinder profile for SDGs)
	Knowledge on climate change	Educators	Lack of understanding of climate change precepts and challenges	Fostering connections between climate education and action	Fostering first-hand experience to encourage students to engage in sustainable behaviours
			Climate change is perceived as a complex concept	Train trainers on climate change	Ongoing skill needs assessment on climate education
			Climate change is not a priority in the curriculum	Encouraging curriculum development on climate change	Integrating applied research into climate curriculum development
			Educators' scepticism on climate change		
			Limited educators' capacity to teach climate change and engaging in innovative methodologies for climate education/action		

(continued)

Table 14.1 (continued)

Type	Factor	Actor	Hindering factors	Fostering factors	Policy recommendations
External factors	Family dynamics	Parents	Lack of role models to foster sustainable behaviours	Strengthening family engagement around sustainable behaviours	Family discussions on sustainable lifestyles and behaviours
			Sustainable behaviours are perceived as expensive		
	Culture	Community	Cultural constrains	Value shift on the youngest generation. 'What is perceived as successful is not necessarily sustainable'	Fostering community engagement though campaigns for sustainable lifestyles and behaviours
			Superimposed unsustainable values	Fostering community engagement on sustainable lifestyles	
			Lack of local engagement to foster sustainable lifestyles and behaviours		
	Policy	Market and governments	Market trends and marketing campaigns on unsustainable fast fashion and use of technology	Market and government policies prioritizing environment over economy	Resource allocation in the form of time and financial to foster sustainable change
					Implementing stronger policies on climate education
					Building on partnerships amongst key actors for climate education to action

Both internal and external factors impact on the transition from climate education to action. As shown in Table 14.1, there are three internal factors that require nuanced attention at education institutions, namely, students' lifestyles and behaviours, classmates and peer pressure and educators' knowledge on climate change. These three factors and actors are also impacted by external factors that emerge outside education institutions: parents' values and family dynamics, community and cultural constrains and policy driven by the market and/or governments. Both external and internal factors have a strong influence on climate education, and so, integrating those at the policy level and curriculum has the potential to contribute to climate action.

Hindering Factors

Hindering factors are those that prevent key actors from engaging in climate action. There is a consensus on resistance to sacrifice unsustainable lifestyles and behaviours at education institutions (see Table 14.1). This is due to the lack of incentives to change behaviours such as diet and clothing consumption patterns. Such resistance is triggered by peer pressure at schools as well as fashion trends, music, and use of technology, amongst others. Whilst students are trapped in an 'unsustainable value system' (Weber and Duderstadt 2012; Adams 2013; Waas et al. 2012 cited by Aleixo et al. 2016), educators try to integrate the global climate agenda into the curriculum with little success. The plethora of definitions and approaches to sustainability-related issues such as climate change (Suleri and Cavagnaro 2016) is interpreted by students in alignment with their own values (Nazir et al. 2011; Sidiropoulos 2014). Climate change is therefore perceived as an abstract theme at education institutions, and limited educators' capacity to teach climate change in a relevant manner has exacerbated the issue (Pfister et al. 2016).

Instead of integrating innovating methodologies in climate education, educators provide students with solved problems. Yet, climate action occurs when asking questions and encouraging students to find their own answers. Likewise, the ambiguity of the notion of climate change itself (Aleixo et al. 2016), lack of funding for sustainability (Steele 2010; Aleixo et al. 2016), limited educator training in climate change (Madueno et al. 2015; Martin and Carter 2015), hierarchy issues to freely discuss climate change matters at schools (Nazir et al. 2011) and overcrowded educators' curriculums are some other hindering factors at education institutions, to name a few.

Outside education institutions, policymakers and key actors involved need to be aware of the lack of role models at home to foster sustainable behaviours, an external factor that needs to be paid nuanced attention. Parents perceive sustainable behaviours as expensive. This barrier is also imposed by market trends and corporate misconceptions on the cost of sustainability. In addition, at the community level, cultural constrains, superimposed values and limited local engagement around sustainable practices diminish the opportunities for climate action.

Fostering Factors

Fostering factors in the transition to climate action vary, and it is all actors' responsibility to facilitate this transformation. Linking climate policy to younger generations' values has the potential to create an identity around sustainable behaviours and reduce peer pressure that impedes students from embracing climate action (see Table 14.1). At education institutions, this should be coupled with a transformation in the education system that challenges traditional structures and practices that do not support climate action (Tilbury 2015). In doing so, resource allocation for educators' capacity building on climate change curriculum in alignment with students' first-hand experiences, across all ages and levels, is essential (Sidiropoulos 2014; Colliver 2017). Outside the education system, this requires stronger family and community engagement around sustainable practices such as clothing swap and communal vegetarian food consumption to enable the value shift amongst the youngest generations. At the policy level, this needs more effective policies and partnerships that prioritize the environment over the economy (Steele 2010).

14.3 Discussion and Results

Based on educators' and students' insights on climate education to action, this chapter focuses on three challenges that prevent both sets of actors from engaging in climate action: students' consumption patterns and behaviours, peer pressure from classmates and educators' lack of knowledge on climate change. Other challenges and factors are listed in Table 14.1. These issues have a strong influence on climate education; thus addressing these explicitly in curriculum has the potential to contribute to climate action. Results presented below were obtained at a workshop led by the corresponding author at the 'First RCE Thematic Conference: Towards Achieving the SDGs'. Nearly 60 educators participated from various geographical locations globally.

Address Unsustainable Consumption Within Education Institutions
There is increasing resistance amongst students to sacrifice unsustainable lifestyles. This is due to the lack of incentives to amend negative behaviour, mainly associated with diet and clothing consumption. Furthermore, the growing gap between student values and the schooling curriculum exacerbates this issue. Educators claim that students do not identify with sustainable lifestyle, which prevents them from translating their sustainable education into the relevant climate action. Below is a short description of factors that impact on students' engagement in the achievement of SDG 13.

As a result, incentives must be created to motivate students to change their unsustainable consumption patterns and behaviours. Whilst education institutions are not responsible for behaviour outside of their walls, they can incentivize behaviours that encourage sustainable consumption. In doing so, educators and other key actors should encourage students to engage in research through first-hand experiences aimed to foster sustainability. It is therefore clear that building students' capacity such that they reconsider their consumption patterns and drift towards a sustainable lifestyle should be imbedded into both the formal and informal schooling curriculum.

Such incentives for students to unleash their potential and entrepreneurial skills for advancing climate action through addressing consumption can include climate leadership campaigns, as well as events including clothing swaps and school markets for sustainable products commercialization. Both of which work to reduce climate pressure and amend unsustainable consumption patterns. Students should also engage in vocational programmes that address climate action in their transition towards higher education or gainful employment in sustainability-related areas.

Peer Pressure

The pressure to follow cultural trends including the unsustainable use of technological hardware is becoming an increasing issue for education institutions. As students are trapped in an 'unsustainable value system' (Aleixo et al. 2016), thus far climate education has been unable to address the peer pressure to follow unsustainable cultural trends, including the desire for the newest phones, computers or clothing.

Creating an identity around sustainable behaviours and integrating innovative methodologies for climate action can assist educators and key actors in addressing these issues with students. For instance, the prevalent use of technology, whilst represented as a social issue, can also serve as an opportunity in the sustainability movement. A case in point is the 'OLIO' (The Food Sharing Revolution 2019), the Tinder-like food sharing app. Through connecting neighbours and friends with local shops and cafes, they promote food sharing and the reduction of food waste. As a result, they are using the power of technology to encourage reduced consumption for the promotion of a more sustainable future ('OLIO – The Food Sharing Revolution' 2019).

Build Capacity to Educate on Climate Change

As climate change has been assigned a plethora of definitions and approaches (Suleri and Cavagnaro 2016), students can often only relate through their own experiences and contexts (Nazir et al. 2011; Sidiropoulos 2014). As a result, climate change is often perceived as an abstract theme, which is too difficult to act upon. This opinion is principally exacerbated by educator's limited ability to effectively teach climate education (Pfister et al. 2016).

Climate action generally occurs through student inquiry regarding their first-hand experiences and encouraging learners to find their own relevant resolutions. As a result, educators should be applying innovative teaching methodologies, which promote these relevant context-driven solutions. In contrast, educators generally provide students with 'one-size-fits-all' solutions, which only widen the gap between climate education and action. Further barriers to effective climate education include the ambiguity of the notion of climate change (Aleixo et al. 2016), lack of funding for both climate and sustainability education (Steele 2010), limited educators' knowledge and training on climate change (Madueno et al. 2015; Martin and Carter 2015), hierarchy issues that prevent free discussion of 'taboo' climate-related matters including family planning (Nazir et al. 2011) and overcrowded curriculums.

Regardless of the context, educators must be equipped with essential knowledge to facilitate climate action integration into the curriculum. This will facilitate efficient student capacity building, developing practical skills to action climate education and forge more sustainable lifestyles. An early step in equipping educators to do this would be to examine existing barriers to climate action within the current curriculum. This is warranted to better understand the challenges and opportunities arising from the interactions between students, curriculum, school policy and other impeding factors in the transition from climate education to action. Other examples of issues and possible ways to overcome existing barriers are presented in Table 14.1.

As education institutions don't exist in a vacuum, educators should be given professional development opportunities to participate in extracurricular climate initiatives that engage with the broader community. This approach can enable the development of climate action curriculum relevant to students' own local climate requirements. For example, exposure to programmes such as Sandwatch will allow students to closely relate to the issues and promote the significance of climate action. Sandwatch is a volunteer network of schools and a community-based organization, which allows students and community members to learn how to monitor beach environments in the local area. This works to develop awareness regarding the fragility of their coastal environment (The Sandwatch Foundation 2018).

Furthermore, it is necessary to couple these essential actions with a complete transformation of the education system, one that challenges traditional teaching structures and practices, which hinder the promotion of climate action (Tilbury 2015). Such transformation requires heavy investment and resource allocation to identify gaps in climate knowledge and capacity building.

The development of the curriculum in line with both current scientific research and students' contextual experiences (Sidiropoulos 2014; Colliver 2017) is paramount to fostering a successful transition from climate education to action. In order to promote locally contextualized climate action, educators' training programmes should work with local public, private and non-profit actors to increase curriculum's impact on students within their specific area.

14.4 Impact Sustainability: Final Remarks

This chapter has explored several significant obstacles in the journey towards achieving SDG 13 and therefore a successful transition to climate action. However, the impact sustainability recommendations below work to facilitate this transition, such that both educators and students can learn to work and live in ways which will sustain a healthy planet for future generations (Steele 2010; Tilbury 2015).

Within education institutions, incentives must be created to help students change their unsustainable lifestyles and behaviours. In doing so, key actors should integrate innovative methodologies. For example, the predominantly unsustainable use of technology could undoubtedly shift from a sustainability concern to an opportunity.

However, the use of new technology and teaching techniques for the promotion of climate education must be supported by ongoing scientific research. Therefore, the importance of connecting educators with academia to develop a relevant climate education curriculum. Likewise, students must also become active participants in relevant research through exposure to first-hand experiences. Additionally, an ongoing assessment regarding existing sustainability skills is required to identify climate-related capacity gaps.

Regarding external stakeholders, family and community engagement will help facilitate a smooth transition towards climate action. Yet, these efforts require the allocation of significant resources, and as such it is the responsibility of policymakers to foster stronger policies promoting climate action. However, businesses and governments are increasingly recognizing the value of contributing to climate action, contributing through building on existing partnerships for investment in education institutions, specifically those in which a lack of resources is distinctly compromising climate education.

The successful integration of climate education into school curriculums is essential in order to inspire the next generation to become active participants in the formation of a more sustainable future. As such, we believe responsibility falls on both educators and students to successfully facilitate this progression.

In order to facilitate this amendment, it is the responsibility of educators to adhere to the following recommendations, and they must be equipped with the relevant knowledge required to integrate climate change into the curriculum and foster participation in extracurricular climate initiatives. Additionally they must enable climate action coordination amongst universities, the private sector and governments. Furthermore, encouraging educators to conduct applied research skills to examine existing barriers to climate action in their respective community would be highly effective.

Additionally, the authors' recommendations relevant to students include building students' capacity, such that they alter their consumption patterns and drift towards more sustainable lifestyles. It is also recommended that incentives be created for students to unleash their potential for acting on climate change; this works to encourage students to engage in climate-oriented vocational programmes in their transition towards higher education, helping them access employment in sustainability-related areas.

References

Aleixo AM, Leal S, Azeiteiro UM (2016) Conceptualization of sustainable higher education institutions, roles, barriers, and challenges for sustainability: an exploratory study in Portugal. J Clean Prod 172:1664. https://doi.org/10.1016/j.jclepro.2016.11.010

Apronti P, Osamu S, Otsuki K, Kranjac-Berisavljevic G (2015) Education for disaster risk reduction (DRR): linking theory with practice in Ghana's basic schools. Sustainability 7(7):9160–9186. https://doi.org/10.3390/su7079160

Brundiers K (2017) Educating for post-disaster sustainability efforts. Int J Disaster Risk Reduct 27:406. https://doi.org/10.1016/j.ijdrr.2017.11.002

Colliver A (2017) Education for climate change and a real-world curriculum. Curric Perspect 37(1):73–78. https://doi.org/10.1007/s41297-017-0012-z

Herrera Cano C (2016) Disaster risk management in business education entrepreneurial formation for corporate sustainability. AD-minister (28):33–48

Madueno J, Cejas M, Pena F (2015) An approach to the implementation of sustainability practices in Spanish universities. J Clean Prod 106:34

Mahat H, Mohamad S, Ngah C (2016) 3R practices among moe preschool pupils through the environmental education curriculum, vol 23. EDP Sciences, Les Ulis

Martin J, Carter L (2015) Preservice teacher agency concerning education for sustainability (EfS): a discursive psychological approach. J Res Sci Teach 52(4):560–573. https://doi.org/10.1002/tea.21217

Naoufal N (2014) Peace and environmental education for climate change: challenges and practices in Lebanon. J Peace Educ 11(3):279–296. https://doi.org/10.1080/17400201.2014.954359

Nazir J, Pedretti E, Wallace J, Montemurro D, Inwood H (2011) Reflections on the Canadian experience with education for climate change and sustainable development. Can J Sci Math Technol Educ 11(4):365–380. https://doi.org/10.1080/14926156.2011.624673

OLIO – The Food Sharing Revolution (2019) Retrieved from https://olioex.com/

Pfister T, Schweighofer M, Reichel A (eds) (2016) Sustainability. Routledge/Taylor & Francis Group, London/New York

Shaw R (ed) (2014) Community practices for disaster risk reduction in Japan. Springer, Tokyo

Sidiropoulos E (2014) Education for sustainability in business education programs: a question of value. J Clean Prod 85:472–487. https://doi.org/10.1016/j.jclepro.2013.10.040

Steele F (2010) Mainstreaming education for sustainability in pre-service teacher education in Australia: enablers and constraints/ [program leader and author of this report Fran Steele; program assistance Rebecca Blair … [et al.]; editing: Jessica North]. Department of the Environment, Water, Heritage and the Arts, Canberra

Suleri J, Cavagnaro E (2016) Promoting pro-environmental printing behavior: the role of ICT barriers and sustainable values. Int J Educ Dev Using Inf Commun Technol 12(2):158–174

The Sandwatch Foundation (2018) Retrieved from http://www.sandwatchfoundation.org/

Tilbury D (2015) Education for sustainability. Foro de Educación 13(19):7–10

Chapter 15
SDG 14 Life Below Water

Introducing Fish Skin as a Sustainable Raw Material for Fashion

Elisa Palomino

Abstract In recent years there has been a growing interest in fish skin – a by-product of the food industry – as an alternative sustainable raw material for fashion. Global production of fish has steadily increased over the last decade, and more than 50% of the total remaining material from fish capture results in 32 million tonnes of waste. A substantial amount of this waste is the skin of the fish; only a small percentage of this skin is processed into leather. While, to date, the European Environment Agency allows seafood processors to dispose of fish skins in marine waters, this is expected to change as the decomposing organic waste can suck up available oxygen from marine species and introduce disease into the local ecosystem. Fish skin leather processing could prevent and significantly reduce marine pollution and sustainably protect marine ecosystems in order to achieve healthy and productive oceans. This paper describes the conditions necessary for the development of fish skin craftsmanship within a Fashion Higher Education sustainable curriculum. In order to enhance the innovation and sustainable design of fish leather products, the author has developed an impactful capacity-building approach connecting fashion students with the Icelandic fish leather industry, which is renowned for sustainable sourcing from Nordic fish farms, promoting the sustainable use of ocean-based resources.

Keywords Fish skin · Food industry by-product · Sustainable management of the oceans · Arctic economic growth of fisheries · Fashion education for sustainability

E. Palomino (✉)
Central Saint Martins, University of the Arts, London, UK
e-mail: e.palomino@csm.arts.ac.uk

15.1 Introduction

This chapter outlines the importance of fish skin, a by-product of the food industry, as an innovative sustainable material for fashion. Fish skins are sourced from the food industry, using waste and applying the principle of circular economy. None of the fish used to make this alternative leather are farmed for their hides. They require no extra land, water, fertilisers or pesticides to produce them, and they have low environmental impact, unlike conventional leather (Jacobs 2018). The processing of fish skin leather avoids throwing the fish skins into the ocean and can significantly reduce marine pollution and sustainably protect marine ecosystems in order to achieve healthy and productive oceans.

The Atlantic Leather tannery, located on the north coast of Iceland, has processed fish leather since 1994, based on the ancient Icelandic tradition of making shoes from the skins of catfish. It supports local economies by sourcing from sustainably managed Nordic fish farming. The manufacturing of fish skin leather works with three aspects of sustainability: the economic benefit of creating value from waste; the social benefit of reconciling sustainability with fashionably exotic fish skin; and the environmental benefit of producing skins without damaging biodiversity or endangering animals.

The results presented in this study are based on the United Nations Sustainable Development Goal 14, Life Below Water, and they highlight the opportunity to develop fish skin leather as a key part of achieving sustainable development of the ocean. SDG 14 deals with the conservation and sustainable management of the oceans, seas and marine resources, and it is strongly connected with other SDGs, in particular SDG 2 (ending hunger and achieving food security, improved nutrition and sustainable agriculture) and SDG 12 (sustainable production and consumption).

Fisheries and aquaculture make a crucial contribution to global food security, nutrition and livelihoods, but overfishing, unsustainable seafood farming practices, ocean pollution and acidification will threaten the future of seafood availability worldwide.

The oceans are recognised as indispensable for addressing many of the global challenges facing the planet, from food security and climate change to the provision of energy and natural resources. The use of the sea and its resources for sustainable economic development (blue economy), contributing to prosperity today and into the future (WWF 2015), is expanding rapidly, but the oceans are under stress. They are already overexploited, polluted and confronted with climate change. As carbon emissions have risen over time, the ocean has absorbed much of the carbon dioxide, leading to acidification. Sea temperatures and sea levels are rising, resulting in loss of biodiversity and habitat and changes in fish stock composition. Future ocean development is threatened by overfishing and depleted fish stocks in many parts of the world (OECD 2016).

Atlantic Leather, an Icelandic fish skin tannery, uses the full potential of the ocean by taking a responsible, sustainable approach during fish skin processing and taking into consideration the wealth of the ocean and its great potential for boosting economic growth, employment and innovation.

This chapter explores the qualities of fish skin leather, a by-product of the food industry and part of the ocean economy: its capacity for creating future employment and innovation and its role in addressing the global challenges facing oceans. Special attention is given to the new technologies used in fish skin production, their potential for innovation and their contribution to addressing challenges such as energy, environment and climate change.

The World Ocean Council (WOC) has been working to advance global ocean business collaboration to develop industry-driven solutions to sustainable development. The Atlantic Leather fish skin tannery is playing an influential role in promoting sustainable practices, producing fish skin leather that implements sustainable concepts, reducing environmental impacts and creating social value. Creating sustainable value chains within ocean and maritime industries is a key priority for the Icelandic fishing sector. Since maritime technology has been developed in Iceland and private companies have set more value on businesses supporting ocean sustainability than before, the concept of the blue economy has received increasing attention and interest (Hansen 2018).

Atlantic Leather works with fisheries with blue technology, which considers the intersection of the economic benefits of the ocean, environmental health and societal value in policies and best practices (Hansen 2018) and exploits the harvested raw material of fish skin to the maximum level within its value chain. By maximising the usage of fish skin, Atlantic Leather has added value within the fishing industry; the company's efforts to utilise 100% of the raw material contribute to maintaining fish stocks at biologically sustainable levels. Using the entire fish adds to the value chain, benefitting fashion buyers as well as the fisheries themselves. Since the benefits of 100% utilisation can be applied when the supplier or fishermen are registered and connected within the value chain, this idea of 100% utilisation helps reduce undocumented fisheries and overfishing.

The activities of Atlantic Leather which are most closely related to the relevant SDG 14 targets are:

1. Reducing ocean acidification: Atlantic Leather reduces CO_2 emissions by ceasing to throw fish skins into the ocean.
2. Regulating harvesting and ending overfishing and illegal unreported fishing practices: 100% utilisation of fish raw materials is applied when the supplier or fishermen are registered, and this helps to decrease undocumented fisheries and overfishing.
3. Increasing economic benefits from the sustainable use of marine resources: Atlantic Leather develops fish skin leather while preserving environmental biodiversity by sourcing from Nordic sustainably managed fish farming.
4. Providing access for small-scale artisanal fisheries to marine resources/markets: Atlantic Leather promotes value-added profits within the value chain by creating new job opportunities for coastal dwellers using a by-product of the fishing industry to produce fish skin leather for the luxury fashion industry.

5. Increasing scientific knowledge, developing research capacity and transferring marine technology: Atlantic Leather uses new technologies for the development of sustainably produced fish skin leather.

This chapter has presented a brief introduction of the development of fish skin use and reflected on the sustainable concepts of fish skin production in Iceland. The study proposes that the sustainable development of fish skin as a by-product could become an innovative sustainable raw material for the fashion industry. After this introduction, Chap. 2 reviews the historical context of fish skin leather. Chapter 3 introduces the fish skin concepts of sustainability, the main contribution of knowledge to this field, and then reviews how fish skin leather and the Atlantic Leather tannery align with the SDG14 Life Below Water. Through the findings presented, this chapter aims to provide insights relevant to policymakers, fish industry stakeholders and academia and to encourage continuous improvement towards more sustainable fashion practices. Chapter 4 introduces methods and methodology for the case study and action research.

Chapter 5 presents a case study of Best Practices in Fashion Higher Education with Arctic students. The study examines a fish skin workshop developed at the Icelandic tannery Atlantic Leather involving fashion students from five Nordic universities, providing new data on cross-collaboration between industry and academia. Such insights will inform industry and academia how fish skin leather, a by-product of the food industry, can better contribute to responsible marine resource use. Chapter 6 focuses on implementing measures and recommendations for both academia and industry, followed by the conclusions of the study.

15.2 Literature Review

15.2.1 *Historical Context:* Fish Skin Through History

Making leather from fish skin is an age-old craft practised by many societies along rivers and coasts around the world, and there is evidence of historical fish skin leather production in Scandinavia, Alaska, Hokkaido, Japan, north east China and Siberia. Before synthetic fibres were invented, people clothed themselves with natural materials available in the surroundings where they lived, including fish skin (Jiao 2012).

Arctic people display a remarkable intelligence in utilising natural resources, reforming natural conditions, adapting to the environment and creating a better life; in the past, this included making clothes from fish skin leather. However, the shortage of raw materials and the omnipresence of modernity have challenged the preservation of the fish skin craft. Better access to the modern world meant that Arctic people were able to access textiles like cotton and silk to create their clothing, leaving fewer people to develop the traditional fish skin craft. There are currently only a few people left who know how to create these fish skin garments (Campbell 2010). Overfishing and water pollution have caused fish stocks to drop, and many Arctic aboriginals have turned to farming and tourism to make a living, abandoning their fish skin skills (Lin 2007).

15.2.2 Iceland's Traditional Knowledge of Fish Skin

For much of their history, Icelanders wore shoes made of fish skins processed using traditional tanning methods. Each shoe was cut from a single piece of fish skin, with a vertical seam at the heel and a seam at the toe. They were soft, supple, flat-soled traditional footwear (Mould 2018). Contemporary accounts of travels around Iceland in the mid- to later eighteenth century describe and illustrate men wearing traditional fish skin shoes (Hald 1972), suggesting that the working man wore them on a daily basis. Icelanders made their shoes from wolf fish leather, and they measured distances by how many pairs of fish skin shoes would be worn out by walking over the path.

15.2.3 Fish Waste: Use of Fish By-Products by Aboriginal Peoples

The use of fish by-products was well-known to aboriginal peoples in Arctic communities (Hardy 1992). The specific material properties of fish skin have been known since ancient times. Some human cultures developed unique techniques to process fish leather from fish skin and used this leather for clothing and shoes (Ehrlich 2015). Icelandic history, right from the settlement of Iceland in the ninth century, has been interwoven with marine resources, and fish have been their main source of food and income (Sigfusson and Arnason 2017a, b). Icelanders are known for reusing everything, and they still have their ancestors' spirit of finding the useful in everything. Improved usage of so-called waste and other by-products could help meet increasing demand for seafood without further stress to the ecosystem. Some "waste" products can have a very high value if they are used. A more efficient use of resources will benefit society, the environment and the industry's bottom line (Bechtel 2003).

The use of fish skin by the Arctic's aboriginal peoples has recently been assimilated as an innovative sustainable material for fashion due to its low environmental impact. Fish skins are sourced from the food industry, using waste, applying the principle of the circular economy (Jacobs 2018).

15.2.4 Protecting Natural and Cultural Resources

For indigenous Arctic people, the relationship with fish, and specifically with salmon, plays an important role in maintaining their identities as distinct cultures. Salmon provides them with more than nutrients. It also plays a role in ceremonial traditions, creating important ties between people and their environment.

The Arctic is undergoing dramatic climate change which threatens indigenous people, impacting their food security and traditional knowledge systems which rely on fishing activities for their physical, cultural and spiritual wellbeing. Coastal indigenous peoples in the Arctic share links to marine environments, mainly through fishing. The relationship with the sea plays an important role in maintaining their identities as distinct cultures, but climate change is threatening indigenous people's ties to oceans and marine resources around the world (Yoshitaka 2017). Fisheries management is a human security issue as well as an environmental issue, and we need to bring social equity into global governance of the oceans, to respect coastal indigenous peoples and their relationship with fish.

Preserving traditional knowledge with regard to fish skin is essential to the Arctic world. This chapter seeks to draw attention to the vital importance of traditional fish leather craft to the Arctic people as the basis of their culture and a component of their identities and to encourage their artisans to reintroduce the skills used by their ancestors as a tool for community development.

15.3 Sustainability Context Icelandic Fish Skin Leather: Concepts of Sustainability. Aligning Fish Skin Leather with the United Nations Sustainable Development Goal 14. Life Below Water

15.3.1 SDG 14.3 Reduce Ocean Acidification

Sustainable Development Goal 14:3 is "Minimise and address the impacts of ocean acidification, including through enhanced scientific cooperation at all levels". Atlantic Leather's production of fish skin leather using fish waste aligns with this goal.

Global production of fish has steadily increased over the last decade, and more than 50% of the total fish catch becomes waste material resulting in 32 million tonnes of waste (Arvanito and Kassaveti 2008). To date, the European Environment Protection agency allows seafood processors to dispose of fish skins in marine waters, but this is expected to change as the decomposing organic waste can suck up available oxygen from marine species and introduces disease into the local ecosystem (EPA 2012). Unlike the EU's fisheries policy, the Icelandic system decrees that nonsaleable fish cannot be tossed back into the ocean but must be brought ashore and counted towards the quota, therefore maintaining fish populations (Deliso 2015).

Fishermen create waste by using fishing methods that are not discriminating enough or by targeting only part of the fish (e.g. roe, fins) and discarding the rest. Fishermen and sea processors are incentivised to discard low-value species or trimmings to help maximise the value of their catch. Moreover, fish waste is typically unsorted and geographically dispersed, which makes it costly to collect and process. The high-value uses of seafood by-products like fish leather make fish skin collection and upcycling more feasible and attractive for the fishing industry (Henning, and Jain 2017). Iceland has also made voluntary commitments to reduce marine litter in its waters and to address acidification by producing an updated climate mitigation strategy by the end of 2017, in line with the Paris Agreement, with obligations of a 40% reduction of greenhouse gas emissions by 2030 (Gunnarsdóttir 2017).

The technology for sustainable processing of fish leather can be of great environmental benefit as well as profit for the global economy. Fish skin leather processing could prevent fish waste ending up in marine waters and significantly reduce marine pollution and sustainably protect marine ecosystems in order to achieve healthy and productive oceans. Before Atlantic Leather started using fish skin to produce leather, fish skins used to be thrown away. Now, they are a source of income for the local people, besides avoiding being turned into biological waste.

15.3.2 SDG14.4 Regulate Harvesting and End Overfishing

Sustainable Development Goal 14:3 is "Conserve and sustainably use the oceans, seas and marine resources for sustainable development".

By 2020, the aim is to effectively regulate harvesting and end overfishing, illegal, unreported and unregulated fishing and destructive fishing practices and implement science-based management plans, in order to restore fish stocks in the shortest feasible time, at least to levels that can produce maximum sustainable yield as determined by their biological characteristics.

15.3.2.1 Marine Biodiversity

Concentrating on improving the sustainable value chain, Atlantic Leather works closely with the fisheries who supply them with fish skin to improve the sustainability of fish resources and to unite the efforts of the world's leading seafood industries to reduce the global extent of illegal, unregulated fisheries. These activities and efforts have linked Icelandic private enterprises in a common movement to save Icelandic marine biodiversity by restricting illegal fishing and undocumented fisheries. The fish catch in Iceland is made in a conscious and nonpredatory way that respects environment laws and procreation periods.

Atlantic Leather uses fish from Nordic government regulated farms with sustainable management, which provides employment for local communities and a sustainable source of food while maintaining fish stocks. Fish is a key part of both food and the local economy in Iceland.

The fishing is carried out in a sustainable way, under the control of government agencies dedicated to the preservation of species and biodiversity. It helps the local communities and respects the Nordic environmental balance. With stock sustainability and the ecological effects of fishing and management systems as core concerns, Iceland has become even more competitive in the global marketplace (Sigfusson and Arnason 2017a, b).

15.3.2.2 Sustainable Management of Arctic Fisheries

As much as 40% of the ocean is heavily affected by depleted fisheries and other human activities (UN 2018). The growth in aquaculture could put stress on the fish industry to meet increased demand, by ignoring fishing quotas imposed by responsible governments. The sustainability role of the fisheries industry is an important issue which has to be taken into consideration as a concern about the future availability of raw material for fish (Bechtel 2003).

Fisheries are the single most important industry in Iceland, and the living marine resources are their most important natural resources, but they are limited, and it is important to utilise these resources in a sustainable way. In 1984, fixed quotas for each vessel were introduced in order to control exploitation of the fish stocks (Valdimarsson 1990). The main objective of the quota legislation was to prevent overfishing and to encourage responsible handling of all catches and exploitation of under-utilised marine life. There is no doubt that the quota system has had a major effect on changing attitudes towards full utilisation of catches. The fishermen and the processing industries are becoming more aware of the possibilities of making marketable products from raw materials that are currently discarded, such as fish leather. Through research and development, publicly funded institutions assist the industry to increase the utilisation of seafoods (Bechtel 2003). Iceland, as an Arctic coastal state, takes part in the ongoing negotiations on a new agreement to prevent unregulated high seas fisheries in the Central Arctic Ocean and has been engaged, within the Arctic Council, in consultations on increased Arctic Marine cooperation (Gunnarsdóttir 2017).

15.3.2.3 Fish Skin as a By-Product of the Food Industry. Resource Efficiency by Generating Value from Waste

Fashion is an extremely wasteful and polluting industry, creating a negative impact on the environment and on people. The fashion industry is currently going through a significant change in its approach towards sustainability (BCG 2017). Therefore, the fashion industry as a whole must strive to change and rethink its raw materials and processes. There is a trend for the adoption of new materials, which have a lower environmental impact than their conventional alternatives (Textile Exchange 2016). Fish skin is an innovative and sustainable alternative material with a lower environmental and social impact than conventional leather. Fish skin as a new raw material for the fashion industry could provide a (partial) solution for aquaculture waste, which the European Union has committed to reduce through actions in the Circular Economy Package.

Almost half of the fish caught for human consumption is discarded before it even reaches our shelves. This represents a significant amount of potential profit that is effectively being thrown away. The average production of 1 tonne of fish fillets results in roughly 40 kilograms of discarded skins. Improved usage of fish by-product waste could help meet increasing demand for seafood without further stress on the ecosystem. More efficient use of resources will benefit society, the environment and the industry's bottom line. Reducing discards and upcycling by-products will likely increase profitability (IOC 2013).

When Atlantic Leather converts the fish skins into leather, it creates new value and far-reaching economic opportunities. Atlantic Leather has been perfecting the fish skin tanning techniques to turn it into high-value products for non-food sectors (fashion) by upcycling fish skin into exquisite fish leather. Recycling fish skin into leather is eco-friendly, cost-effective and sustainable. With an estimated 43% of fish and shellfish resources ending up as wastage, Atlantic Leather is converting once discarded parts of the fish into desirable products and income: fashion and accessories. The tannery is proud that it can reduce waste while sourcing salmon skins from certified sustainable Nordic fisheries. Atlantic Leather is part of a group of pioneering industry experts from Iceland involved in the commercial fishing, aquaculture and processing sector and the creation of value from fish processing by-products. Companies in Iceland, the pioneers of this industry, have developed a wide range of uses for fish waste: enzymes, pharmaceuticals, dietary supplements, cosmetics and leather goods (Sigfusson and Arnason 2017a, b).

15.3.2.4 Chrome-Free Tanned Fish Skin

The processing of leather is most commonly linked with environmental pollution. Many of the chemicals used during tanning are toxic, with substances like mineral salts and chromium routinely used. Environmental protection standards tend to be insufficient in primary leather producing regions, with waste water and solid waste from the tanning process dumped directly into rivers, devasting nearby flora and fauna. Tanning does not just have an environmental cost; a number of the chemicals used to tan leather are carcinogenic, endangering the health of those who labour in tanneries (Shean 2018).

Atlantic Leather commits to SDG 14 Actions for Businesses: Record and disclose information on the chemical usage within products to facilitate closing the loop. Atlantic Leather produces chrome-free fish leather using mimosa bark in a traditional process of vegetable tanning, avoiding chromium salts, which are extremely toxic and polluting. Vegetable tanned fish leather is sustainable, durable and surprisingly strong, even stronger than other kinds of leather. This is due to the alignment of the fibres in the skin: in mammals, these run parallel to each other, but in fish, they are in a cross-hatched pattern, making fish leather much stronger for the same thickness.

15.3.2.5 Harnessing Renewable Energy

Iceland has a unique situation in an era when climate change is making it necessary for countries around the world to implement sustainable energy solutions. Today, almost 100% of the electricity consumed in Iceland comes from renewable energy. The glaciers and rivers of the interior of the country are harnessed to generate 80% of the country's electricity needs through hydropower, while the geothermal fields provide the remaining 20%. Iceland has also focused on sharing its knowledge and technical expertise in geothermal development (Logadóttir 2015). The entire process of producing fish skin at Atlantic Leather relies on the power of nature and is non-impactful on the environment – even in terms of electricity consumption – as geothermal water is used to produce fish skin leather, and their electricity comes from a nearby hydroelectric power station.

The use of geothermal energy for fish by-products is likely to increase in the future. The interest in Iceland is focused on the use of geothermal energy in low-heat regions. It can be expected that the price of oil will increase more than the local energy in the future, and therefore it is worth paying attention to the use of locally available energy sources for the fishing industry (Bechtel 2003).

15.3.3 SDG14.9 Support Artisanal Fishermen

Sustainable Development Goal 14:9 is "Provide access for small-scale artisanal fishermen to marine resources and markets".

In recent history, fisheries and fish processing jobs have been in decline in Iceland. Like many other countries, Iceland has been mindful not to overfish. With stock sustainability and the ecological effects of fishing and management systems as core concerns, Iceland has become even more competitive in the global marketplace by using fish by-products. Iceland has discovered one way of creating value and jobs, especially in remote and rural areas where such opportunities are not taken for granted (Sigfusson 2017a, b).

This approach has been beneficial to all levels of the supply chain, including fishermen in remote areas who have seen the prices of fish triple in recent years due to increased interest in value-added issues. The sustainability of the Icelandic system means that fishermen now rank amongst Iceland's highest-paid workers. The Icelandic model has proved reliable, and this model could be duplicated in seafood industries around the world, creating new opportunities in coastal areas (Sigfusson 2017a, b).

15.3.3.1 Creation of New Job Opportunities for Coastal Dwellers

Since the ninth century, Icelanders have derived vitality and stamina from fish. Atlantic Leather has propelled a Nordic tradition to increase the utilisation and value from fish waste to create fish leather and in so doing create new job opportunities, especially for coastal dwellers.

The Atlantic Leather tannery sits in a thriving community on the North East coast of the island – inhabited by fewer than 3000 people – with fishing grounds located just off-shore. Such proximity to the source means that transportation to the point of manufacture is significantly reduced; it also presents the innovation that the tannery only uses waste fish skins from food consumption.

Atlantic Leather creates blue tech and blue jobs in a remote coastal area promoting a sustainable ocean industry. A key challenge for these coastal areas is to maintain the viability of the fisheries sector and to attract young people to work in it. Atlantic Leather aims to preserve the rich cultural traditions that have been developed within the Icelandic fishing industry when processing their fish leather.

Fish leather is also benefiting other sectors, such as tourism. In 2014, Atlantic Leather – Iceland's last remaining tannery – opened a museum for tourists. The museum recreates the traditional and contemporary tanning process of fish leather and displays historical photos and implements (Deliso 2015).

15.4 Methodology

15.4.1 Study Methodology

The aim of this case study is to explore the link between sustainable materials (fish skin, a by-product of the fish industry, as a new raw material for fashion) and transferring the intangible heritage skills of fish skin craft from Arctic ethnic minorities to higher education fashion students from Nordic universities.

The literature review highlighted that in Iceland, as in many other countries around the world, better utilisation of marine resources is being widely called for. There is well-documented support for the Icelandic commitment to a sustainable seafood sector and a reduction of seafood waste.

To reflect upon the interaction of fish skin using traditional craft techniques, bibliographic and documentary research was initially done:

- Enquiry [Theory]. Following the workshop (see Chap. 5), data was collected through primary and secondary sources to reveal areas of potential development.
- Contextual and visual analysis.
- Making [Practice]. Higher education students produced fish skin samples in collaboration with an Arctic ethnic minority craftsperson. Photographic documentation was used for the illustration and classification of results.

- Sharing [Dissemination]. Feedback has been sought through activities such as conferences, published articles, teaching and communication via the author's website http://www.fishskinlab.com.

15.4.2 Methods

Action research was used during this study. The data was collected through:

- Archival research in museums to study traditional knowledge in fish skin processing.
- Mapping traditional fish skin crafts to validate their technical feasibility.
- Field trip. The field trip covered the area around Sauðárkrókur on the North East coast of Iceland.
- Workshop on fish skin leather craft to test ideas through teaching and learning, observing students' design approaches using fish skin as an alternative material.
- Photographs and video recording.
- A documentary filmed during the workshop, featuring interviews with students, curators and craftsman to observe students' development of fish skin finishes as a form of design research.
- Sketchbook development.
- Literature review.

15.5 Discussion and Results: Best Practices in Fashion Higher Education Increasing the Co-production of Knowledge Sharing Indigenous/Traditional Knowledge with Arctic Higher Education Fashion Students

The Fish Leather Craftsmanship workshop was organised by the author, Elisa Palomino, BA, Fashion Print pathway leader at Central Saint Martins, London, and Katrin Karadottir, Programme Director in Fashion Design at Iceland Academy of the Arts, in collaboration with Atlantic Leather tannery, with the participation of students from Iceland University of the Arts, the Royal Danish Academy of Arts, Boras University, Aalto University and Central Saint Martins.

In order to provide an inspiring environment in Arctic higher education, to enhance student engagement and test a new learning experience, the author designed a workshop encouraging Artic design students to produce fish leather designs using traditional skills built over generations by Arctic indigenous peoples. The aim was to promote the vast set of knowledge and skills on fish skin that the north possesses, developing sustainable design within the Arctic's traditional ways of life in areas with a history of fish skin leather production, such as Iceland, Sweden, Finland and Denmark, preserving and using fish skin cultural heritage and strengthening networking activity.

The workshop took place in Sauðárkrókur, Iceland, and combined learning about traditional knowledge on fish skin tanning with studying the technological progress of the Icelandic tannery Atlantic Leather, which has been turning local fish skin into highly sustainable leather since 1994. A total of ten students from universities in the circumpolar area (Iceland, Denmark, Sweden, Finland) and the UK benefited from the workshop. A Swedish craftsperson from the Sami ethnic minority delivered the workshop. Lotta Rhame shared traditional Sami fish skin tanning methods and passed down the endangered fish skin craft to the next generation of Nordic students as part of a sustainable fashion higher education programme to learn best practices for social change and sustainability. The programme included preparation, implementation, evaluation and a follow-up phase.

The workshop was designed to build community knowledge around material culture and to bring participant voices together to promote understanding of fish skin craft culture, with the aim of improve knowledge of fish skin craft to address the pressing sustainability issues in the current fashion industry and to understand the duty to change fashion systems through education, inculcating fashion students with the values of sustainability. The workshop aimed to develop new fashion practices, taking students out of the classroom and into nature and contributing to the learning experience about fashion sustainability. Another important aim was to improve the awareness and protection of traditional Arctic fish skin culture. Students learnt traditional fish skin handicraft heritage in order to integrate it into their fashion practice. According to Fletcher, participatory design and codesign structures are key to changing fashion systems and to fostering lasting relationships between the makers and the final product (Fletcher 2018). The workshop's main objectives were to:

- Map existing traditional knowledge of fish skin craft from the Sami ethnic minority in the Arctic
- Build an interdisciplinary collaborative network which intersects craftspeople from Arctic ethnic minorities and higher education students to study fish skin ancient traditions
- Preserve and disseminate Arctic cultural heritage connected with fish skin, promoting sustainable development of their unique craft culture
- Help higher education students develop fish skin leather samples as an environmentally responsible alternative material for fashion
- Enhance the visibility and attractiveness of fish skin leather as a new sustainable material for Nordic fashion students
- Bring together sustainable methods from fashion design and traditional crafts to foster international knowledge exchange that will develop the capacity for practice in these fields
- Identify tools about best practice on fish skin leather craft and test the ideas at higher education fashion institutions in the Arctic and internationally, supporting

students to engage in sustainability facilitated through the use of fish skin as an alternative material
- Promote collaboration between industry and education in order to ensure that fashion programmes are meeting industry needs; industry involvement to train fashion graduates on sustainability issues; and attracting Arctic fashion students to the maritime industry
- The workshop was 5 days long and included:
- An introduction to the sustainability background
- Lectures on historical fish skin artefacts at international museums
- A visit to the Atlantic Leather fish skin tannery
- A visit to the local textile museum
- Traditional fish skin tanning
- Sketchbook development

The workshop brought together traditional knowledge holders and community representatives from across the Arctic in order to explore the roots of Nordic fashion and design traditions linked with fish skin, to create space for communities, to share wisdom, skills and techniques around fish skin processing and to co-produce new work using both traditional knowledge and British sustainable design methodologies. The workshop promoted sustainable material engagement through a full-immersive experience in a teaching-in-the field approach, creating a collaborative network for further projects and setting up an international design environment for sharing knowledge.

The workshop methodologies reflected the geographical contrasts of the area. The harshness of the weather, the isolation and the limited availability of materials formed a unique source of creativity and inspiration for the students during the workshop. Fish skin was the only available material, spurring students to think creatively and seek new design possibilities. Eco-consciousness played a fundamental role in the students' designs using remnant materials. Fish leather offers outstanding longevity, one of the most important elements in sustainability, and has the benefit of being a highly biodegradable natural by-product.

The object of the workshop was the preservation and dissemination of the cultural heritage connected with fish skin. In order to achieve this, the collaboration and cooperation amongst different Artic areas, universities and professionals provided a key element in the project. This was a fine example of an innovative way of linking the preservation of traditional knowledge and culture and the development of culturally relevant programmes for students, community involvement and the conservation of resources. The project provided a case study for working across Arctic universities to develop their cultural identities and foster narratives of social sustainability. The cross-disciplinary project has created a new structure to demonstrate how much Arctic communities have in common.

The workshop seeks to inspire fashion lecturers involved in the development of sustainability and craftsmanship within their curriculums to implement this transformative teaching and learning experience in their own practice. Hopefully, the workshop will inspire new ideas across the student and staff communities that were involved, which in turn may contribute to public debate on sustainability issues in the fashion industry (Fletcher and Williams 2010). The Nordic fish skin network has blended the highly qualified skills of a Swedish craftsperson, Lotta Rhame, with British cutting-edge sustainable design education. Development of sustainability within the curriculum has been identified as a high priority for students (Reid 2011), and this project's outputs will inform existing courses naming sustainability, as well as a broad spectrum of design courses.

Through this workshop, the author, as a member of the London College of Fashion, Centre for Sustainable Fashion, has brought its commitment to using fashion to drive change, build a sustainable future and improve the way we live, using human and ecological resilience as a lens for design in fashion's artistic and business practice (CSF 2015). The author has made a contribution to the field of Design for Sustainability (DfS) in fashion, furthering the sense of our interconnections as people and to our natural world. This workshop has specifically supported the following 4 of the 17 United Nations Sustainable Development Goals:

- SDG 4 – Ensure inclusive and equitable quality education and promote lifelong learning opportunities for all.
- SDG 12 – Ensure sustainable consumption and production patterns.
- SDG 13 – Take urgent action to combat climate change and its impacts.
- SDG 14 – Conservation and sustainable management of the oceans, seas and marine resources.

The workshop has developed a collaboration framework between industry and education and has managed to:

- Raise awareness on ocean-related issues and the maritime economy with higher education students
- Create new cooperation between education and maritime industry
- Improve the employability of students thanks to the acquisition of new sustainability and craft skills
- Share resources between different educational institutes and industry at the transnational level

The case study recommends transferable skills for educational models and demonstrates how relevant the indigenous fish skin knowledge in partnership with sustainable design strategies can be to connect people to their culture, communities and the environment. The case study reflects on the dialogue between indigenous craftsmen and Nordic fashion students on common Arctic issues, in particular issues of sustainable development, sustainable material innovation and Arctic environmental protection, in order to restore some of the damage that has already happened to the Arctic's indigenous culture related to fishing rights and fish skin clothing traditions, helping to build resilience amongst the Arctic communities. This project recommends engagement with local communities and traditional fish skin knowledge holders, laying the groundwork for an assessment that is co-produced by both traditional knowledge and British fashion education.

15.6 Impact Sustainability: Final Remarks

The supply of fish in the oceans is not endless, and therefore we need to manage fisheries in a more sustainable way. By developing fish skin leather, we could achieve sustainable ocean development, optimising fisheries management and increasing the value of the catches (Bechtel 2003). The future availability of seafood, however, is threatened by overfishing, unsustainable seafood, farming practices, ocean pollution and acidification. Strategies aimed at increasing the utilisation of fish skin that would otherwise be discarded must be carefully considered. Creating markets for fish skin runs the risk of incentivising bigger catches and creating fishing pressure for species currently viewed as fish skin potential.

The Icelandic fish skin model has proved reliable, and this model can be duplicated in seafood industries around the world, creating new opportunities in coastal areas (Sigfusson 2017a, b). The project could be scaled up by developing a model of fish leather-waste production that can be used by factories in other countries with a big consumption of fish in their diet and countries with a history of using fish skin leather. By doing so, indigenous fishing communities which used to subsist upon, and dress themselves with, fish skin leather items – like the Ainu in Hokkaido, the Nanai in Siberia and Alaska's Inuit – will be able to reach agreements with nearby fishing plants for the supply of fish skins to recover their ancient craft skills of tanning fish skins and develop productions that will boost their economy.

Fish skin leather can be used in wallets, bags and shoes. The process is low-tech and requires little capital, which makes it ideal for small businesses or for setup in developing countries (Henning and Jain 2017). The overall findings align with the Icelandic industry's commitment towards greater sustainability. The study suggests that there is a great financial opportunity to use fish skin as a new raw material for fashion. Countries with both high demand and cultural reliance on fish are potential candidates for the marketing and sale of fish skin leather. It is highly recommended that similar case studies are developed in other fish-producing consumer areas.

The case study has also given new insights into the potential for even greater sustainability actions through the implementation of workshops within higher education fashion. Students have studied how fish by-products are used in the value chain and how it demonstrates positive waste reduction.

Through examining the strategic management of fish skin, this study has outlined the ability for the aquaculture industry to produce more value from the same amount of resources. In conclusion, there are economic and environmental benefits that should be considered in order to develop fish skin further as a new raw material for fashion.

With collaboration between industry and academia, the rise of fish skin as a new by-product raw material for fashion will contribute to the sustainable development and future growth of the aquaculture and fashion industries.

Acknowledgements Professor Dilys Williams, Director Centre for Sustainable Fashion at London College of Fashion, and Simon Thorogood, Senior Research Fellow Digital Anthropology Lab at London College of Fashion, provided overall guidance for this research.

The research has been funded by AHRC LDoc London Doctoral Design Centre Award.The research has been funded by the EU Horizon 2020-MSCA-RISE-2018. Research and Innovation Staff Exchange Marie Sklodowska Curie GRANT NUMBER 823943: FishSkin: Developing Fish Skin as a Sustainable Raw Material for the Fashion Industry.

In addition, this project could not have been completed without support from the Nordic Culture Fund, OPSTART and the Society of Dyers and Colourists' grants to deliver the Nordic Fish Leather craft workshop at the Atlantic Leather tannery.

The author would like to express their gratitude to the research assistant Joseph Boon for his vast contribution to the publication, to the Swedish fish skin craftsperson Lotta Rahme, to Katrin Karadottir, Programme Director in Fashion Design at the Iceland Academy of the Arts, to the Icelandic tannery Atlantic Leather and to the Icelandic Textile Centre. The author would also like to thank all the student participants in the study case workshop and the photographer.

References

Arvanito I, Kassaveti A (2008) Fish industry waste: treatments, environmental impacts, current and potential uses. Int J Food Sci Technol 43:726–745

Bechtel PJ (2003) Advances in seafood byproducts: 2002 conference proceedings. Alaska Sea Grant College program, University of Alaska Fairbanks, Fairbanks. 566 pp

Campbell J (2010) Why do the Hezhe people rarely make clothing out of fish skin anymore? Textile Museum of Canada

CSF (2015) Centre for sustainable fashion vision, Plans, Actions and Change Strategy 2015–2020

Deliso, C. (2015) Iceland, preserving the bounty of the sea. November 30, 2015. Retrieved October 16, 2018, from http://www.the-report.com/reports/iceland/a-strong-yet-sustainable-recovery/preserving-the-bounty-of-the-sea/

Ehrlich H (2015) Fish skin: from clothing to tissue engineering. In: Biological materials of Marine origin. biologically-inspired systems, vol 4. Springer, Dordrecht

EPA. Environmental Protection Agency (2012) Ocean Disposal of Fish Wastes. The 2012 guidelines. Retrieved October 16, 2018, from https://www.epa.gov/ocean-dumping/ocean-disposal-fish-wastes

Fletcher K (2018) The fashion land ethic: localism. In: Clothing activity and Macclesfield. http://ualresearchonline.arts.ac.uk/id/eprint/13641

Fletcher K, Williams D (2010) Shared talent: an exploration of the potential of the shared talent collaborative and hands-on educational experience for enhancing learning around sustainability in fashion practice. In: Lens conference: sustainability in design NOW! Bangalore, India

Gunnarsdóttir K (2017) The permanent Mission of Iceland to the United Nations statement by H.E. Thorgerdur Katrín Gunnarsdóttir, Minister of Fisheries and Agriculture High-level United Nations Conference to Support the Implementation of Sustainable Development Goal 14: Conserve and sustainably use the oceans, seas and marine resources for sustainable development General Debate 6 June 2017 www.iceland.org/un/nyc

Hald M (1972) Primitive shoes. An archaeological-ethnological study based upon shoe finds from the Jutland Peninsula, Publications of the National Museum, Archaeological-Historical Series I Volume XIII. Copenhagen: The National Museum of Denmark

Hansen E (2018) Ocean/maritime clusters: leadership and collaboration for ocean sustainable development and implementing the sustainable development goals World Ocean council white paper

Hardy RW (1992) Fish processing by-products and their reclamation. In: Pearson AM, Dutson TR (eds) Inedible meat by-products, Advances in meat research series, vol 8. Springer, Dordrecht

Henning J, Jain M (2017) Seafood waste and by-products. *Fish 2.0*. Investment Insights

IOC, Iceland Ocean Cluster (2013) IOC Analysis: Double value for 40% of the catch

Jacobs B (2018) The future of leather is plant-based. HowNow Magazine https://www.hownow-magazine.com/innovation/2018/6/22/the-future-of-leather-is-plant-based

Jiao F (2012) Keeping the legend of the fish skin tribe alive. http://www.chinatoday.com.cn/ctenglish/se/txt/2012-02/02/content_423289.htm

Lin, Q (2007) Look who's tipping the scales in favour of skinny suits. http://www.chinadaily.com.cn/cndy/2007-07/10/content_5422759.htm

Logadóttir H (2015) Iceland's sustainable energy story: a model for the world? *UN Chronicle*. December 2015

Mould Q (2018) Male footwear in 17th and 18th century Iceland: traditional dress and imported fashion. Archaeological leather group Retrieved October 16, 2018, from www.archleathgrp.org.uk

OECD (2016) The ocean economy in 2030. OECD Publishing. PAME Retrieved October 16, 2018 from https://www.pame.is/index.php/projects/marine-protected-areas

Reid C (2011) Education for sustainable development (ESD) and the professional curriculum. University of Dundee. The Higher Education academy (HEA)

Shean A (2018) Shedding leather, not necessary a sustainable solution. Sancroft team May 11, 2018. Retrieved October 16, 2018, from https://sancroft.com/2018/05/11/shedding-leather-not-necessarily-a-sustainable-solution/

Sigfusson T (2017a) Intro to Iceland Ocean cluster. Rep Np: np, nd Web 2 Oct 2018

Sigfusson T (2017b) Icelandic fisheries conference. http://www.icefishconference.com/the-conference/conference-programme

Sigfusson T, Arnason R (2017a) A new utilization movement

Sigfusson T, Arnason R (2017b) Iceland Ocean cluster concept paper. Rep Np: np, n.d. Web. 2 June 2017

Textile Exchange (2016) "Preferred Fibers & Benchmark – Sector Report 2016" by Textile Exchange http://textileexchange.org/downloads/preferred-fiber-materials-benchmark-sector-report-2016/

The Boston Consulting Group (BCG) and Global Fashion Agenda (GFA) (2017) Pulse of the fashion industry. Global Fashion Agenda & The Boston Consulting Group

UN (2018) About the United Nations Sustainable Development Goals. https://www.un.org/sustainabledevelopment/sustainable-development-goals/2018, September 5

Valdimarsson G (1990) Utilization of selected fish by-products in Iceland: past, present, and future. In: Keller S (ed) Internat. Conf. On fish by-products, Fairbanks, AK

WWF (2015) What a blue economy really is. Retrieved October 16, 2018, from http://wwf.panda.org/homepage.cfm?249111/

Yoshitaka O (2017) For indigenous communities, fish mean much more than food. The Conversation January 30, 2017. Retrieved October 16, 2018, from https://theconversation.com/for-indigenous-communities-fish-mean-much-more-than-food-70129

Chapter 16
SDG 15 Life on Land

A Review of Sustainable Fashion Design Processes: Upcycling Waste Organic Yarns

Claudia Arana, Isabel B. Franco, Anuska Joshi, and Jyoti Sedhai

Abstract The fashion industry has had a significant impact on the environment and overall global sustainability. Evidence shows it is the most polluting industry and the largest consumer of water, accounting for 20% of global water wastage detrimentally affecting both life on land and underwater. As such a few key stakeholders in the fashion industry have begun undertaking key preventative measures. These include but are not limited to the use of organic cotton crops, reduction of water use throughout the production chain, the implementation of a zero-waste patternmaking technique, second-hand shops, recycling of production materials, recycling discarded fishing nets into nylon fibre and increasing the use of biodegradable fibres, crop's waste fibres, bio-based fibres and bio-textile processes and renewable sources like bamboo and hemp. The review presented in this chapter examines the fashion production cycle, the use of alternative organic materials and recycling processes for the sustainable production of yarns whilst exploring the connections between the Sustainable Development Goals (SDGs) 15 Life on Land and 14 Life below Water.

Keywords Sustainable fashion · Waste · Organic yarns · SDG 15 Life on Land · SDG 14 Life below Water · Sustainability

---

C. Arana (✉)
Bunka Gakuen University, Fashion Global Concentration, Shibuya, Japan
e-mail: arana@bunka.ac.jp

I. B. Franco
Institute for the Advanced Study of Sustainability, United Nations University Shibuya-ku, Tokyo, Japan

Australian Institute for Business and Economics, The University of Queensland, Brisbane, Australia
e-mail: connect@drisabelfranco.com

A. Joshi · J. Sedhai
United Nations University, Institute for the Advanced Study of Sustainability, Tokyo, Japan
e-mail: a.joshi@student.unu.edu

© Springer Nature Singapore Pte Ltd. 2020
I. B. Franco et al. (eds.), *Actioning the Global Goals for Local Impact*, Science for Sustainable Societies, https://doi.org/10.1007/978-981-32-9927-6_16

16.1 Introduction

From sourcing to post-consumption disposal and waste management, the fashion industry compromises several forms of life on land and therefore compromises the achievement of the Sustainable Development Goal 15 (SDG 15) and SDG 14 Life below Water (FAO, ITPS 2015). Major issues confronting the industry range from use of raw materials, waste generation and pollution of land and water. Such issues have profound impacts on life and are often difficult to estimate due to their complexity and global reach (Šajn 2019). Introducing more efficient production processes in the fashion industry would contribute to reducing waste and production of greenhouse emissions and promote sustainable resource management. Whilst producing more environmentally friendly apparel can contribute to the achievement of SDG 15, the industry should commit to closing the whole cycle production in a sustainable and responsible manner. This requires the participation of key stakeholders in the industry, namely, designers, factory workers, material suppliers and consumers, contributing in the development of new practices, behaviours, technologies and processes for recovering and reusing waste.

In this context, this chapter presents a review of fashion design processes for upcycling organic yarn through exploring its connection to the promotion of SDGs 14 and 15. The fashion industry has grown rapidly since garments and textiles first began mass production in the nineteenth century. It has also become the second largest industrial consumer of water (UNECE 2019). Additionally, evidence shows it is responsible for the production of more than 1.2 billion tonnes of greenhouse gas emissions (UNFCCC 2018), causing severe environmental damage (Masson-Delmotte et al. 2018). In response, stakeholders in the fashion industry have come up with sustainability-driven innovations to address such industrial issues. Some of these actions include but are not limited to textile production from recycled materials, the use of innovative technologies in clothing production and the use of zero-waste pattern techniques, among others. Although these practices have gained popularity among consumers, the fashion industry as a whole is far from becoming sustainable. This chapter presents a review of material sourcing and waste management processes throughout the design process. The review will cover four main stages of the fashion cycle as well as case studies examining sustainable fashion practices and techniques. This will hopefully increase our understanding of the development of impact sustainability solutions towards the promotion of SDG 15 and SDG 14 in the fashion industry. Thus, this review is useful for educators and researchers working to promote sustainable fashion innovations. The chapter begins presenting connections between the sustainability of the fashion industry, SDG 15 and SDG 14. It then provides a review of material sourcing, which consists of the separation of materials into two groups: agriculture-sourced or natural fibre-sourced and synthetic fibre-sourced. This is followed by a review of the production stage, which transforms raw materials into fibres, yarns, fabrics, etc. This section also highlights case studies exemplifying global sustainable practices, specifically reviewing waste management practices, and finishes with a few recommendations for impact sustainability research to further promote sustainable innovation within the fashion industry.

16.2 Towards Sustainable Fashion: A Literature and Practice Review

16.2.1 Introduction

The fashion industry has a clearly negative implication on land and water ecosystems, through the release of both microfibres and pollutants. The International Union for the Conservation of Nature (IUCN) has calculated that 34.8% of microplastics released into the oceans are due to the laundry of synthetic textiles (Boucher and Friot 2017). Evidence also shows that for each item of clothing washed, the amount of microfibres released into the water stream is as high as 700,000; these microplastics then make their way through the food chain (Napper and Thompson 2016). Fashion is also the second highest industrial polluter – after the oil industry. The pollutants released from the industry are often released in the affluent which can consist of toxic chemicals like lead, arsenic and mercury, as well as the release of chemical fertilizers used in the production of fibres to supply the industry. Some other factors compromising the overall sustainability of the fashion industry will be explored in detail in the following sections.

16.2.2 The Fashion Industry and Sustainability: Making the Links

This review presents the links between the fashion industry and both SDG 14 Life below Water and SDG 15 Life on Land. The Sustainable Development Goal report 2018 has shown that SDGs 14 and 15 are in a dire state due to increases in the exploitation of nature, levels of pollution and the acidity of water sources (UN 2019). Marine acidity has increased by about 26% since the industrial revolution and now faces conditions that are entirely unprecedented. Also, with the increased rise of pollution, it has been estimated that coastal eutrophication will rise by 20% by 2050 (UN 2019). Additionally, the increase of pollutant and microfibre levels in the ocean has a detrimental impact on the aquatic life, including a significant impact on relevant bioaccumulation processes. Most of these pollutants are non-biodegradable and highly bioaccumulative (the concentration of these toxics increases as it passes through the food chain). The fashion industry has also been shown to emit high amounts of toxic bioaccumulators such as mercury.

Consumers' awareness has recently increased (Kim and Damhorst 1998; Gam 2011), as consumers look to engage in sustainable lifestyles and opt for green products (Diamantopoulos et al. 2003; Zimmer et al. 1994). The fashion industry is mostly influenced by seasonal trends linked to eco-friendly purchasing behaviour (Kunz 2005: 4). Therefore, companies are introducing sustainability as part of their business strategy, making green products increasingly available in the marketplace (Fraj and Martinez, 2006; Gam 2011). However, this business strategy in isolation is not sufficient. Scholars have documented some other setbacks facing the industry in its journey towards sustainability. Findings show that the main issues preventing the complete integration of sustainable principles into the fashion industry include the lack of education regarding sustainable or "green" fashion, the perception that green fashion lacks glamour and style, the prevalence of unsustainable production practices across the supply chain, the short life cycle of products and the limited corporate disclosure in the fashion industry. Notable exceptions include Bono's Eden, the British Stella McCartney or Stewart -Brown which are perceived as sustainable fashion brands (Cervellon and Wenerfelt 2012).

16.2.3 The Fashion Industry: Production Cycle

The fashion industries' main contribution to the deterioration of both aquatic life and life on land is in the form of water consumption and waste generation, with huge amounts of pollutants released in the form of factory effluents. The completion of a clothing item has a complex life cycle consisting of different phases including resource production, fibre manufacturing, apparel assembly, transportation, consumer use and finally recycling or ultimate disposal. Each of these processes can cause environmental impacts such as resource depletion, fossil fuel emission, water wastage and solid waste, seriously compromising the sustainability of the fashion industry in the long term. Thus, this section presents a review of the sourcing of materials, the production cycle and several case studies showcasing sustainability practices in the industry.

16.2.3.1 Sourcing Materials in the Fashion Industry: A Review of Fibres

Textile production in the fashion industry also involves sewing, cutting and assembling, which require a large workforce (UNECE 2019). One of the major concerns of the fashion industry is waste production throughout the supply chain from crop yields, yarns and textiles to pattern cutting, and the post-production stages, "about 15 percent of fabric intended for clothing ends up on the cutting room floor" (Rosenbloom 2010). Findings also show that discarded clothes more often than not end up as landfill. Most of the clothing that ends in landfills creates polluting gas and heavy metal releases as well as additive discharges into soil and groundwater causing soil degradation (Choudhury 2017).

Statistics show that 70% of the apparel fibres production is synthetic and non-biodegradable and the other 30% of natural fibres are often mixed with synthetics in the textile production phase. Blended fibres are challenging to recycle. To date, there is no specific analytical technique to identify the type of fibre in the garment, and chemical separation is often difficult as different types of fibre require their own specifications (Peters et al. 2014). Likewise, current recycling methods lack efficiency. For example, only 20% of the fibres in a pair of jeans can be recycled and polyester fabrics, also difficult to recycle. Moreover, recycling requires the separation of garments into different colours which is often a very labour-intensive process (Walker 2017).

Fibres can be natural or synthetic. Textile production is possible due to the manufacturing of fibres, which can be turned into yarn for knitting or weaving. Natural fibres are either sourced from agriculture or the production of synthetic (non-cellulosic) fibres from petroleum through chemical synthesis. However, the production of natural fibres through agriculture consumes a significant amount of total freshwater available for human consumption (Radhakrishnan 2017). Despite the obvious benefits of biodegradable fibres, those demand fertilizers and pesticides, which reduce soil fertility and consequently result in the biodiversity loss (ITPS 2015: 127). Data shows that cotton itself represents 82.7% of the total natural fibre production for the apparel sector whilst other fibres, which are more beneficial for the environment, represent a minority. For instance, wool represents 5.3%, flax 2.5% and cellulosic fibres 9.9% (FAO-ICAC 2013). Among all the natural fibres, cotton has the most adverse environmental impacts, followed by polyester, acrylic, elastane and nylon (Karthik and Rathinamoorthy 2017). In 2010, cotton represented 32.9% of the total world apparel fibre consumption (FAO-ICAC 2013). Evidence also shows that 350 million people are engaged in cotton farming, and it is manufactured in 100 countries (Radhakrishnan 2017). According to UNECE (2019), 2700 litres of water are needed to produce an average cotton shirt. Additionally, cotton farming is responsible for 24 percent of insecticides and 11 percent of pesticides despite using only 3 percent of the world's productive land (UNFCCC 2018). As for natural fibres like cotton, the recycling process consists of cutting clothes or textile waste and making them small enough, through stripping machines to pull them apart into fibres and finally spin into yarns. These shorter fibres then need to be blended with virgin cotton. Although natural fibres seem more beneficial to the environment, they can create negative environmental impacts due to unsustainable production practices and poor natural resource management.

On the other hand, synthetic fibres depend on the extraction of raw materials, such as petroleum, coal and limestone, adversely impacting soil and water supplies. Since Nylon became popular in the 1930s, its demand kept on increasing as it became a substitute for silk due to the scarcity of the latter in World War II. Nylon was mainly used for the production of military products like parachutes. However, it also opened the door for other synthesized polymers derived from petroleum. By 2010 synthetic (non-cellulosic) fibres represented 70% of the world's apparel fibre consumption (FAO-ICAC 2013). This is a major sustainability issue, as most polymers are non-biodegradable, and their manufacturing produces many harmful chemicals and emits greenhouse gases, fostering global warming.

Both natural and synthetic processes for the conversion of fibres into textiles use natural resources in the wet-treatment phases, including bleach, dye, print and final finishing to name a few. The mix of water and chemicals resulting from these processes contains large amounts of dyes, diluents, bleaches, detergents, optical brighteners and heavy metals among others. Evidence shows that these chemicals pollute the environment due to their high levels of acidity or alkalinity (pH) (Choudhury 2017).

16.2.3.1.1 Natural Fibres

Cotton
Cotton represents almost 80% of natural fibres for apparel, yet it is an excessively water-consuming crop, representing a problem for waste management. It also involves many issues regarding pesticide control; as with the higher use of pesticides, the insects' resistance to pesticides increases, demanding more aggressive doses resulting in a greater threat to surrounding biodiversity and human health (Choudhury 2017). The environmental pressure of cotton industry is huge, and as pointed out by WWF, even bringing cotton production to an acceptable environmental standard is a very challenging task (WWF 2019). Monsanto, the company that dominates the market of genetically modified (GM) seed, embarked into the reduction of the use of pesticides. Yet, the Pesticide Action Network (PAN) states that GM is not the appropriate solution as Bt cotton contains the toxin from the bacterium *Bacillus thuringiensis* (Bt) which kills some natural pests, such as the cotton bollworm (Black 2012) (not sure if this is correct).

However, cotton production cannot solely rely on this practice as there are other processes affecting yields, including rain distribution, soil conditions and farming practices. Moreover, whilst some pests are controlled by Bt cotton, others will replace them resulting in the so-called pesticide treadmill (Black 2012). Long-term solutions lie in the use of organic fertilization, avoiding synthetic chemicals, crop rotation and farming practices that are aligned with sustainable parameters for both nature and human wellbeing. Farmers and supply chain members have the support from organizations such as Better Cotton Initiative BCI, a not-for-profit organization which offers information and training programmes on this subject.

Hemp
In 2009, at the International Year of Natural Fibres symposium, the use of hemp was acknowledged as a natural fibre which should be used widely. Hemp's ecological footprint is less harmful. It grows quickly and densely and requires low-level water consumption. It is resistant to ultraviolet radiation, is naturally anti-bacterial and dyes well (FAO 2009). To balance the production of natural fibres, sources such as hemp should be considered more seriously by the textile industry. In Japan, for example, hemp has been used since the Jōmon Period for weaving clothing and baskets (Clark and Merlin 2013). Hemp thrives with average rainfall, and its naturally long fibres allow efficient spinning, using less energy during production,

reducing carbon emissions (Choudhury 2017). Unfortunately, hemp's links with marijuana makes its commercialization difficult. Despite this, it is by far the most environmentally friendly fibre, and its popularity is growing. China, for example, has become the leading producer and exporter of hemp fibres (FAO 2009).

Producers more often engage in the production of softer textured textiles through mixing hemp fibres with cotton. CRAiLAR (2018), for example, uses an enzyme-based process to transform blast fibres such as Hemp into softer fibres. This process can give hemp fibres a similar feeling and appearance to cotton-based fibres. Although there is not an exclusive process for blast fibres, these fibres including hemp, flax, jute and kenaf are suitable for cold weather and do not require irrigation. Blast fibres are competitive in the market, and there is no need to blend them with others. Furthermore, this process could be mono-material or multi-material; blast fibres used in this enzyme process are all biodegradable, becoming biological nutrients during decomposition.

Mulberry

Another fibre experiencing rapid growth with high adaptability to poor soil and climate change is mulberry. Mulberry pulp has been used to make paper in Asia, particularly in Korea, Japan, China, Thailand and the Philippines (Muthu and Gardetti 2016). Mulberry yarns were first used in the late twenty-first century in Korea where manufactures and research institutions developed yarns as a ply-twisted or filament yarn. These can be used for weaving and knitting textiles and in the production of cooling clothing (Muthu and Gardetti 2016). Mulberry is made using a rotary slitter and twisted. Water is added during the twisting process, so the yarn is softer and has a high tensile strength (Park and Lee 2014). North Face Korea and Youngone Outdoor – both multinational outdoor companies – have already used mulberry in their products. However, Mulberry yarns, much the same as hemp, are stiff and rough, making the textile process difficult.

These case studies clearly show there are many sustainable materials available to the fashion industry. These in conjunction with sustainable processes could contribute towards a more sustainable fashion industry. Such practices include disuse of synthetic chemicals that may contaminate water supplies and promotion of low-energy textile manufacturing. Tencel fibres, known as Lyocell, are fibres produced by the international group Lenzing, which produces wood-based viscose fibres, modal fibres, lyocell fibres and filament yarn. Lyocell fibres are produced through regenerating cellulose in an organic solvent, N-methylmorpholine-N-oxide hydrate. This procedure makes textiles softer, with greater absorption than cotton (TENCEL 2018). These fibres are also "derived from sustainable wood sources – natural forests and sustainably managed plantations. Wood and pulp used by the Lenzing Group are harvested from certified and controlled sources" in Austria and neighbouring countries (TENCEL 2018). Lenzing fibres are manufactured in a closed-loop, as cellulosic fibres are fully compostable. "Cellulose disintegrates into its native substances and prepares the ground for new plants to grow" (Lenzing 2018). These textiles are created through the use of nanotechnology in an award-winning solvent-spinning process which recycles water and reuses the solvent at a recovery rate of "more than 99%" (TENCEL 2018). Lenzing products including lyocell and modal fibres are the closest example of harvesting, producing and recovering materials with less than 0.1% of negative impact in their full-life cycle. Lenzing exemplifies sustainable practice in the fashion industry and should thus be replicated by other corporations.

Another case in point is soybean fibres produced by spinning the protein distilled from the soybean. Like hemp, this fibre was used to replace cotton in the World War II, but due to technical difficulties, its popularity quickly diminished (Fletcher 2014). In recent years, this fibre has gained strength in the textile market in China. However, there are major concerns regarding the soybean crops intensive tilling that require consideration.

The use of natural fibres in the fashion industry is not holistically sustainable, and therefore complementary fibres are needed throughout the production process, such as synthetic fibres. Below is a review of cases using synthetic materials and how those could be integrated into the fashion industry in a more sustainable manner.

16.2.3.1.2 Synthetic Fibres

The main issue regarding the production of synthetic fibres is energy consumption. Synthetics are made of petroleum feedstock containing chemicals harmful to the environment. They are also the major contributors of pollutants, resulting in clearing of forests for resources as well as emitting huge amount of GHGs (Superego 2018). Thus, synthetic fibres are not sustainable.

The use of antimony, for example, in the production of polyester turns into wastewater, releasing high amounts of greenhouse gases and volatile organic compounds (VOCs). "Over 70 billion barrels of oil per year are used to make polyester" (Karthik and Rathinamoorthy 2017). Synthetic fibres, if recycled, are downcycling, a process in which the value of a material decreases during recycling or the material ends up in landfills. Despite being unsustainable, synthetics are essential for some technological developments such as the design of space suits. Polyester is also used in fibres, bottles, containers and photographic films to name a few (Karthik and Rathinamoorthy 2017). Although polyester waste management often results in environmental problems, if "design begins at the molecular level, synthetic products can be conceived as technical nutrients, which are materials specifically designed to feed or be returned to, industrial systems without any harmful effects" (McDonough and Braungart 2002). In 2001, Victor Textiles introduced Eco-Intelligent™ Polyester. The company affirms to be the first dyed polyester fibre made of environmentally safe ingredients, such as dyestuff, auxiliary chemicals and titanium- and silica-based catalysts that replace the metalloid antimony (McDonough and Braungart 2002). Eco-Intelligent™ Polyester is designed to have a high performance and durability, with no limitations on colour choices, and can be woven in any jacquard pattern. Additionally, it is a fibre-to-fibre product, safely recycled to produce the same high-quality fibre as the original (Victor 2018). This system in partnership with recycling technology from Unifi, a yarn manufacturer, and

Designtex, a company specializing in the design and manufacturing of electronics, offers a solution for the environmental problems derived from the use of polyester-related textile products.

Biopolymers and Mixed Polymers
Biopolymers do not have the same biodegradability issues as petrochemical-based polymers. Produced from biological renewable sources including lignocellulosic biomass, fatty acids and organic waste, these polymers are biodegradable fibres derived from living organisms. The first biopolymers came from carbohydrate sources including corn, potatoes and other agricultural feedstock. These are synthesized as cellulose, starch, polylactide, chitin and collagen formed in the natural environment during the growth cycles of organisms. They can be produced through bacterial fermentation processes by synthesizing the relevant building blocks from these sources (Karthik and Rathinamoorthy 2017). Poly- (lactic acid) or polylactide (PLA) fibres are made of corn starch and sugar cane and can be used for textiles in the fashion industry. Ingeo™ fibres are produced by NatureWorks, a bioplastic-polymer manufacturing company. The company affirms that their textile products can be used for clothing as well as household products. They can also be knitted or weaved and resist steaming in temperatures of 80 °C for approximately 20 min. Products can also be blended with wool (NatureWorks 2018). The company also uses an Eco-Profile, a tool which analyses input and output data from the manufacturing process, such as "water to grow feedstocks, CO2 sequestered by plants, energy to produce fertilizers, and greenhouse gases" (NatureWorks 2018). This data is then used to measure environmental impacts including greenhouse gas emissions and use of non-renewable energy.

Another case in point is Yulex Pure™, which is a company that produces a 100% plant-based rubber. This fibre replaces Neoprene or polychloroprene, which is a petrochemical substance made by chlorinating and polymerizing butadiene. The company affirms that 99.9% of the harmful impurities (including proteins) are removed from hevea becoming free from toxic chemicals. This has resulted in a biodegradable, non-sensitizing natural elastomer that is suitable for over 40,000 product applications in the fashion industry (Yulex 2018).

Since the early 1900s, DuPont has developed polymers for the fashion industry. One of such products is Sorona®, a fibre made of TPA (terephthalic acid) and PDO (1,3-Propanediol). Bio-PDO is created through the fermentation of the crops glucose and chemical synthesis. Evidence shows that 37% of the polymer is made using plant-based ingredients. DuPont states that compared to nylon 6, Sorona® uses 30% less energy and releases 63% fewer greenhouse gas emissions (DuPont 2018). However, this fibre is neither biodegradable nor compostable because although Sorona® is 37% plant-based, the other 63% is petroleum-based. However, it can be considered as an option for increasing the use of renewable sources and energy saving in the fashion industry.

16.2.3.2 Production Process in the Fashion Industry

16.2.3.2.1 Wet-Treatment

Chemical use and water consumption in the production process are two of the most harmful environmental impacts of the fashion industry. Evidence shows that for approximately every tonne of textile produced, 200 tonnes of water are needed for dyeing, washing, printing, desizing, scouring, bleaching, mercerizing and the fabric finishing processes. All these processes comprise the wet-treatment phase (Choudhury 2017). For example, the production of a t-shirt requires approximately 2650 litres of water. Nearly 20% of this water is used for dyeing process (Lakshmanan and Raghavendran 2017). In order to reduce the water footprint of textile products, the industry has engaged in various technologies throughout the wet-treatment process, such as the use of plasma, ultrasonic-assisted dyeing and the use of supercritical carbon dioxide. These innovations reduce water use significantly and save energy in the wet-treatment process.

Plasma technology, for example, works at a low temperature and reduces the use of chemicals. After going through plasma polymerization, the chemicals lay on the textile surface. This process allows the fabric's internal structure to remain unaffected (Lakshmanan and Raghavendran 2017). Dyeing and finishing processes can be done by plasma gas particles, facilitating the modification of the fabric and improving textile characteristics and functionality. Plasma technology and ultrasonic-assisted dyeing technology can be applied to wool, cotton and polyester fabrics during the dry stage. These methods significantly diminish the use of huge quantities of sodium chloride, decreasing the discharges of dyestuff and other chemicals into wastewater (Lakshmanan and Raghavendran 2017).

The supercritical carbon dioxide method for dyeing synthetic fibres also reduces consumption of both water and energy. Through increasing temperature and pressure, it generates a liquid-gaseous state that produces supercritical fluids. Fabric or fibres are put inside an autoclave with dye powder, then purged with carbon dioxide (CO_2) and finally preheated. Carbon dioxide is non-toxic and non-flammable and can be recycled in a closed system (Lakshmanan and Raghavendran 2017). A case in point is the DyeCoo textile, the "world's first water-free and chemical-free dyeing solution". The CO_2 is reclaimed and has a recycling efficiency of 95% (DyeCoo 2018).

16.2.3.2.2 Green Chemistry

The discharge of harmful substances during the wet-treatment process is another major problem confronting the fashion industry. Some corporations have engaged in solutions to tackle this issue, including Adidas, Gap, H&M, Nike, Puma, Taiwanese dyestuffs makers and Everlight Chemical, which committed to Zero Discharge of Hazardous Chemicals by 2020 (ZDHC) (Choudhury 2017). Two processes can reduce the discharge of chemicals into water streams: recycling and use of enzymes. Enzymes can be used as substitutes for chemicals in the fashion industry. They can be obtained from three primary sources: animal tissue, plants and microbes. Enzymes are safely dischargeable after use, are biodegradable, consume low energy and produce lower greenhouse gas emissions. A case in point is lotus effect, a finishing treatment that alters the surface of the textile by biometric technology making the textile stain and soil-resistant (Kapsali 2012).

16.2.3.3 Product Production: Manufacturing Process

16.2.3.3.1 Design and Patternmaking Techniques

The design process also requires significant sustainability consciousness: "When talking about sustainability-oriented design, it is quite consolidated the fact that designers can play an important role especially in the early stages of design ... where 80% of impacts have been determined" (Marseglia 2017: 4). Design practices, such as zero-waste patternmaking, digital printing, recycling and upcycling, have proven to be of great value to create garments and fashion goods.

A case in point is the zero-waste design. It is a technique in which garments are designed and produced without creating textile waste by fitting all the parts of the pattern in one piece of the fabric. Instead of "forcing" patterns to fit together, zero-waste design requires new approaches to patternmaking. Some designers are applying traditional pattern techniques, using the whole piece of cloth, whereas others consider ways in which clothes should "wrap" the body rather than fit it. The zero-waste movement has been embraced internationally and includes designers such as Mark Liu, Julian Roberts, Holly McQuillan, Yeohlee Teng and Timo Rissanen.

16.2.3.3.2 Sewing Construction Garments

In her book *Sustainable Fashion & Textiles*, Kate Fletcher (2014) states that the most common approach to tackle waste arising from the textile life cycle is to implement waste management strategies (widely known as the 3Rs: reduce, reuse, recycle) (Fletcher 2014). In other words, "everything is a source for something else" (McDonough and Braungart 2002).

There are two types of recycling: downcycling, a process in which the recovered materials are processed into lower value products, and upcycling, in which the recovered materials are transformed into better quality products. Some companies have started using waste for product manufacturing. A case in point is Patagonia, a company committed to reduce its environmental impact by producing the most eco-sound way they can afford to. This philosophy of environmental conservation has been used not only for manufacturing but also for branding purposes. Patagonia uses 100% hemp textiles or hemp blended with recycled fibre textiles for manufacturing. The company also uses organic certified cotton, recycled wool and nylon, among others. Patagonia has also used polyester fabrics recycled from PET soda bottles since 1993, being the first outdoor company to transform this material into fleece (Patagonia 2018). The company also makes use of recycled blended fabrics such as Refibra™ Lyocell made of 80% wood and 20% recycled cotton. Their products are fair trade-certified by bluesign®, a Swiss-based company that certifies textile manufacturing by measuring energy consumption and CO_2 emissions (Power 2012). Patagonia also works with bluesign® technologies since 2000 to approve chemicals, processes, materials and products that are safe for the environment, throughout each step in the textile supply chain (Patagonia 2018).

Another case in point is the German company Vaude. In 2018, it received the GreenTec Award for its sustainable Green Shape Core Collection made with bio-based textiles, recycled or natural materials (Vaude 2018). Some of the bio-based textiles include Tencel, PrimaLoft® and Silver Insulation Natural Blend, made of 30% kapok tree and 70% fibres. Two-thirds of recycled materials used come from post-consumer recycled PET bottles. The company also uses the recycled material Econyl®, a yarn made of fishing nets collected from the ocean. Vaude (2018) also uses organic certified cotton and certified Terracare® leather which is also sustainably tanned.

16.2.3.3.3 Seamless Garments, Shima Seiki Wholegarment

In clothing production, machinery has evolved considerably over the years. Hand knitting has been industrialized by machinery capable of high speed and precise techniques that enable mass production. Every day the evolution of more automatic, faster, cleaner and advanced software has expanded its use in the fashion world. Nowadays, knitting has proven to be capable of manufacturing almost any product on the fashion market, which nobody would have otherwise thought to be possible. For example, running shoes brands like Nike and Adidas are now using knitting in mass production processes.

Knitting technology has reduced the processes in the production chain of clothing, avoiding, for example, weaving, cutting and, in some cases, sewing (Power 2012), especially with full garment or seamless garments machinery known as Wholegarment technology. This technology produces complete pieces ready to wear without requiring processes such as assembly, which results in the reduction of post-production labour as well as cutting down production time. Power (2012) explains that enhancements in knitting production due to technological developments demonstrate improvements in the product's quality, raising productivity and reducing cost "providing opportunities for new and modified products and techniques through innovation; and finally, reducing environmental impact of industrialized production" (Power 2012: 3).

Shima Seiki, a Japanese company, was the first to develop a fully automated glove-knitting machine, and since then, Shima Seiki and the German company Stoll have developed the most advanced technology on computerized machines for knitting. Shima Seiki was the first to introduce whole garment machines that work with CAD software, SDS-ONE APEX 3. This software provides a 3D simulation with final product images and material textures. Among other advantages of this technology, seamless production only requires the minimum amount of yarn needed, meaning no material loss is generated. "The knitting industry can make a significant contribution globally to savings in terms of energy consumption and waste through a number of avenues including tightening internal efficiency (reducing downtime), using indirect technology (needles and oils), implementing innovative technologies to change the manufacturing process (less waste, reduced post knitting operations) and reduced transport costs (complete garment for warp knitting, flat-bed knitting and circular weft knitting)" (Power 2012: 7).

The German company Groz-Beckert, manufacturer of the circular knitting machine needles, created litespeed® to promote environmental protection by changing the needle weight and size. This procedure reduces friction during the knitting process, resulting in energy consumption reduction of up to 20% whilst reducing CO_2 emissions (Power 2012). However, the complexities of the knitted garment construction and the difficulty of visualizing all the variables of the design process have overwhelmed knitting designers as well as technicians. In knitting design, the pattern shape and size are directly related to the material. The pattern is made by the software, and variables such as the roller tension, loop size, stitch type, quantity of threads by carrier, etc. need to be tested and modified for each type of yarn according to its thickness and other characteristics. This design process is far more complex than sewing and cutting which may influence the number of knitted products in fashion collections.

16.2.3.4 Waste and Disposal

Textiles can end up in landfills and can leak chemicals thereby poisoning groundwater, streams and rivers. Urgent actions on waste management are therefore necessary to reverse land degradation and halt biodiversity loss. Recycling is a sustainable practice adopted by various companies, yet the major problem confronting the fashion industry is related to post-consumer goods ending up at the landfill site.

16.2.3.4.1 Polyester Recycling

There are two methods of recycling in the fashion industry, namely, the mechanical and chemical methods. In the former method only single material clothes can be recycled. The process involves cutting them to small pieces, then ripping, spinning and blending them with virgin fibre. After the process is completed, these products are unlikely to be recycled for a second time. Chemical methods are often used for recycling synthetic materials such as polyethylene terephthalate PET, but post-consumed PET fibres are usually not recycled a second time due to various technical difficulties. Methods of recycling PET are often carried out through solvolysis or pyrolysis which can provide environmental solutions. The first is the degradation of waste material by solvents including water. In the second, the degradation process uses heat in vacuum. The challenge lies in recycling polyester fibre-to-fibre in a process that maintains the material quality. During this process, it is important to ensure that the chemical substances utilized in this process and released to the environment can biodegrade safely, thus, reducing energy consumption in the recycling processes of PET bottles (Al-Sabagh et al. 2015).

The Teijin ECOCIRCLE® system "turns polyester products into raw materials, and raw materials into products, in a never-ending ring of recycling resources. It is used in Fibres, PET bottles and other products. This system consumes 84% less energy and emits 77% less CO_2 than the production of polyester fibres from petroleum" (Teijin 2006). Recycling polyester is fundamental to reducing the waste generated by petrochemical feedstock as it lacks biodegradability and also has high-energy consumption rates. The difficulty of recycling polyester textile is that these textiles are usually blended with different materials. One of the most common blends is with cotton, which makes the process of recycling difficult. In order to facilitate this process, the garment should be made entirely of polyester, buttons, zippers, labels, etc. This is the idea of DuPont, Apexa® fibre, made of a biodegradable polymer that can be spun with natural fibres such as cotton, wool, hemp and silk. Labels, zips, tape, etc. can also be made of this polymer (Kapsali 2012).

16.2.3.4.2 Cotton Recycling

Refibra™ is a technology for upcycling cotton scraps from garment production. The cotton is blended with wood pulp and spun to produce new virgin Tencel Lyocell fibres to make fabrics and garments in a closed-loop production (Tencel Refibra 2018). "The leftovers from 16 virgin cotton shirts can be turned into one reclaimed cotton shirt" (Patagonia 2018). The mechanical method used for recycling cotton faces problems in material separation and produces lower-quality fibres during the recycling process. The mechanical method involves carding machines that tear the fabric apart. In this process, the fibres are shortened and blended with virgin material (Fletcher 2014). On the other hand, the chemical method dissolves the cotton into glucose synthetic fibre. The Hong Kong Research Institute of Textiles and Apparel (HKRITA 2018) developed a bioprocess that by enzymatic hydrolysis natural fibres such as cotton are degraded into glucose. This process offers a solution to the natural base materials (e.g. cotton) by increasing the efficiency on the recycling process as well as improving the fibre-to-fibre life cycle. This could also be a solution for blended textiles, which is one of the major difficulties in post-consumer waste recycling.

16.2.3.4.3 Blended Textiles Recycling

According to HKRITA (2018), their textile waste recycling process uses a biological pretreatment method and textile separation by hydrothermal treatment, offering a solution for recycling cotton-polyester blended fabrics. The pretreatment method modifies the structure of the textile waste with reusable chemicals to make them more susceptible to the subsequent hydrolysis. Enzymes, which are required for hydrolysis, are produced through fungal cultivation; these fungi grow in the surface of the fabric in 28 °C for 7 days. *Aspergillus niger* CKB is then recovered from the hydrolysis, in which the enzyme solution is blended with the pretreatment textile waste undergoing hydrolysis in a bioreactor. "This process hydrolyses the cotton into soluble glucose, while the non-biodegradable material (e.g. polyester) remains intact and is separated as a solid form by filtration" (HKRITA 2018). The polyester is then re-spun into yarn, whilst the glucose can be converted into bio-based products. This process provides a successful fibre-to-fibre method and is commercially viable. As such, used textiles discarded in landfills can be converted into new high-quality products, reducing the production of raw material whilst saving energy and resources.

Fabrics such as Climatex Lifecycle are biodegradable fibres (wool and ramie blend), coloured with nonharmful chemicals and manufactured without releasing carcinogens, persistent toxic chemicals, heavy metals or other toxic substances (Fletcher 2014). For example, "Worn Again", a project that by 2021 expects to launch an industrial plant which aims to separate, "decontaminate and extract polyester polymers, and cellulose from cotton, from non-reusable textiles and PET bottles… turning them back into new textile raw materials as part of a continual cycle… this is the first chemical recycling technology to be Cradle to Cradle (C2C) certified" (Worn Again 2018).

A review of recycling processes in the fashion industry suggests that clothing must be well designed to ensure that products can turn into either biological nutrients or technical nutrients. When biodegradation is not possible and products need to be reutilized or recycled for the production of new products, they are called technical nutrients. Furthermore, the aim must be to limit downcycling where possible, as downcycled products cannot be recycled the second time the cycle is interrupted, creating waste. The major challenge to turn textiles into biological nutrients is that many natural fibres are blended with synthetic fibres, which cannot return safely to the soil. In order to address this issue, the industry should align with sustainability standards which ensure the completion of a full cycle.

Fashion textiles often have a mix of different fabrics, making them difficult to recycle and as such contributing to its dumping as a waste material after use. As Tierra points out, a t-shirt composed by 99% cotton and 1% of spandex could not be recycled today and therefore would end up in a landfill or burnt in thermal power station (Tierra 2017). Mono-materials or fabrics made only from one material could solve the problem of recycling blended textiles. However, evidence shows that natural fibres (biodegradable) are blended with synthetics to improve quality, create textures, colour effects, etc. (Fletcher 2014). Mixed materials are also essential in labels, fasteners and elastic bands, etc. Production of yarn and synthetic manufacturing are also the most energy-consuming processes in the industry (Karthik and Rathinamoorthy 2017). In addition, the dyeing process is a major water pollutant, with 40% of globally used colourants containing carcinogens, making textile effluent one of the most significant causes of environmental degradation (Kant 2012). Therefore, there is a need to reduce waste and convert it into a completely reusable system.

16.3 Impact Sustainability: Final Remarks

There are strong connections between research and innovation in the fashion industry and SDG 15 Life on Land and SDG 14 Life below Water. This review clearly showed that major issues confronting the industry pertain to water consumption, impact on ecosystem services, water pollution and the industries' clear contribution to global warming, deforestation and environmental degradation (UN 2019).

The production of raw materials, crop management and production of fibres from petrochemicals feedstock drain natural resources in an unsustainable manner. The fashion industry demands great amounts of water and energy whilst contaminating inland water sources and oceans. Moreover, clothing production increased by "fast fashion" trends heightens soil degradation due to the use of chemically hazardous components. Therefore, the fashion industry needs to embrace solutions which promote the achievement of the SDG 15 and SDG 14 and commit to various socially responsible practices linked to production, gender, labour, poverty issues (UNECE 2019) as well as promoting the use of alternative materials such as blast fibres like hemp, Lyocell fibres Tencel and Yulexis. Sustainable resource management techniques such as reducing the use of pesticides and fertilizers are also essential for biodiversity conservation and consequently soil improvement – thereby preventing soil degradation.

Additionally, the use of technologies for animal and biodiversity protection through the use of alternative animal-like materials such as Zoa™ (bioleather) is necessary to achieve SDG target 15.6: "promoting fair and equitable sharing of the benefits arising from the utilization of genetic resources and promote appropriate access to such resources, as internationally agreed". Likewise, the use of eco-friendly techniques such as finishing plasma, ultrasonic-assisted dyeing and supercritical carbon dioxide methods, whilst promoting effective fibre-to-fibre recycle techniques, such as HKRITA's technologies, can contribute to achieve SDG target 15.3, "restore degraded land and soil". Promoting deployment of these technologies and encouraging the sustainable use of natural resources – from design to consumption– the fashion industry could lessen the impact on the life of land and below water, achieving overall sustainability.

References

Al-Sabagh AM, Yehia FZ, Eshaq G, Rabie AM, ElMetwally AE (2015) Greener routes for recycling of polyethylene. Egypt J Pet 25(1):53–64. https://doi.org/10.1016/j.ejpe.2015.03.00

Choudhury AKR (2017) Sustainable chemical technologies for textile production. In MS (ed) Sustainable fibres and textiles, The Textile Institute Book Series, Pages 267-322

Black S (ed) (2012) The sustainable fashion handbook. London, pp:278–283. isbn: 978-0500290569

Boucher J, Friot D (2017) Primary microplastics in the oceans: a global evaluation of sources. IUCN, Gland, Switzerland. 43pp

Cervellon M, Wernerfelt A (2012) Knowledge sharing among green fashion communities online: lessons for the sustainable supply chain. J Fash Mark Manag 16(2):176–192. https://doi.org/10.1108/13612021211222860

Clark R, Merlin M (2013) Cannabis evolution and ethnobotany. University of California Press. isbn: 978-0-520-27048-0

CRAiLAR (2018) CRAiLAR Fibre Technologies International Inc. Retrieved from https://www.crailar-fti.com/what-is-crailar-fibre-2/

Diamantopoulos A, Schlegelmilch BB, Sinkovics RR, Bohlen GM (2003) Can socio-demographic still play a role in profiling green consumer? A review of the evidence and an empirical investigation. J Bus Res 56: 465–480 (Crossref, ISI, Google Scholar)

DuPont S (2018) DuPont, Sorona®. Retrieved from http://sorona.com/apparel

DyeCoo (2018) DyeCoo. Retrieved from http://www.dyecoo.com/co2-dyeing/

FAO (2009) FAO. Retrieved from FAO Natural Fibres Ancient Fabrics, High-Tech Geotextiles : http://www.naturalfibres2009.org/en/fibres/

FAO, ICAC (2013) World apparel fibre consumption survey. FAO, ICAC. isbn: 9780979390395

FAO, ITPS (n.d.) (2015) Status of the World's Soil Resources (SWSR)- main report. Food and Agriculture Organization of the United Nations; Intergovernmental Technical Panel of Soils, Rome, Italy

Fletcher K (2014) Sustainable fashion and textiles second edition. Routledge. isbn: 978-0415644563

Fraj E, Martinez E (2006) Environmental values and lifestyles as determining factors of ecological consumer behavior: an empirical analysis. J Consum Mark 23(3): 133–144. (Link, Google Scholar, Infotrieve)

Gam HJ (2011) Are fashion-conscious consumers more likely to adopt eco-friendly clothing? J Fash Mark Manag 15(2):178–193. https://doi.org/10.1108/13612021111132627

HKRITA (2018) HKRITA, April 14. Retrieved from HKRITA Garners Admirable Accolades in the International Exhibitions of Inventions of Geneva: http://www.hkrita.com/marketing/PressRelease/20180419_HKRITA_46th_Invention_Award_Press_Brief/Award_projects_eng.pdf

Kant R (2012) Textile dyeing industry an environmental hazard. Nat Sci:22–26

Kapsali V (2012) Nature as a paradigm for sustainability in the textile and apparel industry. In: Black S (ed) The Sustainable fashion handbook. London, pp 278–283. isbn: 978-0500290569

Karthik R, Rathinamoorthy R (2017) Sustainable synthetic fibre production. In MS (ed) Sustainable fibres and textiles, The Textile Institute Book Series, pp 191–240

Kim H, Damhorst MR (1998) Environmental concern and apparel consumption. Cloth Text Res J 16(3): 126–133 (Crossref, Google Scholar, Infotrieve)

Kunz G (2005) Merchandising theory, principles, and practice, 2nd edn. Fairchild Publications, New York (Google Scholar)

Lakshmanan S, Raghavendran G (2017) In: SS (ed) Low water-consumption technologies for textile production, The textile institute book series, pp 243–265

Lenzing AG (2018) Retrieved from https://www.lenzing.com/en/sustainability/

Marseglia M (2017) Design process and sustainability. Method and tools. Des J 20:S1725–S1737

Masson-Delmotte V, Zhai P, Pörtner H-O, Roberts D, Skea J, Shukla PR (2018) Global warming of 1.5 degree celcius: summary for policymakers. IPCC

McDonough W, Braungart M (2002) Cradle to cradle: remaking the way we make things (a Melcher Mediabook). North Point Press, New York

Muthu SS, Gardetti MA (2016) Green fashion volume 1. Springer, Singapore. isbn: 978-981-10-0109-3

Napper IE, Thompson RC (2016) Release of synthetic microplastic plastic fibres from domestic washing machines: effects of fabric type and washing conditions. Mar Pollut Bull 112(1–2):39–45

NatureWorks (2018) NatureWorks. Retrieved from https://www.natureworksllc.com/

Park TY, Lee S (2014) Fibres polymer. 14(311). https://doi.org/10.1007/s12221-013-0311-4

Patagonia (2018) Patagonia. Retrieved from https://www.patagonia.com/materials-tech.html

Peters G, Granberg H, Sweet S (2014) The role of science and technology for sustainable fashion. In: Fletcher K, Tham M (eds) The handbook of sustainable fashion. Routledge, London

Power E (2012) Sustainable developments in knitting. Int J Bus Glob 9(1):1–11

Radhakrishnan S (2017) In: SS (ed) Sustainable cotton production, The textile institute book series, pp 21–67

Rosenbloom S (2010) Fashion tries on zero waste design, August 13. New York Times

Roy AK (2017) Sustainable chemical technologies for textile production. In SS (ed) Sustainable fibres and textiles, The Textile Institute Book Series, pp 267–322

Šajn N (2019) Environmental impact of the textile. European Parliamentary Research Service

Superego (2018) Superego, February. Retrieved from The Dangers of Synthetic Fibers and Fabrics on the Environment: https://superegoworld.com/blogs/the-world/the-dangers-of-synthetic-fibers-and-fabrics-on-the-environment

TEIJIN (2006) TEIJIN CSR Report 2006. Retrieved from https://www.teijin.com/csr/report/pdf/csr_06_en_all.pdf

TENCEL (2018) TENCEL. Retrieved from https://www.tencel.com

TENCEL REFRIBRA (2018) TENCEL. Retrieved from REFRIBRA: https://www.tencel.com/refibra

Tierra (2017) Tierra. Retrieved from Mono-Materials: https://tierra.com/environmental-work-2/mono-materials-recycling/

UN (2019) UN Sustainable Development Goals. Retrieved from Life on Land: Why does it Matter?: https://www.un.org/sustainabledevelopment/wp-content/uploads/2016/08/15_Why-it-Matters_Goal15__Life-on-Land_3p.pdf

UNECE (2019) Fashion and the SDGs: what role for the UN? https://www.unece.org/fileadmin/DAM/RCM_Website/RFSD_2018_Side_event_sustainable_fashion.pdf. Retrieved 9 April 2019

UNFCCC (2018) United Nations Climate Change News, 22 January. Retrieved from https://www.unece.org/fileadmin/DAM/RCM_Website/RFSD_2018_Side_event_sustainable_fashion.pdf

Vaude (2018) Vaude. Retrieved from https://www.vaude.com/en-INT/Company/Project-campaigns/Green-Shape-Core-Collection/

VICTOR Textiles (2018) VICTOR Textiles. Retrieved from https://www.victortextiles.com/en/at-a-glance/#case-study

Walker A (2017) Patent Application Publication. Worn Again Footwear and Accessories Limited

Worn Again (2018) Worn Again. Retrieved from Technologies http://wornagain.co.uk/

WWF (2019) WWF. Retrieved from Sustainable Agriculture: https://www.worldwildlife.org/industries/cotton

Yulex Corporation (2018) Yulex Corporation Products. Retrieved from http://yulex.com/products/fsc-hevea/

Zimmer MR, Stafford TF, Stafford MR (1994) Green issues: dimensions of environmental concern. J Bus Res 30(1):63–74 (Crossref, ISI, Google Scholar, Infotrieve)

Chapter 17
SDG 16 Peace, Justice and Strong Institutions

The Untapped Potential of Women for Sustainable Peace in Resource Regions

Isabel B. Franco and Ellen Derbyshire

Abstract This chapter makes a strong contribution to SDG 16 as it examines the role of women and livelihood options in fostering sustainable peace. Research was conducted in rural communities in the vicinity of Risaralda, Colombia, a resource-rich region. The results obtained are important as they provide insights into the heterogeneous composition of communities, particularly women and their identities, which explains contrasting perceptions towards project development. Secondly, outcomes entail a practical dimension which suggests that in the process of assessment, development and management of resources, differences related to community identity, religion and context-based factors must be recognized and taken into account to foster sustainable peace and overall regional sustainability.

Keywords SDG 16 Peace, Justice and Strong Institutions · Women · Sustainable livelihoods · Community · Resource regions

---

I. B. Franco (✉)
Institute for the Advanced Study of Sustainability, United Nations University Shibuya-ku, Tokyo, Japan

Australian Institute for Business and Economics, The University of Queensland, Brisbane, Australia
e-mail: connect@drisabelfranco.com

E. Derbyshire
Faculty of Business, Economics and Law, Business School, The University of Queensland, Brisbane, QLD, Australia
e-mail: ellen.derbyshire@uq.net.au

17.1 Introduction

The notions of sustainable livelihoods and gender are taking centre stage in the context of developing resource-rich regions, particularly those in a situation of conflict. Evidence shows that top-down approaches to development, which neglect women's development aspirations regarding their livelihood options, are consistently implemented. This is not only detrimental for women but also for companies operating in resource regions as it will escalate community discontent and resentment towards the industry (Franco 2014). A situation which increases political instability, particularly in locations with a history of armed conflict. Instead, fostering the participation of women can open new paths for sustainable peace and foster overall sustainability in resource regions.

Evidence shows that the representation of women through gender equality initiatives presented in the peace deal with FARC has been pivotal in the government's awareness of the importance of women. This recent agreement exemplifies the significant role that gender has to play in the development of sustainable livelihoods in Colombia's mining regions. A 2012 United Nations analysis of 31 peace processes between 1992 and 2011 shows that women make up 4% of signatories, 2.4% of chief mediators, 3.7% of witnesses and 9% of negotiators (UN Women 2012). What this means for the future of gender inclusion and the representation of women in resource-rich regions is significant. The agreement in the Colombian case seeks to atone destruction and limitations imposed by conflict whilst addressing inherent inequalities for women in Colombia. As evident in Clause 5 of the agreement, "... the sub-commission on gender to conclude the work it has been engaged in as promptly as possible, consisting of the revision of the agreements 1, 2 and 4 of the General Agreement with a gender-based approach" (Alto Comisionado para la Paz 2016). The first agreement is comprised by plenipotentiary members of the delegations, to which other advisors decided by each party may be added, in addition to the advisors responsible for each topic, in order to facilitate rapprochements at the strategic level, make the warranted decisions to streamline the drafting of agreements and supervise the work of the delegations. The inclusion of women in the governmental bureaucratic processes and discussions further represents the growing importance in women as actors in the public sphere. This change is a significant step towards legitimizing the role of women in these mining regions. Furthermore, it empowers women to become primary economic actors of socio-developmental change within resource-rich regions.

The clear increase in resource development has severely impacted women and their opportunities to fashion sustainable livelihoods. The extractive industry operating in these locations has experienced an economic boom, significant for the global economy with revenues from the sector accounting for US$ 863 million in 2009 in countries like Colombia (Molina-Escobar and Restrepo-Baena 2010; Ponce 2010). However, evidence shows that communities have not benefited from these economic gains (Franco 2014), a situation that causes resentment and tension at the community level, compromising overall sustainability and the achievement of the Sustainable Goal 16 (SDG 16), Peace, Justice and Strong Institutions.

Integrating livelihoods of women in the extractive industry or other sectors relevant for the local economy will create value for both the resource industry and women themselves. However, stakeholders (particularly the extractive industry) meet difficulties in pursuing such integration and building resilient assets. Women's livelihoods have undergone substantial transformations led by local dynamic change. For example, violent times in Risaralda, Colombia, diminished community opportunities to develop livelihood options other than mining. During times of armed conflict in Colombia, illegal groups tapped on revenues coming from natural resource extraction and very often forced community members, particularly women, to engage in illegal mining. The low income that some families received from mining was inefficiently spent, and gender and power dynamics amongst small miners elevated internal conflict in the community. In addition, "low educational level impeded women miners effectively administering the income they got from gold extraction" (Community Leader, Interview, November 2012). Informal miners did not know how to handle the benefits derived from informal mining, neither did they realize the magnitude of the social and environmental problems triggered by poor mining practices. This situation led to some local authorities partnering with community members and other local stakeholders to provide capacity building in the form of education (e.g. a vocational programme on making jewellery), providing women with sustainable human capital and therefore alternative employment and livelihood options.

Communities, particularly women, adjacent to extractive operations more commonly experience loss of livelihood options. Hence, there is a need to engage the extractive industry in helping women fashion sustainable livelihoods in the regions they operate (Franco 2014). This can create added value to the industry and local communities, reduce tension in resource-rich regions and foster paths for sustainable peace through the creation of new sustainable livelihood options. Moreover, there is a recognition that natural resources, particularly minerals and metals, will not last for long and that local communities, mostly women, shouldn't solely rely on natural resource extraction in order to advance towards sustainability. The discussion in this chapter is organized as follows: Sect. 17.1 introduces the literature review, followed by the methodology and presentation of Risaralda, Columbia, the case study area. The paper finishes with the discussion and data analysis and final remarks on impact sustainability.

17.2 Literature Review

17.2.1 *Sustainable Livelihoods*

The world has experienced rapid resource development, particularly in developing and less industrialized economies. The total net output of the resource sector in the world grew by 1.7% in 2013 (UNIDO 2014). However, increased growth of the sector has given rise to complexities within the industry, leading to unsustainable resource development at locations where the industry operates with implications for women and sustainable livelihood options. Poor resource development practice in rural areas poses significant caveats to social welfare services and discourages the participation of women in local political dialogues, with educational inequalities further perpetuating the rural land tenure. Due to gender gaps, it is often the case that rural female household heads have limited access to land, training, financing and other benefits that are generally available to men. Thus, public emphasis on support and investment aimed at helping women are vital for increasing productivity and improving livelihoods for women in resource-rich regions.

Concepts like sustainable livelihoods are abound in the present-day sustainable development (SD) literature. Whilst the notion of SD owes its origins to environmental activists in the nineteenth century, in the contemporary era, SD is seen as a broad term encompassing a wide range of social, economic, environmental and political elements (Dresner 2008). Global organizations such as the World Bank and the United Nations have embraced the Brundtland Commission's definition of SD, one that states it is "Meet(ing) the needs of the present without compromising the ability of future generations to meet their own needs" (World Commission on Environment and Development 1987, p. 31).

Forging sustainable livelihoods necessitates placing people, particularly women, at the centre of the resource industry. This approach transcends the notion of sustainability and therefore has the potential to overcome community development limitations (often caused by community natural resource dependence). Scholarship in sustainability, corporate social responsibility and business ethics argues that communities receiving paternalistic treatment by corporations encounter limitations in achieving sustainable development aspirations (Cornelius et al. 2008; Jenkins 2004; Veiga et al. 2001). Jenkins (2004, p. 26) calls this "false dependency": a scenario in which corporations act as providers of services and business for communities. In reality, it only worsens the situation as communities become "mere puppets in the regeneration game played out by large national, regional and local agencies" (Cornelius et al. 2008, p. 358). Yet, resource industries can however significantly contribute to helping communities, particularly women, forge more sustainable livelihoods through integrating them into the industry or into other sectors relevant for the local economy. This approach also has the potential to assist the industry in fostering sustainable resource practices, creating diversified livelihood options and reducing tensions emerging from unequal distribution of resources.

17.2.2 Women's Sustainable Livelihoods

Forging sustainable livelihoods in resource regions can assist women in becoming active participants in shaping their life plans and achieving their sustainable development aspirations. It can also foster paths for sustainable peace in resource regions commonly impacted by conflict. Recent resource development has positioned the extractive industry as one of the major economic activities in the region. In the Colombian case, operations have been undertaken by Canadian and domestic exploration and mining companies. Metal production in Risaralda represented 6.71% of Colombia's total production in 2005 (UPME 2006). The extractive industry plays a strong role in enhancing women's human capital assets, an approach that is leading to more sustainable livelihoods in Risaralda, Colombia. Research shows that active engagement and empowerment of women, state government's public policies and effective governance arrangements are factors that have contributed to forge sustainable women's livelihoods (Franco 2014).

Risaralda's government authorities agree that the extractive industry is essential to meet regional competition standards and achieve the sustainable development goals, particularly SDG 16, Peace, Justice and Strong Institutions. It is important to note that companies are made accountable for meeting acceptable sustainable resource development and are asked to make a strong commitment to regional development. There are three key approaches taken by the local government to forge sustainable livelihoods; these include education, knowledge transfers and entrepreneurship (Franco 2014). These three pillars, as constituent components of Risaralda's governance framework, are considered as fostering factors to develop less resource-dependent communities and aid in creating more sustainable livelihood options. However, we can see the transferal of economic dependency from local capacity-building initiatives to short-term dependency on resource revenue. For areas such as Risaralda, localization of corporate policies will provide an economic boost to grassroots capacity building and upskilling initiatives. Thus, the involvement of women in the extractive industry and other local economic sectors is paramount to forging sustainable resource regions and for the progress towards sustainable peace (Franco 2013). Business literature therefore recognizes the role of women participation in the local economy, particularly in running business in nontraditional sectors such as mining, construction and manufacturing. In addition, scholarly debates address concerns regarding the lack of empowerment and poor decision-making skills in integrating both the community and women in the local and global economy (Christensen and Grant 2007; Mabudafhasi 2002; Muthuri 2007; Said-Allsopp and Tallontire 2015).

One of the most significant research findings in the Risaralda case study has to do with active women engagement. Women have played a strong role in the formation of local development agendas, becoming active participants in achieving their own development aspirations (Franco 2014). Women have also benefited from socially responsible practices intended to help them cope with livelihood transformations induced by the extractive industry. Such initiatives comprise activities, like agribusiness, dressmaking, jewellery manufacture, coffee production and agriculture. For example, former informal miners and miners' wives have been provided by the company with dressmaking training to create alternative livelihoods and realize their entrepreneurial potential. These actions have been implemented either as a result of effective government-corporation partnerships or as a response to women's requests.

17.3 Methodology

This chapter is based on empirical research in Risaralda, Colombia. The selection of the case study was based on three criteria: (1) a developing resource region, (2) evidence of women starved for sustainable livelihood options and (3) a region which displays both complexities and opportunities in integrating livelihoods of female entrepreneurs in extractive industry or other sectors in the local economy. This section also provides a more detailed description of the Colombian case study.

Colombia is a Latin American country located in the north of South America that has experienced an escalating resource growth over the last three decades. Mining along with oil extraction represented 4.6% of the Colombian GDP in 2005 (UPME 2006). The resource industry is also becoming one of the most representative economic sectors in Risaralda. Only 3.6% of Colombian municipalities are both rural and show a high human development index. Thus, many local governments in these municipalities have succeeded in fostering sustainable resource development through effective fiscal management and efficient public institutions. Although some others have failed due to external factors such as armed conflict and political instability, there are other factors which heavily influence the quality of women's livelihoods such as economic strength, human capital accumulation and the emergence of a larger middle class of "rural" landowners. The gender gap and rurality are key caveats to the empowerment of females in Colombia. More specifically, the key limitations we see in resource-rich regions are evident in the monopoly of a state-centric approach to sustainable community development.

Risaralda is located in the Colombian Andes mountain range. Mining and exploration projects operating in Risaralda have impact on women and their livelihoods in both urban and non-urban areas. According to the last census register (DANE 2005), Risaralda hosted a population of 859,666 people in 2005. Out of the total population, 665,104 people inhabited urban areas, whereas 194,562 were located in peri-urban and rural areas. 51.3% of the population was female and 48.7% male. Statistics also show that Risaralda hosted 230,532 households in urban and non-urban areas. According to the "Mission to Link Employment, Poverty and Inequality Surveys" (MESEP), an individual is considered poor if he or she earns less than US$103.30/month. Using this income threshold, almost half of the country's population lives below the poverty line. The income of the average poor Colombian is below the poverty line by $48.10/month. The gap is even wider for the rural poor at $51.88/month; and more significantly, rural women's incomes are farthest below the poverty line at $53/month. As women were the most susceptible to this "poverty trap," it increases the complexity of finding livelihood options and opportunities to aid their transition out of poverty (CIAT 2013), a situation that also compromises political instability, hence demonstrating the importance of providing recommendations that assist the stakeholders in resource regions in forging sustainable livelihoods and paths to sustainable peace.

17.4 Discussion

This section aims to discuss existing approaches to help women in developing resource regions forge sustainable livelihoods and open new pathways towards sustainable peace. It is argued that women's livelihoods are very often threatened due to lack of sustainable assets to realize their full potential, armed conflict and inequality (Franco 2014). The extractive industry, particularly the mining sector, has been pivotal in protecting women's livelihoods. However, there is still room for improvement in helping women enhance their assets to cope with mining-induced changes. Mining will eventually affect the livelihood of women in farming, jewellery design, artisanal mining, large-scale mining and other economic activities. Women farmers, for example, are choosing artisanal mining over agriculture which might displace traditional livelihood options in the long term. However, the minimization of the extraction industry's detrimental effects on the community and the fostering of opportunities for sustainable peace are mainly due to the active participation of women and the enhancement of women's assets. Active participation of women in income generation has also helped them proactively embark on enterprises, request the required resources and skills to positively transform their lives and help them cope with potential mining impacts. Existing approaches have also helped women achieve their development goals and forge sustainable livelihoods. This has also assisted stakeholders in responding to conflicting demands in resource locations.

Based on women's expectations, the extractive industry and governments have embarked on sustainable development agendas and mechanisms to foster peaceful arrangements at the community level. This has had positive impacts on locals, as these practices are more in tune with local circumstances. Female consultation in relation to needs and expectations is highly appreciated in the development of these initiatives, resulting in immediate benefits. For example, local coffee and jewellery producers, as well as female entrepreneurs, have already experienced the benefits of these initiatives:

> We have been trained in jewellery design ... the company has also provided us with some financial assistance to attend international fairs so that we can promote and sell our products. (Community Members, Interview, October, 2012)

There is a close relationship between good governance and sustainable livelihoods for women. Collaboration for forging sustainable livelihoods of women has been a driver to enhance resilient assets. Governance is not restricted just to the role of government but also involves other parties such as the extractive industry and civil society (Davies 2005). It is part of the role of these three parties to aid women in developing asset-based adaptation strategies to help women forge sustainable livelihoods. Furthermore, understanding the political and economic agendas of international companies involved in operations is necessary to gage the impact and success of sustainable livelihood development and capacity building initiatives. Exploration projects currently undertaken by Canadian companies occupy large areas of land that host important reserves of gold. Informal miners who previously worked in this area are involved in ongoing negotiations with the Canadian company to explore possibilities of relocation and opportunities for their livelihood transformations. A group of informal miners has obtained a concession to keep mining the land where the Canadian company operates. Current dialogues between small-scale informal miners and multinational corporations have a strong focus on employment generation. A dialogue tha is necessary to enforce the necessity to localize capacity building in a gender inclusive manner.

Lessons can be learnt from the Risaralda case and applied to other developing resource locations. Sustainable livelihoods for women need to be further developed according to their expectations and needs. This bottom-up economic development approach is already having a significant effect on many women in Risaralda, Colombia, and equipping them with more resilient assets that will help them sustain their livelihoods in the long term.

In addition to a bottom-up approach to development, collaboration has been essential. In 2008, government authorities extended the impact of this initiative and agreed with communities that they were going to add sustainable value to raw materials extracted in the region. This was the beginning of a successful multi-stakeholder collaboration in which "governments and civil society actors partnered to create a jewellery association—a community-based organization" (Community Leader, Jewellery CBO, Interview, November, 2012). The association is one of a number of income generators at the local level. At the present time, the organization is self-sustainable and employs locals. The arrival of multinational companies in the region has further strengthened the work of CBOs in Risaralda. A Canadian company operating in the region strongly supports this initiative as they are keen to "promote social sustainability in other industries relevant for the community... [Similarly, the international corporation] is in a permanent dialogue with State actors to know more about community's needs and be able to assist women and local communities increase income" (Corporate Representative, Interview, October, 2012).

Collaborative approaches for enhancing women's assets have resulted in positive outcomes for the women involved. For example, "the company helped the municipality to open a plant for waste collection which has the potential for generating income for the local community" (Community Leader, Jewellery CBO, Interview, November, 2012). Ongoing collaboration between companies, educational institutions and civil society organizations has also helped women create asset-based strategies to increase local income: "Universities located in the region are helping us undertake market analysis to improve cost production and increase income ... this is very important for us because we are a key stakeholder for [Risaralda's agriculture sector]" (Community Leader, Blackberry CBO, November, 2012).

Fashioning sustainable livelihood options for women has also been driven by the need for economic diversification. This has led to some positive livelihood transformations, as women have been able to generate income from industries other than mining. In addition, supported by private and government organizations, community associations, in which women actively participate, are not only increasing income but also fostering livelihood opportunities in jewellery design and sales. This has reduced women's dependency on mining and helps sustain female livelihoods.

17.5 Impact Sustainability: Final Remarks

A community's pre-existing vulnerabilities, as well as the complexities of the extractive industry, are likely to result in more adverse impacts on women and their livelihoods. Successful integration of livelihood options in the local economy faces a number of complexities that might increase or diminish the extractive industry's ability to help women forge more sustainable livelihood options and respond to conflicting demands from local actors. This chapter identified effective approaches to sustain women livelihoods and integrate them in the local economy. It also indicated that priority livelihood areas which are the most valuable for women are those that help them achieve their own development aspirations. Because of this, women-based approaches are more likely to create value for female leaders than corporate and/or government top-down approaches.

In these cases, in which companies have embraced community-oriented and women-driven agendas, women claim to have become more resilient to cope with mining-induced transformations, including conflict. This approach has served as a means to reduce disputes, achieve SDG 16 and sustain women's livelihood overtime. Women who have been properly consulted regarding livelihood priorities have been able to strengthen key assets, becoming more capable of coping with mining-induced changes. However, such approaches need to be included as a constituent component of both corporate and government policies. It is not solely the resource industry's responsibility to make these approaches valuable for women as governments also need to share responsibilities with companies in this regard. Governments also need to consult women and get them involved in decision-making about matters such as priorities for the allocation of funds and asset transformation. Furthermore, these consultation processes should specifically include female miners and women entrepreneurs. Similarly, major attention must be paid to the potential growth of the existing agriculture community associations, in helping women become active suppliers for the extractive industry or other industries relevant to the local economy. These lessons drawn from the Colombian experience can be applied to other Latin American regions and elsewhere.

References

Alto Comisionado para la Paz (2016) Acuerdo Final para la Terminación del Conflicto y la Construcción de Una Paz Estable y Duradera: Alto Comisionado para la Paz

Buitrago Franco I (2013) Mining, capacity-building and social license: making the links. Paper presented at the World Mining Congress. Montreal, Canada. Peer reviewed paper.

Buitrago Franco I (2014) Building sustainable communities: enhancing human capital in resource regions. PhD Dissertation, The University of Queensland, Brisbane

Christensen J, Grant M (2007) How political change paved the way for indigenous knowledge: the Mackenzie Valley Resource Management Act. Arctic 60(2):115–123

Cornelius N, Todres M, Janjuha-Jivraj S, Woods A, Wallace J (2008) Corporate social responsibility and the social Enterprise. J Bus Ethics 81(2):355–370

DANE (Departamento Administrativo Nacional de Estadísticas) (2005) Perfil Risaralda Bogota: DANE
Davies JS (2005) Local governance and the dialectics of hierarchy, market and network. Policy Studies 26(3–4):311–335
Dresner S (2008) Principles of sustainability. Earthscan, London
International Centre for Tropical Agriculture (CIAT) (2013) Policies for bridging the urban-rural gap in Colombia: CIAT
Jenkins H (2004) Corporate social responsibility and the mining industry: conflicts and constructs. Corp Soc Responsib Environ Manag 11(1):23–34
Mabudafhasi R (2002) The role of knowledge management and information sharing in capacity building for sustainable development--an example from South Africa. Ocean Coast Manag 45(9–10):695–707
Molina-Escobar JM, Restrepo-Baena OJ (2010) Colombian mining sustainability. Dyna-Colombia 77(161):149–151
Muthuri JN (2007) Corporate citizenship and sustainable development – fostering multi-sector collaboration in Kenya. J Corp Citizsh 28(12):73–84
Ponce A (2010) Panorama del Sector Minero. Unidad de Planeación Minero Energética, Bogota
Said-Allsopp M, Tallontire A (2015) Pathways to empowerment? Dynamics of women's participation in global value chains. J Clean Prod 107:114–121
UN Women (2012) Women's participation in peace negotiations: connections between presence and influence: UN Women
UNIDO (United Nations Industrial Development Organization) (2014) Mining and utilities. Industrial Development Organization, Vienna
UPME (Unidad de Planeación Minero Energética) (2006) Colombia País Minero 2019. Unidad de Planeación Minero Energética, Bogota
Veiga M, Scoble M, McAllister ML (2001) Mining with communities. Nat Res Forum 25(3):191–202
WCED (World Commission on Environment and Development) (1987) Report of the world commission on environment and development: our common future. WCED, Oslo

Chapter 18
SDG 17 Partnerships for the Goals

Global Business Networks: Accounting for Sustainability

Isabel B. Franco and Masato Abe

Abstract Corporate accountability has expanded rapidly in recent years as a tool that business can employ to respond for corporate performance. Nevertheless, the exercise of corporate accountability is challenging as it requires stronger corporate capacity and commitment to respond to external stakeholders in alignment with voluntary regulatory norms. In response, corporate agendas are being significantly shaped by internal and external stakeholders that are employees, suppliers, and customers. Yet, an examination of the current status of accountability both as a historical trend and in current corporate agendas in selected cases shows an increasing gap between policy and impact. Based on a case study method and theory building, the research reported in this article shows the status of corporate accountability of companies operating in Asia-Pacific and adhered to the Global Compact Network. It also provides some conceptual and practical tools toward enhanced accountability, contributing to the achievement of the Sustainable Development Goal 17 (SDG 17) and its targets *Multi-stakeholder partnerships* (17.16 and 17.17) and *Data, monitoring, and accountability*.

Keywords Corporations · SDG 17 · Sustainability · Reporting · Accountability · Disclosure · Global Compact Network

---

I. B. Franco (✉)
Institute for the Advanced Study of Sustainability, United Nations University Shibuya-ku, Tokyo, Japan

Australian Institute for Business and Economics, The University of Queensland, Brisbane, Australia
e-mail: connect@drisabelfranco.com

M. Abe
United Nations Economic and Social Commission for Asia and the Pacific, Macroeconomic Policy and Financing for Development Division, Tokyo, Japan
e-mail: abem@un.org

18.1 Introduction

There is a general consensus that corporations need to become more accountable to both their shareholders and stakeholders. Limited understanding of how businesses account for their sustainability performance and to what extent they are being truly accountable to external stakeholders is an area that remains underinvestigated. Corporations are more often tasked to become more responsible to external stakeholders and the environment where they operate. This need for greater external accountability is increasing, yet companies still hold limited capacity to duly report on corporate performance. There is also an increasing concern among scholars and practitioners about how businesses can become more accountable both in policy and practice (Post 2013; Parsons 2017; Schembera 2018). While corporations are increasingly pressured to respond for social, economic, and environmental sustainability matters, there is a dearth of knowledge about the practice of accountability, both internal and external, and to what extent it compromises stakeholders' ability to report successfully on sustainability performance and in alignment with voluntary regulatory frameworks and global accountability networks (Abe and Ruanglikhitkul 2013; Franco 2014).

This chapter is based on three cases examined in the light of a Global Compact Network supporting ten universal principles in the areas of human rights, labor standards, environmental protection, and anticorruption. The network is typically complex, reciprocal, and accountability based (Gilbert and Behnam 2013). There are cases in which these interactions are inexistent compromising members' compliance with the network. Despite the relevance of accountability for successful reporting, the literature on the global reporting networks remains relatively silent when it comes to investigate the extent to which companies are being truly accountable and in line with the network's principles. The conditions and mechanisms that contribute to the constitution of accountability in the network are poorly understood which is reflected in limited corporate capacity to respond to external stakeholders. The authors argue that enhanced external accountability from signatory members and partners of the Global Compact to external stakeholders is a precondition to foster overall sustainability. Therefore, this research links to the Sustainable Development Goal 17 as it contributes to the achievement of targets on *Multi-stakeholder partnerships* (17.16 and 17.17) and *Data, monitoring, and accountability* (Table 18.1). Against this background, the aim of this chapter is to examine current forms of accountability of a selected group of signatory members of the examined network. A thorough investigation of the connection between accountability and the network itself may help identify concrete measures for increasing accountability to external stakeholders and increasing not only our understanding but also the implementation of the network's ten principles in a sustainable manner.

Based on a case study method and theory-building methodology, this manuscript reports research on selected cases, namely, Republic of Korea, Myanmar, and Sri Lanka. Findings also show that corporations need take accountability seriously to truly respond for sustainability performance. Interestingly, evidence indicates that

Table 18.1 Sustainable Development Goal 17: targets

Sustainable Development Goal (SDG 17) – targets
Multi-stakeholder partnerships
17.16 Enhance the global partnership for sustainable development, complemented by multi-stakeholder partnerships that mobilize and share knowledge, expertise, technology, and financial resources, to support the achievement of the sustainable development goals in all countries, in particular developing countries
Multi-stakeholder partnerships
17.17 Encourage and promote effective public, public-private and civil society partnerships, building on the experience and resourcing strategies of partnerships
Data, monitoring, and accountability
17.18 By 2020, enhance capacity-building support to developing countries, including for least developed countries and small island developing States, to increase significantly the availability of high-quality, timely and reliable data disaggregated by income, gender, age, race, ethnicity, migratory status, disability, geographic location and other characteristics relevant in national contexts

accountability in the three cases is mainly internal. In addition, in those cases in which external accountability occurs, it is not fully exercised as companies do not necessarily comply against the network's principles nor display accurate accountability mechanisms, such as sustainability reports. The study shows that accountability is a premature practice and further development in this area is needed in the selected cases.

These and other research findings are reported as follows: the first section in this manuscript provides an extensive review of the notion accountability, its evolution, and current debates in the corporate sector. The second section of this manuscript presents and discusses a comparative analysis of three country cases, namely, the Republic of Korea, Sri Lanka, and Myanmar. These three countries were selected according to the corporate engagement and alignment with the network, stages of their socioeconomic development, and geographical location in the Asia-Pacific region. This manuscript finishes with conclusions and recommendations for further research in this field.

18.2 Literature Review

The concept of accountability is not recent, yet it has evolved rapidly as a business practice. Nevertheless, how accountability is interpreted and operationalized by the corporate sector is an issue that deserves further investigation. Accountability is a concept that has multiple interpretations that makes it a complex (Mulgan 2000) and chameleon-like term (Sinclair 1995). Debates on accountability first took place around 700 B.C.E in archaic literature and later on played a key role during the ascendancy of the Athenian democracy, in medieval times before evolving rapidly as a governance practice both at the private and public spheres.

A segment of scholars in law and ethics agree that accountability is a logical consequence of Athenian democracy (Roberts 1982). According to Von Dornum (1997), legal institutions that established the roots of accountability date to around 700 B.C.E. Accountability mechanisms first appeared in archaic Greek literature, in the form of poems that encouraged the regulation of the polis', citizens', and officials' behaviors. Nowadays, this characteristic can be found in accountability practices as a legal and technical tool to assess performance. Hesiod's poems, written in 700 B.C.E, first introduced the notion of euthyna, a form of vertical accountability to follow gods' will. Solon's literature pieces, written in 620–590 B.C.E, highlighted the importance of citizens' participation in political systems to make public decisions more transparent. Pindar's victory odes in 500 B.C.E suggested the idea of common cultural values to make citizens and officials more accountable for their behavior. Euthyna was based on the principles of rectitude as it was applied to make officials accountable to polis' citizens (Von Dornum 1997). Euthyna consisted of auditing of financial and nonfinancial dealings to guarantee citizens' rights. It also punished officials who behaved illegally by way of multiple accountability mechanisms such as fines, imprisonment, detention, partial exclusion or expulsion, corporal punishment, or death (Foxhall and Lewis 1996).

This form of accountability was based on the principles of rectitude as it was applied to assess official performance and procedure to make officials' conduct more visible for polis' citizens. Euthyna consisted of the auditing of financial and nonfinancial dealings in order to guarantee citizens' rights. It also punished officials who behaved illegally by way of multiple accountability mechanisms such as fines, imprisonment, detention, partial exclusion or expulsion, corporal punishment, or death (Foxhall and A.D.E.Lewis 1996, pp. 74–83). The contributions from the notion euthyna not only constructed the concept of accountability but also supported the ascendancy and development of the Athenian democracy. The notion of accountability in the Athenian democracy became a legal mechanism to preserve the exercise of authority as well as a partially inclusive mechanism that allowed polis' citizens – except women – to get involved in public decisions. This framework set in Classical Greece became the basis of the roots of medieval and more contemporary accountability.

In Medieval England, accountability was conceived as a mode of financial regulatory mechanism (Godfrey and Hooper 1996). It was also applied to economic relationships to explain the landowners' obligation in exchange of workers' social protection. Likewise, it was used to explain the monarchy's concern for strengthening financial governance and decision-making regarding land ownership (Galbraith 1961; Godfrey and Hooper 1996). Contemporary forms of accountability in the form of financial accountability to shareholders are more common approaches than external accountability approaches.

Contemporary approaches to accountability began with "Accountingization" processes that facilitated interactions and trust relationships among actors (Hood 1991; Gray 1997; Power and Laughlin 2003). These approaches emerged along with public management practices such as privatization. In 1996, the World Bank (1996) stated that "public provision must become the exception rather than the rule."

Privatization emerged as a new public management practice that led to public cuts and gained status with the implementation of government reforms under the administrations of Margaret Thatcher and Bill Clinton Hughes 2003). Privatization encouraged corporations to become more responsive to external stakeholders, that is, customers, suppliers, governments, and other civil society groups and individuals.

Accountability can be internal or external (Mulgan 2000; Stewart 1999). Internal accountability occurs between the CEO and board chair account for financial reporting and guarantees profitable revenues to shareholders. This form of accountability became the norm in the corporate sector after the publication of the Cadbury's Report in 1992 (Brennan and Solomon 2008). The report embraces accountability as a corporate principle to improve corporate effectiveness (Committee on the Financial Aspects of Corporate Governance 1992). The notion of internal accountability was reinforced by Friedman (1970) and other economists, promoters of neoliberalism between the1960s and 1970s (Eisenhardt 1989a; Wilson 1968). They argued that the responsibility of a business is to increase profits and that corporations are not accountable to external stakeholders but only to shareholders. Internal accountability mechanisms are presented in the form of reports from board chairs to shareholders (Bovens 1998) and the engagement of external auditors that prove the accuracy of these reports (Brennan and Solomon 2008).

External forms of accountability, instead, are based on the premise that society is an active rather than a passive stakeholder entitled to request accountability by corporations (Gray et al. 1997). Freeman's (1984) work on stakeholder theory challenges conventional internal accountability practices, adding democratic, social, and sustainability attributes to this notion. External accountability, therefore, contests old business practices (Cooper 1992; Gray et al. 1997; Mulgan 2000). Under this form of accountability, corporations are encouraged to engage with external stakeholders, be more inclusive and sustainable, and comply against voluntary sustainability accountability norms (Eisenhardt 1989a; Mulgan 2000; Cummings 2001; Fontrodona and Sison 2006). Lack of external accountability compromises corporations' ability to respond for their performance (Boiral and Henri 2017) resulting in adverse impacts on the environment, local communities, and corporations themselves (Amer 2015).

18.3 Methodology

A multiple case study approach was used because it facilitates a comparative examination of the three regions of interest. The selection of the three case studies was based on three criteria: (1) regions where companies adhere to the network's principles, (2) locations where businesses display various forms of accountability, and (3) cases that display accountability issues. The examination of country cases first consisted of a review of business accountability in the form of sustainability reporting and performance of companies adhered to the global network. Theory building was also applied to draw inferences from the literature, and policy reviews are also applied to examine the selected cases. Policy analysis on corporate data was obtained from 80 companies adhered to the network.

A case study methodology was combined with theory building. The qualitative research strategy sought to inductively build new theory concerning accountability and reporting. As such, our data analysis followed procedures for inductive theory building from each country as prescribed by Braun and Clarke's (2006) thematic analysis and Yin's (2009) chain of evidence method to analyze data (also Eisenhardt 1989b). The chain of evidence method calls for clear links in moving from data to theory. Theory building is a process of data reduction based on cause-effect evidence. The process was to first group data into countries based on (1) accountability status, (2) accountability mechanism, and (3) focus area. From these three sets of themes, researchers then induced theoretical inferences of cause-effect between features and available information. Tables 18.2, 18.3, and 18.4 record the data from this inductive theory-building process. After obtaining results from separate countries, a comparative analysis was undertaken to identify commonalities and differences. Findings were validated whenever possible with literature and policy reviews and presented in the discussion and in the conclusion sections.

18.4 Discussion

Accountability is an essential component of corporate governance and overall sustainability. The need to make corporations more responsible for their sustainability performance has led to the proliferation of voluntary sustainability regulations. The examined network is one of such and establishes minimum requirements that ensure active members are making changes that are socially and environmentally responsible. The framework ensures that business practices remain transparent and accessible by all shareholders and stakeholders. Based on a review of three cases, namely, Republic of Korea, Sri Lanka, and Myanmar, this section explores to what extent corporations aligned with the network's principles are externally or internally accountable and how accountability is operationalized.

The study showed that despite the existence of voluntary regulatory frameworks, internal accountability prevails upon external forms of accountability. Three types of external accountability mechanisms were identified, namely, sustainability policies, sustainability performance/activities, and sustainability reporting of both policy and performance. The latter concerns to corporate disclosure through sustainability reports often aligned with the network's principles. Sustainability policy relates to policies or corporate statements that show corporate commitment to sustainability. Finally, evidence indicates, however, that companies applying a sustainability performance approach, display a plethora of scattered initiatives in the form of corporate philanthropy reflecting a lack of a cohesive sustainability strategy or framework.

18.4.1 Republic of Korea Case Study

The interest of Korean businesses in being seen as accountable has grown since recent years. However, how they are exercising such an accountability is an issue that deserves nuance attention (Sok and Whang 2014). For the purpose of this analysis, a sample of 30 companies that claim to be adhered to the network's principles was selected across different sectors in RoK. The status of accountability in selected companies is shown in Fig. 18.1.

Overall, the analysis indicated that accountability is mainly internal and top-down as companies are being mainly accountable to shareholders (Mulgan 2000; Ospina et al. 2002) and in some cases to the government (Ministry of Legislation, 1998). Interestingly, the current form of top-down accountability, adopted by selected companies, seems to be the norm. Yet, it is not sustainable (Abe and Franco 2017). The RoK case shows a growing gap between accountability and sustainability performance which may compromise social viability and economic growth within these companies.

The study found that 63% of businesses do not display any accountability mechanism. Eleven (11) per cent only account for sustainability performance. Twenty-three (23) per cent of companies reviewed only present a sustainability policy or corporate sustainability statement. Interestingly, just 3% of companies present a sustainability report accounting for both sustainability policy and performance. Thus, the analysis shows that only a limited number of companies take full accountability to external stakeholders.

Evidence shows that businesses still lack effective accountability capacity to respond for their sustainability performance in an adequate manner. Findings show

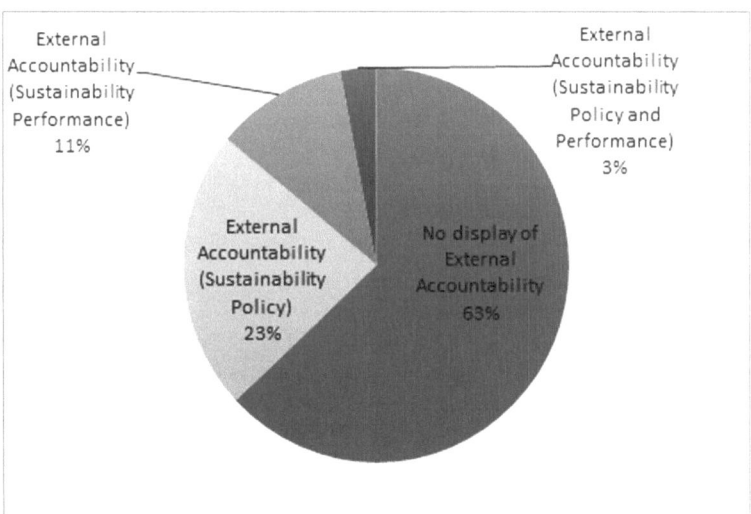

Fig. 18.1 Accountability in the Republic of Korea. (The author 2018)

also that, very often, businesses embrace the notion of accountability to rebuild their reputation and brand image. Accountability is therefore perceived as value-enhancer, rather than a genuine practice to truly account for sustainability performance.

Data analysis also indicates that most of the selected companies do not have accountability mechanisms in place. Yet, in those cases in which corporations do display accountability mechanisms, those do not comply effectively against the network's principles. Instead, companies report on a plethora of scattered activities that in most cases take the form of corporate philanthropy.

18.4.2 Myanmar Case Study

Research findings show that businesses operating in Myanmar experience major difficulties to remain accountable for their sustainability performance. Societal and organizational factors are two of the greatest challenges that need to be overcome to enable a more ethical business culture. These challenges emerge due to weak legal enforcement systems that are binding and high rates of the violation of codes of conduct by organizations and individuals. Despite the strong regulatory framework, however, enforcement lacks, resulting in poor sustainability knowledge and practice. A case in point is the Central Bank Law that mandates board directors to release reports on monetary and financial stability (State Law and Order Restoration Council 2013).

Surprisingly, however, a review of 35 companies shows limited capacity to account externally for sustainability performance. Findings also show there are two issues, businesses need to pay more attention to stakeholder engagement and transparency on revenue management and reporting. Figure 18.2 illustrates the status of accountability of selected corporations in RoK. The analysis showed that 46 percent of businesses are not accountable externally. Accountability within this group of companies is mainly internal. Further, 29% of the companies only display their sustainability policy but fail in duly accounting externally for sustainability performance. Information is reported mainly on social media channels (e.g., Facebook) and concerns to the business itself rather than to corporate sustainability performance. Evidence also indicates that 14% of corporations provide a list of philanthropic actions to account externally for their sustainability performance. However, businesses in this range do not show a comprehensive sustainability strategy that backs up their performance. The analysis also revealed that only 11 percent of businesses issue sustainability reports containing both sustainability policy and performance. Some of the areas in which companies claim to be accountable for include but are not limited to reliefs to refugees and natural disaster victims, tree planting, community healthcare, and education. Yet, compliance with the network's principles remains to be an issue.

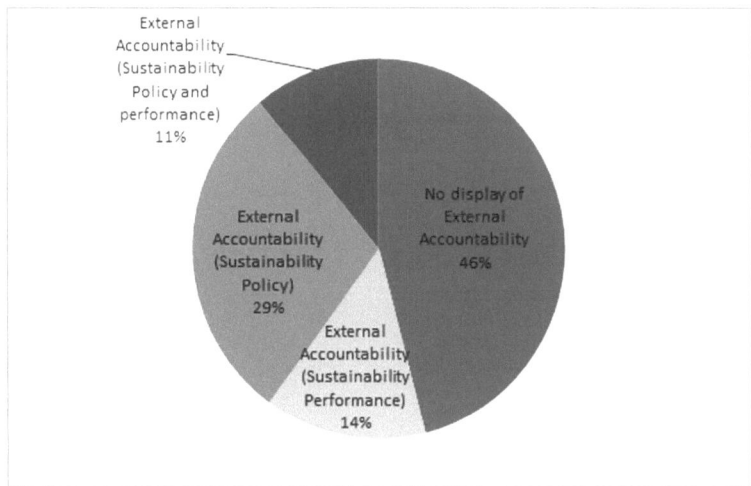

Fig. 18.2 Sustainability accountability in Myanmar
Source: The author

18.4.3 Sri Lanka Case Study

Research findings revealed that companies need to become duly accountable to external stakeholders and the environment they operate within. A sample of 20 companies was selected across different sectors in Sri Lanka. The analysis shows that 42% of the companies do not exercise external accountability. Twenty-five (25) per cent of the companies account for sustainability policy but fail in translating it into practice. Approximately 33% of companies account for sustainability performance usually in the form of philanthropy. It is surprising that none of the businesses selected fully account for their sustainability performance. These results show that internal accountability is the preferred approach to respond for corporate performance. However, in those cases in which external accountability is exercised, research findings showed that corporations hold limited capacity to be duly accountable. Figure 18.3 presents the status of accountability of companies adhered to the network in Sri Lanka.

Evidence indicates that accountability in Sri Lanka is often used to gain political advantage, rather than as a business strategy to truly account for corporate performance. In response to this, the government is more often playing a stronger role to encourage businesses to boost their capacity on business ethics (Bedewella and Fairbrass 2016).

Stronger government intervention facilitates the prevention, detection, investigation, and prosecution of the offences of money laundering and the financing of terrorism, respectively. In the Sri Lankan case, government institutions undertake due diligence measures to combat money laundering and the financing of terrorism. A case in point is the Ministry of Finance and Planning, whose responsibility is to

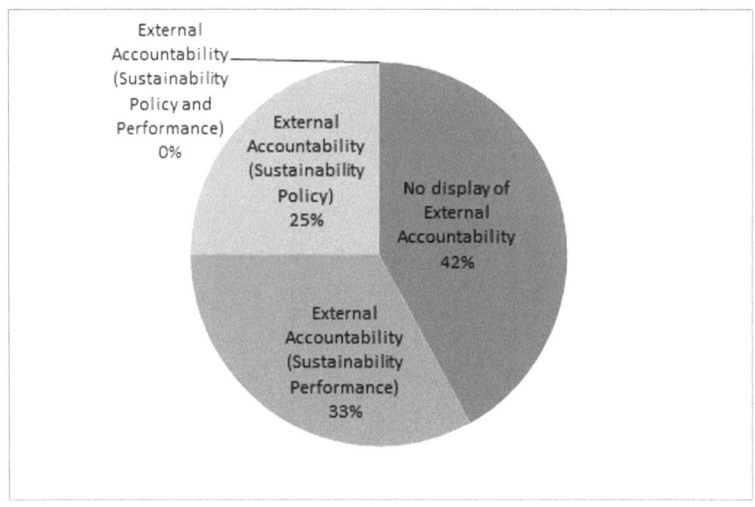

Fig. 18.3 Sustainability accountability in Sri Lanka
Source: The authors

facilitate the prevention, detection, investigation, and prosecution of the offences related to terrorist financing. This body also ensures that corporations remain legally accountable for their conduct.

18.4.4 Commonalities and Differences

The analysis shows there is limited corporate understanding of external accountability as a mechanism to respond for impacts on the environment and the communities. Accountability is a major issue in selected cases, with 63%, 46%, and 42% of companies not displaying external accountability mechanisms in RoK, Myanmar, and Sri Lanka, respectively. Internal accountability is therefore the most predominant approach among reviewed companies. External accountability is poorly exercised mainly through social media (e.g., Facebook, websites). Companies operating in RoK seem to be the most critical case with 63% of companies not exercising fully external accountability.

A limited number of companies, however, issue sustainability reports in compliance with the network's principles. Selected companies operating in the Sri Lankan case seem to face the most critical issues as evidence shows that none of the reviewed companies have a sustainability mechanism in place or report any sustainability policies or strategies. However, the Sri Lankan case shows dissimilar outcomes in regard to sustainability performance with 33% of companies fully accounting externally. Yet, the nature of reported initiatives deserves further investigation as companies are adopting corporate philanthropy as an approach to external accountability.

Similarly, reported initiatives, in most cases, do not comply with the network's principles.

Concerning sustainability policy, the analysis shows even results among cases. The investigation, however, found a high number of companies are not accountable for either sustainability policy or performance. A situation that may compromise overall corporate sustainability.

18.5 Impact Sustainability: Final Remarks

This study outlined the evolution of accountability theory and current practice in alignment with a global network for corporate accountability. It highlighted the current state of accountability and approaches to corporate accountability applied to day-to-day operations in the selected cases in the Asia and Pacific region. This research is essential for businesses and governments interested in improving corporate accountability practices. Hopefully the conclusions and recommendations below assist stakeholders in developing business accountability practices that truly respond to impacts on the environment and the community. The authors also hope this research provides some conceptual and practical instruments to fully exercise accountability, contributing to the achievement of the Sustainable Development Goal 17 (SDG 17) and its targets *Multi-stakeholder partnerships* (17.16 and 17.17) and *Data, monitoring and accountability*. Although companies have a primary responsibility to their shareholders, they are increasingly being tasked – through international agreements and guidelines and in-country obligations with social and environmental responsibilities to external stakeholders. In this context, companies need to contribute to ensuring the effective development, provision, and assessment of corporate accountability practices, which must be a constituent component of the business strategy enabling them to become more accountable to external stakeholders. Although this research focused on selected cases in Asia and the Pacific region, the conclusions are highly likely to be applicable to other locations in the region and elsewhere.

If the fostering of business accountability is identified as a corporate goal, then companies need to be more active in support of this and to be more accountable to external stakeholders. While corporate accountability agendas are vital to achieve development aspirations and protect the environment, in practice they have serious constraints. Lack of government support to encourage corporate accountability and limited corporate capacity for external accountability are two of the main obstacles that hinder business accountability practices. It is recommended that the international community plays an active role in advising and building corporate and government capacity on corporate accountability. It is equally important that both governments and companies and global business networks further engage in collaborative approaches to corporate accountability, especially where it has been identified as being of high priority, thereby increasing opportunities to overall sustainability.

The study also unveiled that in doing so, governments need to engage in developing more effective regulatory frameworks to help companies comply with international sustainability mandates such on sustainability reporting. Although companies operating in the three cases have proactively adhere to global sustainability mandates, they are not obliged to be fully accountable for and report on sustainability performance. Disconnection between sustainability practices and sustainability mandates remains to be a critical issue. The study suggests that if business accountability is identified as a business strategy to account for sustainability performance, businesses need to be both more active in support of it and more responsive to external stakeholders and the environment they operate within. As such, there is a strong need to enhance both businesses' and governments' capacity, so that businesses can go beyond business as usual and become truly accountable for the impact of their business operations.

Appendix (Tables 18.2, 18.3, and 18.4)

Table 18.2 Corporate accountability status in Republic of Korea

Company	Accountable	Accountability mechanism	Focus area
Doosan Engine Co., Ltd.	Yes	Website (strategy and report)	Code of conduct EHS (environment, health, and safety) Community involvement and development
Onemount	Yes	Website	None

NAVER Corporation	Yes	Website (business ethics)	Costumer
			Value for costumers
			Respect for costumers
			Costumer protection
			Stockholder
			Protect shareholder's interests
			Disclose management information
			Partners
			A win-win internet environment
			Protection of partners information
			Society
			   Abide by legal regulations and social norms
			   Contribute to social development and the creation of a healthy culture
			   Retain political neutrality
			Employees
			Respect for employees
			Fair and responsible treatment

(continued)

Table 18.2 (continued)

Company	Accountable	Accountability mechanism	Focus area
CJ CheilJedang	Yes	Website	CEO's statement: "As illustrated by our campaign of Farmers and CJ Happy Together,' we will strengthen CSV (Creating Shared Value) to apply CJ CheilJedang 's key competitiveness to fostering the food industry ecosystem, taking further steps from our current corporate social responsibility activities like donations and volunteer works" There are no reports available to account for existing responsible business practices
Daewoo Engineering & C	Yes	Website (sustainability)	Sustainability management report Social contribution Quality management Environmental management Safety management
LOGOS PIT	No	N/A	N/A
Macoll Communication C	Yes	Website	None
LG Chem, Ltd.	Yes	Website (sustainability)	Sustainability management Environment, safety, and health Social partnership Sustainability report
Bio Focus Co., Ltd	Yes	Website (community)	None
G.S. ACE Industry Co	Yes	Website	Policy statement Human respect and customer satisfaction
Woo Young Logistics Co	Yes	Website (corporate value)	"Focus on talent cultivation to build a sustainable development logistics service company"
Panko Corporation	Yes	Website	None
Miraeroedu Co., Ltd	No	N/A	N/A
Gimm-Young Publishers	Yes	Website	N/A
Hankook Tire	No	N/A	N/A
Korea Transportation S	No	N/A	N/A
The Basic House Co., Ltd	Yes	Website	CEO's message "Good Company" is the main management philosophy of The Basic House, and with this philosophy we believe it is possible to continuously develop and contribute to our society

(continued)

Table 18.2 (continued)

Company	Accountable	Accountability mechanism	Focus area
Bucheon City Facilities	Yes	No	N/A
Korea Exchange Bank	Yes	Website	None
JW Medical Corporation	Yes	Website	Ethical management: JW Group strives to become a trustworthy corporate that is responsible for the healthy future of citizens based on transparent and fair management
KIRD	No	N/A	N/A
KEOSAN Machinery Co, Ltd	Yes	Website	Company's statement: Our company is a promising venture business that makes incessant efforts to develop new technologies to become a business focused on respecting environment and humankind and taking a responsibility for the development of our nation
Saturn Bath Co., Ltd	Yes	Website	None
Mediana, Co, Ltd	Yes	Website	Corporate statement: We seek to continuously optimize our manufacturing and business processes. We promote fair and the best business practices. Our ultimate goal is to earn the trust of our customers by using our imagination and skills to continuously offer them better medical solutions
HMDC Co, Ltd	No	N/A	N/A
Anyang Public Amenities	No	N/A	None
Asia Seed Co., Ltd.	Yes	Website	None
Flotron Corporation Li	Yes	Facebook	None
Line Reclamation Corporation	No	N/A	N/A
Lexcode Inc.	Yes	Website	N/A

Table 18.3 Corporate accountability status in Myanmar

Company	Accountable	Accountability mechanism	Focus area
Octagon International Services Co., Ltd	Yes	Facebook	None
Ayeyarwady Bank	Yes	Website	Community participation
Asia Green Development Bank L	Yes	Website	None
Diamond Mercury Group of Companies	No	–	–
United Paints Group Co., Ltd.	Yes	Facebook	Mission statement
Han Investment Group Co., LTD	Yes	Facebook	None
YKKO Group of Companies Limited	Yes	Website	Mission statement
Myanmar Culinary Holdings Company	Yes	Facebook	None
Shwe Yaung Pya Argo Co Ltd	No	–	–
Best Industrial Company Limited	Yes	Website	–
City Mart Holding Co., Ltd	Yes	Website	Flood victims Tree planting No plastic bag campaign Mobile soup kitchen City Love and Hope Foundation
Asia Royal Hospital	Yes	Website	None
Myanmar Marketing Research Services	Yes	Website	None
Fortune International	Yes	Website	Mission statement Refugees of ethnic conflicts and victims of natural disaster
Blue Ocean Operating Management Co. Ltd.	Yes	Website	None
Device Services Company Limited	Yes	Facebook	None

Parami Energy Group Of Companies	Yes	Website	Sustainable business practices:
			   Transparency policy
			   Anti-corruption policy
			   Human rights policy
			   Policy action plan
			   Action plan to be released

(continued)

Table 18.3 (continued)

Company	Accountable	Accountability mechanism	Focus area
Medi Myanmar Group Ltd.	Yes	Website	None
Max Energy Co., Ltd.	Yes	Website	None
FAME Pharmaceutical Industry (2014)	Yes	Website	Mission statement on CSR
Ocean Emerald Pearl Group Co., Ltd	Yes	Website	Vision statement
Dagon International Limited	Yes	Website	Report available on healthcare and education donations
Serge Pun & Associates (Myanmar) Limited	Yes	Website	Community Myanmar Business Coalition on AIDS Cape Negrais Relief and Recovery Committee Anti-malaria Program No reports available
Shwe Taung Group of Companies	Yes	Website	Policy statement
Scipio Services Co., Ltd.	Yes	Website	None
Myanmar Computer Co., Ltd	Yes	Website	None
Myanmar Information Technology Pte. Ltd.	No	–	–
KMD Company Limited	Yes	Website	Online report on compliance with the principles of the Global Compact
Information Matrix Co. Ltd.	Yes	Website	Report on compliance with the principles of the Global Compact
Mandalay Technology	Yes	Website	None
Today Top Star General Services Company Limited	No	–	–
First Ocean Breeze	No	–	–
K.L.S Partners Co.,	Yes	Website	None
Access Spectrum Company Limited	Yes	Facebook	Code of ethics

Source: The authors

Table 18.4 Corporate accountability status in Sri Lanka

Company	Accountable	Accountability mechanism	Focus area
Growrite Substrates Pvt.	Yes	Website	Strategies and activities on SRB shown on the website
Global links international	No	–	–
Thread Works (Pvt) Ltd.	Yes	Website	People agenda Solutions for aging population
Biogrow Lanka (Pvt)	No	–	–
Diesel & Motor Engineering (2015)	No	–	–
Access Engineering PLC	Yes	Website	Sustainability statement
Tropicoir Lanka (Pvt)	Yes	Website	Program: "Children of today"
Cargills (Ceylon) PLC	Yes	Website	Sustainability strategy
Chemanex PLC	Yes	Website	CSR statement
Ceylon Asset Management (2015)	No	–	–
Aspirations Education	No	–	–
Lanka ORIX Leasing Com	No	–	–
M/s. Eswaran Brothers	No	–	–
Hayleys PLC	Yes	Website	Sustainability statement
Dialog Axiata PLC	No	–	–
Brandix Lanka Ltd.	Yes	Website	CSR strategy
Kelani Valley Plantation	Yes	Website	Ethical business Environment
Mabroc Teas Ltd.	Yes	Website	CSR programs
MAS Holdings Ltd.	Yes	Website	CSR statement

References

Abe M, Franco I (2017) Socially responsible business: a model for a sustainable future. United Nations Economic and Social Commission for Asia and The Pacific, Bangkok. Retrieved from http://www.unescap.org/publications/socially-responsible-business-model-sustainable-future-studies-trade-investment-and

Abe M, Ruanglikhitkul W (2013) Integrating sustainability reporting into global supply chains in Asia and the Pacific, Chapter II. From corporate social responsibility to corporate sustainability: moving the agenda forward in Asia and the Pacific, studies in trade and investment no. 77. United Nations Economic and Social Commission for Asia and the Pacific (ESCAP), Bangkok

Amer E (2015) The penalization of non-communicating UN Global Compact's companies by investors and its implications for this initiative's effectiveness. Bus Soc 0007650315609303

Bedewella E, Fairbrass J (2016) Seeking legitimacy through CSR: institutional pressures and corporate responses of multinationals in Sri Lanka. J Bus Ethics 136:503–522

Boiral O, Henri JF (2017) Is sustainability performance comparable? A study of GRI reports of mining organizations. Bus Soc 56(2):283–317

Bovens M (1998) The quest for responsibility. Accountability and citizenship in complex organisations. Cambridge University Press, Cambridge

Braun V, Clarke V (2006) Using thematic analysis in psychology. Qual Res Psychol 3(2):77–101

Brennan NM, Solomon J (2008) Corporate governance, accountability and mechanisms of accountability: an overview. Account Audit Account J 21(7):885–906

Committee on the Financial Aspects of Corporate Governance (1992) Financial aspects of corporate governance report with code of best practices. Gee (A division of Professional Publishing, London, p 90

Cooper C (1992) The non and nom of accounting for (M)other nature accounting. Audit Account J 5(3):23

Cummings J (2001) Engaging stakeholders in corporate accountability programmes: a cross-sectoral analysis of UK and transnational experience. Bus Ethics Eur Rev 10(1):45–52. https://doi.org/10.1111/1467-8608.00211

Eisenhardt KM (1989a) Agency theory: an assessment and review. Acad Manag Rev 14(1):57–74

Eisenhardt KM (1989b) Building theories from case study research. Acad Manag Rev 14(4):532–550

Fontrodona J, Sison A (2006) The nature of the firm, agency theory and shareholder theory: a critique from philosophical anthropology. J Bus Ethics 66(1):33–42. https://doi.org/10.1007/s10551-006-9052-2

Foxhall L, Lewis A (1996) Greek law in its political setting justifications not justice. Clarendon Press Oxford, New York

Franco I (2014) Building sustainable communities: enhancing human capital in resource regions – Colombian case. The University of Queensland, Brisbane

Freeman RE (1984) Strategic management: a stakeholder approach. Cambridge University Press, New York

Friedman M (1970) A Friedman doctrine: The social responsibility of business is to increase its profits. The New York Times Magazine, 13(1970): 32–33

Galbraith VH (1961) The making of Domesday Book. Clarendon Press, Oxford

Gilbert DU, Behnam M (2013) Trust and the United Nations global compact: a network theory perspective. Bus Soc 52(1):135–169

Godfrey A, Hooper K (1996) Accountability and decision-making in feudal England: Domesday Book revisited. Account Hist 1(1):35–54. https://doi.org/10.1177/103237329600100103

Gray R, Owen D, Evans R, Zadek S (1997) Struggling with the praxis of social accounting. Stakeholders, accountability, audits and procedures. Account Audit Account J 10(3):325–364

Hood C (1991) A public management for all Seasons? Public Adm 69(1):3–19. https://doi.org/10.1111/j.1467-9299.1991.tb00779.x

Hughes OE (2003) Public management and administration, vol VIII, 3rd edn. Palgrave Macmillan, New York

Ministry of Legislation (1998) Foreign Investment Promotion Act. Ministry of Legislation of ROK

Mulgan R (2000) Accountability: an ever-expanding concept? Public Adm 78(3):555–573

Ospina S, Diaz W, O'Sullivan JF (2002) Negotiating accountability: managerial lessons from identity-based nonprofit organizations. Nonprofit Volunt Sect Q 31(1):5–31

Parsons C (2017) The (in) effectiveness of voluntarily produced transparency reports. Bus Soc 0007650317717957

Post JE (2013) The United Nations global compact: a CSR milestone. Bus Soc 52(1):53–63

Power M, Laughlin R (2003) Critical theory and accounting (Alvenson Willmont ed.). Sage, London

Roberts JT (1982) Accountability in Athenian Government: books on demand

Schembera S (2018) Implementing corporate social responsibility: empirical insights on the impact of the UN Global Compact on its business participants. Bus Soc 57(5):783–825

Sinclair A (1995) The chameleon of accountability: forms and discourses. Acc Organ Soc 20(2–3):219–237. https://doi.org/10.1016/0361-3682(93)e0003-y

Sok K, Whang (2014) Corporate social responsibility implementation in South Korea: lessons from American and British CSR policies. J. Int. Area Stud 12(2):99–118

Stewart R (1999) Public policy (Robyn Flemming ed.). Sydney MacMillan Publishers Australia PTY

The State Law and Order Restoration Council (2013) Central bank law. The State law and Order Restoration Council

Von Dornum D (1997) The straight and the crooked: legal accountability in ancient Greece. Columbia Law Rev 97(5):1483–1518

Wilson R (1968) The theory of syndicates. Econometrica 36(1):119–132

World Bank (1996) Environment matters. World Bank, Washington

Yin RK (2009) Case study: research design and methods, vol 5, 4th edn. Sage, Los Angeles

Chapter 19
Impact Sustainability: Conclusions and Lessons Learned

Ellen Derbyshire, Isabel B. Franco, Tathagata Chatterji, and James Tracey

This book deconstructed and re-evaluated the key issues that sustainability leaders, namely, educators, scientists, governments, practitioners and policymakers, face when achieving the sustainable development goals (SDGs). The overarching aims of this book were to provide a coverage of results of research conducted in accordance with the sustainable development goals and to better understand the integration of the SDGs as an integral part of impact research, curriculum and community capacity-building for sustainability. The impacts of climate change, unsustainable resource development, widening of gaps between socioeconomic groups and social conflict are pressuring sustainability leaders to collaborate and reconstruct normative approaches to developing more consistent and impactful sustainability agendas. This stems from an inability to turn knowledge and theory into impactful outcomes. However, quantified targets do not always mean quantified solutions, and these outcomes are often impeded by homogenized interpretations of globalized targets.

E. Derbyshire (✉)
Faculty of Business, Economics and Law, Business School, The University of Queensland, Brisbane, QLD, Australia
e-mail: ellen.derbyshire@uq.net.au

I. B. Franco
Institute for the Advanced Study of Sustainability, United Nations University Shibuya-ku, Tokyo, Japan

Australian Institute for Business and Economics, The University of Queensland, Brisbane, Australia
e-mail: connect@drisabelfranco.com

T. Chatterji
Xavier School of Human Settlements, Xavier University Bhubaneswar, Kakudia, Odisha, India
e-mail: tathagata@xub.edu.in

J. Tracey
Faculty of Engineering, University of New South Wales, Sydney, Australia

Thus, these 18 chapters broke down these traditional approaches and presented innovative and agile strategies to action the SDG targets.

This material provides an original contribution to sustainability science, education, policy and practice through breaking down and contextualizing theories, policies and strategies for SDG operationalization and impact sustainability. In Chap. 1, Franco and Minnery addressed SDG 1 by proposing the use of the sustainable livelihood framework (SLF) as a powerful conceptual approach for research aimed at understanding the interaction between global investment, local livelihoods and poverty reduction in resource regions. Research presented in this manuscript with regard to SDG 2, Creegan and Flynn argued that by imitating the productivity of natural ecosystems and returning carbon-based materials to the soil source, biomass utilization promotes reduced reliance on water pollution and often prohibitively expensive synthetic fertilizers. The case studies presented show the effectiveness of curriculum and education tools in addressing the targets for SDG 2. This research highlights the need for more financial investment in education about biomass utilization and carbon source availabilities in emerging circular economies. Thus, providing greater insight into alternative strategies that address the targets set by SDG 2 Zero Hunger.

An innovative assessment of healthcare and sustainable development by Belen Federico in connection with SDG 3 explored the effects of ultraviolet radiation on human DNA from the point of view of sustainable healthcare. Franco and Derbyshire analysed SDG 4 through the lens of stakeholder partnerships and collaboration for the implementation of Education for Sustainable Development. Research argues that greater collaboration is required in order to meet the targets set by SDG 4. Later the piece presents investigation by Franco, Salinas-Meruane and Derbyshire linked to SDG 5. Based on a qualitative assessment of current leadership and organizational discourse, their research explores the role of women in male-dominated and gender-segregated fields, such as the extractive industry. The chapter builds knowledge and understanding of the limiting factors and barriers that prevent women from embarking on a sustainable leadership pathway in the early stages of their career. SDG 6 is analysed based on the efficient use of energy and water resources in the mining sector. This research by Cano, Velasco, Garcia and Franco identified key environmental impacts associated with the use of water and energy resources within the mining industry. Franco, Power and Whereat explored the linkages with other SDGs which showed how women cope with rising complexities in resource regions. Through the cases of Japan and Colombia, this chapter argued that key stakeholders in the energy sector, both renewables and fossil fuels, need to further engage in the enhancement of women's assets and capacities towards the achievement of SDG 7. An examination of linkages with SDG 8 presented key lessons to be learned from fair trade certified small producers in attaining economic growth whilst maintaining decent work conditions.

Whilst targeting SDG 9, another study reported in this manuscript presented an innovative roadmap that addressed indicators that can enhance the ability of small-scale miners to cope with the demands of sustainability issues over time. The study showed that there are essential community capacity-building areas that can foster sustainable community development in mining regions. Power examined SDG 10 through reduced inequalities through the lens of Australia's superannuation system. This review examined how SDG 10 can assist with mitigating the burden of financial insecurity that women experience in retirement in three central parts. The results highlight key systemic biases experienced by women in Australia's superannuation system. An examination of SDG 11 argued that the issues of global sustainability cannot be addressed, without strongly addressing sustainability at the urban scale. This study by Vadiya and Chatterji focused on SDG 11 as the analytical framework to explore how the transformative force of urbanization represents opportunity and challenge to meet several other sustainability challenges. Based on a multidimensional approach to entrepreneurship for wellbeing and linkages with SDG 12, Franco and Newey's research reported in this manuscript also examined the role of corporations in fostering sustainable community development. Later, SDG 13 was presented by Franco, Tapia and Tracey who explored the challenges of climate education addressing remedial measures to stakeholders involved towards the achievement of SDG 13.

The book also provided recommendations on education and research for sustainable fashion whilst connecting with SDGs 14 and 15. Palomino merges the fashion and food industry together by presenting the importance of fish skin as an alternative and sustainable material source. Fish skins are sourced from the food industry, using waste, applying the principle of a circular economy. Arana, Franco, et al. explored the production cycle in the fashion industry and provided a critical review of the unsustainability of existing materials and procedures and linkages with SDG 15. This review is essential for practitioners and students in the fashion industry. Research here reported also made a strong contribution to SDG 16 by Franco and Derbyshire who examined the role of women and livelihood options in fostering sustainable peace. Conducted in rural communities in Colombia, the results obtained are important as they provide insights into the heterogeneous composition of communities, particularly women and their identities, which explains contrasting perceptions towards project development. This investigation also provided recommendations for further research as well as policy recommendations to be considered in the pursuit of peace and sustainability in developing resource regions. The book finishes showcasing research that links to SDG 17. Franco and Abe examined the incidence of global business networks for corporate sustainability. This investigation provides recommendations for sustainability managers and practitioners in Asia and the Pacific on how to better account for sustainability performance.

This compelling analysis of how sustainability science, education, policy and practice presents key contributions that address the paradox between lack of knowledge on SDG operationalization and increased need for impact sustainability practices. At the heart of this publication lies the imperative to bridge the gaps between dialogue and constructive action. Rapid population growth, increasing resource demand and exacerbation of socio-economic inequalities all place pressure on policymakers, higher education institutions, governments, business and institutions to address these challenges. By 2030, the global world of work, political dynamics and the global environment will be vastly different from its current state. Thus, the purpose of this book is to provide sustainability leaders with analyses and tools that will help them remain prepared and agile to meet the challenges of the post 2030 world.

Acknowledgements Dr. Isabel B. Franco, the lead editor of this manuscript, is thankful to the Japan Society for the Promotion of Science (JSPS). This work was supported by JSPS (Grant No. JP17777), Project: "Stakeholder Networks for Impact Sustainability Research and Education" led by Dr. Franco under the JSPS-UNU Fellowship.